# Human Metabolism of Alcohol

## Volume III

### Metabolic and Physiological Effects of Alcohol

Editors

**Kathryn E. Crow, Ph.D.**
Research Officer
Department of Chemistry and Biochemistry
Massey University
Palmerston North
New Zealand

**Richard D. Batt, Ph.D., D.Phil.**
Professor and Head
Department of Chemistry and Biochemistry
Massey University
Palmerston North
New Zealand

CRC Press, Inc.
Boca Raton, Florida

**Library of Congress Cataloging-in-Publication Data**

Human metabolism of alcohol / editors, Kathryn E. Crow, Richard D.
   Batt
      p.  cm.
   Includes bibliographies and indexes.
   Contents: v. 1. Pharmacokinetics, medicolegal aspects, and general
interest — v. 2. Regulation, enzymology, and metabolites of ethanol
— v. 3. Metabolic and physiological effects of alcohol.
   ISBN 0-8493-4523-5 (v. 3)
   1. Alcohol—Metabolism. I. Crow, Kathryn E. II. Batt, Richard
D. (Richard Dean), 1923-
QP801.A3H85 1989
615′.7828—dc19                                               88-24238
                                                                  CIP

# PREFACE

## Volume III

This volume is not a comprehensive coverage of the effects of alcohol on human subjects; to attempt that would have required far more space than that available. In selecting topics for inclusion in this volume, attention has been directed to the breadth of interest in the topics discussed. It should be noted that, because of their general interest, two topics that could have been included in this volume (alcohol-nutrition interactions and the fetal alcohol syndrome) are covered in Volume I, Section III).

# EDITORS

**Kathryn Crow, Ph.D.,** is a Senior Research Officer with the Alcohol Research Group in the Department of Chemistry and Biochemistry at Massey University, Palmerston North, New Zealand. Dr. Crow graduated with a B.Sc. (Honors) degree (First Class) in biochemistry from Massey University in 1972, and continued with Ph.D. studies in the same department. After completing her Ph.D. she spent two years as a Post-Doctoral Fellow in the Laboratory of Metabolism at the National Institute on Alcohol Abuse and Alcoholism in Washington, D.C. Following this, she returned to New Zealand and joined the Alcohol Research Group at Massey University in 1977.

Dr. Crow is a member of the Research Society on Alcoholism, the International Society for Biomedical Research on Alcoholism, the Biochemical Society (British), the New Zealand Biochemical Society, and the New Zealand Institute of Chemistry. She was the recipient, in 1975, of a Fogarty International Post-Doctoral Fellowship, and in 1987 of the Watson Victor award for biochemical research in New Zealand.

Dr. Crow has published 30 papers in areas related to the metabolism of alcohol. Her primary research interest is in the regulation of alcohol metabolism in mammalian liver.

**Richard D. Batt, Ph.D., D.Phil.,** is Head of the Department of Chemistry and Biochemistry and Professor of Biochemistry at Massey University in Palmerston North, New Zealand.

Dr. Batt trained initially as an organic chemist at the University of Otago in Dunedin, New Zealand, obtaining a B.Sc. degree in 1944, a M.Sc. in 1945, and a Ph.D. in 1947. He was appointed to a Nuffield Dominions Demonstratorship at Oxford University in 1948 and completed there a B.A. (Honors) degree in Physiology in 1950 and a D.Phil degree with Professor D. D. Woods in 1952. From 1953 to 1964 he was successively a Lecturer, Senior Lecturer, and Associate Professor in Biochemistry at the University of Otago Medical School in Dunedin. In 1964, he was appointed to the Foundation Chair in Biochemistry in the newly established Faculty of Science at Massey University and, since his appointment, has been Head of the combined Department of Chemistry and Biochemistry. For 16 years he was also Dean of the Faculty of Science.

Dr. Batt was one of the foundation members of the New Zealand Biochemical Society and, for a number of years, Chairman of the Executive Committee. He was the President of the New Zealand Institute of Chemistry in 1984 and the Visiting Lecturer to Australia in that year under the jointly sponsored arrangements of the Royal Australian Chemical Institute and the New Zealand Institute of Chemistry. He holds fellowships from the Royal Society of New Zealand, the New Zealand Institute of Chemistry, and the Royal Society of Chemistry. In 1984, he was appointed an Honorary Fellow of the Royal Australian Chemical Institute and in 1976 he was awarded his M.B.E. for ''services to research''. He is currently a member of the Australian and New Zealand Association for the Advancement of Science Council in Australia and has been a member of the Royal Society Council in New Zealand.

Dr. Batt initiated the involvement of the Department of Chemistry and Biochemistry at Massey University in alcohol research, and he has now been Director of the Alcohol Research Group for 15 years. He has been associated with many of the 102 publications produced by the group during that time. His major interests have been in the areas of acetaldehyde removal from the human body and medicolegal aspects of alcohol metabolism.

# CONTRIBUTORS

## Volume III

**Richard D. Batt, D.Phil**
Department of Chemistry and
  Biochemistry
Massey University
Palmerston North, New Zealand

**Egil Bodd, Ph.D.**
Senior Assistant Physician
National Institute of Forensic Toxicology
Oslo, Norway

**Kathryn E. Crow, Ph.D.**
Senior Research Officer
Department of Chemistry and
  Biochemistry
Massey University
Palmerston North, New Zealand

**John Dich, M.D.**
Associate Professor
Department of Biochemistry A
University of Copenhagen
Copenhagen, Denmark

**Graeme Eisenhofer, Ph.D.**
International Research Fellow
National Institutes of Health
Bethesda, Maryland

**Robin Fraser, M.D., Ph.D.**
Associate Professor
Department of Pathology
Christchurch School of Medicine
Christchurch, New Zealand

**Gaut Gadeholt, Ph.D.**
Senior Assistant Physician
National Institute of Forensic Toxicology
Oslo, Norway

**Judith S. Gavaler, Ph.D.**
Associate Professor
Department of Medicine
University of Pittsburgh
Pittsburgh, Pennsylvania

**Robert M. Greenway, Ph.D.**
Reader
Department of Chemistry and
  Biochemistry
Massey University
Palmerston North, New Zealand

**Niels Grunnet, Dr.Phil.**
Associate Professor
Department of Biochemistry A
University of Copenhagen
Copenhagen, Denmark

**Yedy Israel, Ph.D.**
Professor
Departments of Pharmacology and
  Medicine
University of Toronto, and
  Addiction Research Foundation
Toronto, Ontario, Canada

**Ralph H. Johnson, M.D., D.Phil.**
Director
Postgraduate Medical Education
University of Oxford
Oxford, United Kingdom

**Jens Kondrup, M.D., Ph.D.**
Senior Registrar
Medical Department A
Rigshospitalet University Hospital
Copenhagen, Denmark

**Markku Kupari, M.D.**
Consultant Cardiologist
Cardiovascular Laboratory
Helsinki University Central Hospital
Helsinki, Finland

**David G. Lambie, Ph.D.**
Senior Advisory Officer
Department of Scientific and Industrial
  Research
Wellington, New Zealand

**Richard Laverty, Ph.D.**
Professor and Chairman
Department of Pharmacology
University of Otago Medical School
Dunedin, New Zealand

**John M. Littleton, Ph.D.**
Professor
Department of Pharmacology
Kings College
London, England

**Jorg Mørland, Ph.D.**
Director
National Institute of Forensic Toxicology
Oslo, Norway

**Robin John Olds, M.B. Ch.B.**
Training Fellow
Department of Pathology
University of Otago
Dunedin, New Zealand

**Hector Orrego**
Departments of Pharmacology and
  Medicine
University of Toronto
  and Addiction Research Foundation
Toronto, Ontario, Canada

**J. N. Santamaria, M.D.**
Department of Community Medicine
St. Vincent's Hospital
Melbourne, Victoria, Australia

**Leslie Owen Simpson, Ph.D.**
Senior Research Officer
Department of Pathology
University of Otago
Dunedin, New Zealand

**J. G. T. Sneyd, Ph.D.**
Professor and Chairman
Department of Clinical Biochemistry
University of Otago
Dunedin, New Zealand

**Hernan Speisky**
Department of Pharmacology
University of Toronto
Toronto, Ontario, Canada

**Antti Suokas, M.D.**
Research Fellow
Research Unit of Alcohol Diseases
Helsinki University Central Hospital
Helsinki, Finland

**David H. Van Thiel, M.D.**
Professor
Department of Medicine
University of Pittsburgh
Pittsburgh, Pennsylvania

# ACKNOWLEDGMENTS

We would like to thank all the contributors to these volumes for their interest and enthusiasm for the project, for the high quality of manuscripts we received, and for the many helpful suggestions and comments that were provided.

We are particularly grateful to the following local contributors for their help and advice: Trevor Kitson, Mike Hardman, Bob Greenway, Patsy Watson, Paul Buckley, Len Blackwell, Alistair MacGibbon, Allan Stowell, Leena Stowell, and Dick Laverty. We would also like to thank our co-workers in the laboratory, Kathryn Stowell, Rachel Page, and Duncan McKay, for their help and support throughout the project.

We would also like to acknowledge the interest and support shown by many members of the Department of Chemistry and Biochemistry at Massey University during the preparation of these volumes.

**Dr. Kathryn Crow**
**Professor R. D. Batt**

# HUMAN METABOLISM OF ALCOHOL

## Volume I

## PHARMACOKINETICS, MEDICOLEGAL ASPECTS, AND GENERAL INTEREST

## Volume II

## REGULATION, ENZYMOLOGY, AND METABOLITES OF ETHANOL

**Volume III**

## METABOLIC AND PHYSIOLOGICAL EFFECTS OF ALCOHOL

# TABLE OF CONTENTS

## Volume III

# SECTION I
## *Effects of Alcohol and Its Metabolism on Specific Organs and Tissues*

Chapter 1

# THE METABOLIC EFFECTS OF ALCOHOL ON THE LIVER

**Kathryn E. Crow and Robert M. Greenway**

## TABLE OF CONTENTS

# I. INTRODUCTION

Many aspects of the metabolic effects of ethanol in the liver are dealt with in detail elsewhere in this volume (Chapters 7, 8, and 9). The effects of ethanol on the liver have also been reviewed several times in recent years.[1-4] The purpose of this chapter is to provide an overall picture of the metabolic events which occur as a result of the presence of ethanol in the liver on either an acute or a chronic basis and to cover in more detail some of the metabolic effects that have not been discussed elsewhere in this book.

# II. OVERVIEW

## A. Acute Ethanol

### 1. Immediate Effects

When the liver is presented with ethanol in the bloodstream, the oxidation of this substrate produces some immediate metabolic changes. Acetaldehyde is produced by the enzyme alcohol dehydrogenase (ADH) (see Volume II, Chapters 4 and 5) and other minor oxidative pathways (see Volume II, Chapter 10). Acetaldehyde does not normally accumulate to high concentrations (see Volume II, Chapter 13) as it is rapidly oxidized to acetate by the enzyme aldehyde dehydrogenase (see Volume II, Chapters 6, 7, and 8). Much of the acetate formed during ethanol metabolism is exported from the liver to be metabolized by peripheral tissues.[5] Within 25 min of an acute ethanol dose, hepatic venous acetate concentrations rise from about 0.1 m$M$ to about 1.0 m$M$ and remain at about this level until all the ethanol has been metabolized.[6] The increase in hepatic venous acetate concentration with an acute ethanol dose is higher in alcoholics than in control subjects, probably reflecting higher ethanol clearance rates in the alcoholics.[6] Blood acetate has been suggested as a marker for metabolic tolerance to alcohol, indicative of chronic alcohol intake.[7,8]

Within a few minutes of ethanol reaching the liver, the ratio of free [NAD$^+$]/[NADH] in the cytoplasmic and mitochondrial compartments is reduced.[9] This occurs because both ADH- and aldehyde dehydrogenase-catalyzed reactions require NAD$^+$ and produce NADH. This NADH must be reoxidized for oxidative metabolism to continue. Initially, the liver does not reoxidize NADH as fast as it is being produced, hence, the decrease in [NAD$^+$]/[NADH] ratio occurs. However, the ratio stabilizes quickly, indicating that a new steady state has been reached in which the rate of NADH reoxidation has risen to match the rate of NADH production by ADH and aldehyde dehydrogenase. The reoxidation of NADH produced during ethanol metabolism can account for as much as 60 to 80% of the oxygen utilized by the liver.[10,11] Under some circumstances this is apparently achieved with no net change in oxygen uptake by the liver, although sometimes an increase is observed (see further discussion in Section IV.A).

Thus, in summary, the immediate metabolic effects of an acute ethanol dose on the liver are production of acetaldehyde and acetate, decrease of the free cytosolic and mitochondrial [NAD$^+$]/[NADH] ratios, altered oxygen utilization and, in some instances, increased net oxygen uptake. These immediate effects have mainly been defined in experimental animals, since, for obvious reasons, the direct monitoring of metabolic changes over a short time-scale in human liver is difficult.

### 2. Secondary Effects

As the liver continues to metabolize ethanol, secondary effects of the above changes become apparent. Although acetaldehyde does not reach high concentrations, its reaction with aldehyde dehydrogenase may inhibit the metabolism of biogenic aldehydes, which may then accumulate to higher levels than normal. Acetaldehyde may also undergo reactions with biogenic amines and peptides. The extent and physiological significance of such reactions are still being assessed (for example, see papers in Collins[12]). Acetaldehyde may also bind to liver proteins, and the possible extent and consequences of such binding are also still being clarified (see Volume II, Chapter 13).

Although much of the acetate formed during ethanol metabolism is released from the liver, some is fully oxidized to $CO_2$ or incorporated into lipids after conversion to acetyl CoA.[2,13] Increased lipid content in liver may be observed after a few hours of ethanol metabolism, although the mechanisms by which this occurs are still being defined.[14] (For more details on interactions of ethanol and lipid metabolism, see this volume, Chapter 7.)

The decreased $[NAD^+]/[NADH]$ redox state in the liver has the potential to affect flux through other metabolic pathways in which $NAD^+$ and NADH are involved. The citric acid cycle is inhibited, and a net decrease in $CO_2$ output by the liver results.[10,13] This inhibition explains why NADH produced during ethanol metabolism can be reoxidized without necessarily producing a net increase in liver oxygen uptake. Effects of the decreased $[NAD^+]/[NADH]$ on pathways of glucose and lipid metabolism are discussed in this volume, Chapters 8 and 7, respectively. Ethanol-induced redox effects on steroid metabolism have been discussed recently by Andersson et al.[15] and are referred to in this volume, Chapter 6. Protein synthesis may be inhibited to some extent by the reduced $[NAD^+]/[NADH]$ *in vitro,* although it is not clear whether this occurs *in vivo* (see Section V.A).

Increased oxygen uptake by the liver during ethanol metabolism could lead to localized hypoxia in the pericentral region, although it has been suggested that this probably occurs only in the presence of preexisting liver damage.[16] In normal liver the effect of increased oxygen uptake in the presence of an acute ethanol dose is probably offset by increased blood flow, so that there is no net change in hepatic oxygen tension.[1]

Thus, in summary, the metabolism of an acute dose of ethanol by the liver leads to changes in many of the major metabolic pathways, such as the citric acid cycle and those involved in oxygen utilization and glucose and lipid metabolism. Such acute changes are, however, fully reversible. Again, most have been defined *in vitro* or using experimental animals.

## B. Chronic Alcohol

Long-term changes that have been observed in liver following chronic alcohol consumption include: increased fat accumulation; increased oxygen uptake; increased protein and water content; increased hepatocyte size; disruption of the microtubule system; increased collagen synthesis and fibrosis; proliferation of microsomal membranes, with increase in activities of microsomal enzymes and functions leading to increased free radical formation; mitochondrial damage; and increased immunoreactivity of hepatocyte elements.[1-4] Many of the tests for detection of chronic alcohol intake are based on changes observed in blood samples as a result of altered liver function.[8,17,18] However, the relationship between long-term changes and the acute metabolic effects described above has yet to be clearly defined. For example, there is no clear indication as to how much of the gross lipid accumulation that occurs with chronic alcohol intake can be explained by the acute effects of ethanol on lipid metabolism in the liver (see discussion in this volume, Chapter 7). Also, while it can be presumed that metabolic changes seen with chronic alcohol intake are related to the structural changes observed in cirrhotic liver (see the following chapter), cause-effect relationships have yet to be clearly established. At one time it appeared that chronic ethanol consumption led to an ordered progression from increased liver lipid content, through fatty liver degeneration, to the final stages of cirrhosis with gross liver malformation and malfunction. Such a progression has not been proven, however, and is not consistent with current evidence. As discussed in the following chapter and in Volume I, Chapter 10, liver cirrhosis occurs in only about 30% of alcoholics and there are probably genetic factors predisposing towards its occurrence, whereas fatty liver appears to develop in all cases of chronic alcohol intake.

Possible connections between alcoholic liver damage and the effects of chronic ethanol consumption on oxygen uptake have been suggested, but are not yet fully defined. It has been suggested that increased oxygen uptake may lead to hypoxia and, hence, to cellular damage in the pericentral region of the liver,[4] but this mechanism has been questioned and changes in redox state leading to effects on fatty acid oxidation and protein synthesis have been suggested as an alternative mechanism for initiation of damage.[1] Individual differences in oxygen uptake with chronic ethanol consumption could well account for the variation in susceptibility to liver damage, but this has yet to be shown.

Determining the relationship between metabolic changes caused by ethanol and end organ damage with chronic alcohol intake is complicated by the effects of ethanol on nutritional factors, leading to secondary malnutrition.[1,19] The development of liver damage caused directly by chronic alcohol intake probably involves the interaction of many metabolic factors such as increased hepatic oxygen uptake,[4] increased lipid content,[14] increased collagen deposition,[3,4] decreased protein export and increased water content[3,4] (leading to enlargement of hepatocytes and portal hypertension), and increased microsomal activity[1,20,21] (leading to increased free radical generation and oxidative damage). Factors such as anemia,[4] mitochondrial damage,[3,22] acetaldehyde reactions,[1,3] and decreased glutathione concentrations[23] (see also this volume, Chapter 9) may also play a role. Alterations in the immune system occur in patients with alcoholic liver disease and could be a significant factor in the development of hepatic damage.[24] Differences in diet may lead to different rates of cirrhosis development in separate populations and it has been suggested that diets high in polyunsaturated fats may predispose towards alcoholic cirrhosis.[25] Some or all of the above factors may be important in any individual and the likelihood of irreversible damage may be increased by additional stresses such as poor nutrition,[19,26] the presence of other hepatotoxins,[27] and respiratory dysfunction or viral hepatitis. With such a complex picture it is improbable that a single, well-defined route to the development of alcoholic liver disease will ever be found.

The final elucidation of links between the metabolic effects of alcohol and liver cirrhosis has also been hampered because the experimental animals used for most metabolic studies, such as rats and mice, do not readily develop a condition analogous to liver cirrhosis. Recently, however, progress has been made towards solving this research problem via a number of different approaches. It has been shown that rats fed a high-fat diet with alcohol will develop centrilobular fibrosis resembling that seen in man.[28] Administration of an ADH inhibitor, 4-methylpyrazole, with chronic ethanol treatment produced increased signs of liver degeneration in rats.[29] Alternative small animal models in which liver damage similar to that seen in humans can be produced have been developed in ferrets[30] and rabbits (this volume, Chapter 2).

Baboons have been extensively used by the group of Lieber et al.[1-3] as a model for human alcohol-induced liver damage. Finally, noninvasive techniques of liver study[31,32] present the promise that many more data may be obtainable through the direct study of human subjects in the future.

The areas of development and of metabolism in fatty liver (this volume, Chapter 7), liver cirrhosis (following chapter), effects of ethanol on liver carbohydrate metabolism (this volume, Chapter 8), interaction with glutathione metabolism (this volume, Chapter 9), and alcohol-nutrition interactions (Volume I, Chapter 11) are discussed elsewhere and will not be covered in any more detail here. Effects of ethanol on the immune system[24] and microsomal activity in liver[2,20,21] have been well covered in recent literature. The effects of ethanol on liver redox state, oxygen uptake, and protein synthesis and export will be discussed in more detail in the following sections.

## III. LIVER [NAD$^+$]/[NADH] REDOX STATE

### A. Acute Effects

It was initially observed in the 1960s that ethanol decreased the ratio of total [NAD$^+$]/

[NADH] in rat liver.[33,34] Subsequently, however, it was recognized that the ratios of most interest metabolically are those of free [NAD$^+$]/[NADH] estimated separately for the cytosolic and mitochondrial compartments. Free [NAD$^+$]/[NADH] cannot be measured directly, but must be estimated using ratios of metabolites participating in near-equilibrium reactions[35] (for reviews see References 9 and 36). It was then established that ethanol increased the ratio of [lactate]/[pyruvate] (used as an indicator of cytosolic free [NAD$^+$]/[NADH]) and of [β-hydroxybutyrate]/[acetoacetate] (used as an indicator of mitochondrial-free [NAD$^+$]/[NADH]) in rat liver.[37-39] The cytosolic effect results directly from the production of NADH in the ADH and aldehyde dehydrogenase reactions. The mitochondrial effect is caused by transport of reducing equivalents from the cytosol to the mitochondria via shuttle systems[40-42] and, probably more significantly, by the production of NADH by mitochondrial aldehyde dehydrogenase.[43] The ratio of free [NADP$^+$]/[NADPH] in the cytoplasm is also reduced, although only transiently and to a lesser extent than the [NAD$^+$]/[NADH] ratio.[37]

Changes in cytosolic and mitochondrial redox state may be linked to changes in cytosolic phosphorylation potential[44,45] and attempts have been made to assess changes in this parameter with acute ethanol treatment. Based on total liver ATP, ADP, and Pi measurements, the phosphorylation potential either increased[37] or did not change[43] with ethanol, although in the second study, a decrease was observed with ethanol and cyanamide together. The latter effect was thought to be mediated by high acetaldehyde concentrations, although the mechanism was not clear.

While changes in free [NAD$^+$]/[NADH] ratio can be estimated directly from metabolite ratios in animal liver, this is not possible in humans. It has been inferred, however, that similar changes in free [NAD$^+$]/[NADH] ratio occur in human liver as the ratio of [lactate]/[pyruvate] is altered in hepatic venous blood from humans metabolizing alcohol.[46,47] In baboons, [lactate]/[pyruvate] ratios during ethanol metabolism have been measured simultaneously in liver biopsy tissue and arterial and posthepatic venous blood.[48,49] The increase in [lactate]/[pyruvate] with ethanol was greatest in posthepatic venous blood, probably indicting a greater change in cytosolic redox state in the perivenular (pericentral) liver area. This area is less well oxygenated and it has been shown that low oxygen tension increases the cytosolic redox state shift in isolated hepatocytes metabolizing alcohol.[48] Alternatively, the increased reduction of the free cytosolic [NAD$^+$]/[NADH] couple could reflect increased ADH activity in the pericentral region. Reports of the distribution of ADH activity across liver lobules have varied, with highest activity claimed to be in the periportal zone,[50] in pericentral regions,[51] or with equivalent activity in both regions.[52] In isolated hepatocytes, rates of ethanol oxidation were the same in cells from periportal and perivenous regions.[53] These conflicting reports are perhaps explained by a recent publication which shows that hormonal interactions govern the distribution of ADH across the liver lobule.[54] In the latter report, the tendency was for ADH activity to be highest in the pericentral regions. Further investigation of the influence of hormones on intrahepatic distribution of ADH and ethanol metabolism, using digitonin perfusion techniques for separation of enzymes[55] and hepatocytes[56,57] from periportal and pericentral liver regions, could prove to be interesting.

An increased redox gradient in pericentral regions has also been observed in studies using redox scanning with two- or three-dimensional recordings of redox state[58,59] and by surface fluorescence and transmittance in perfused rat livers.[60] It has been suggested that ethanol-induced liver damage might be initiated by the enhanced redox shift, since damage begins in the pericentral regions.[1] As stated in Section II, the changes in cytosolic and mitochondrial [NAD$^+$]/[NADH] ratio during ethanol metabolism may play a role in changing the flux through other pathways in which NAD$^+$ and NADH are involved.[1,2,61] When assessing the potential significance of such changes, however, the actual changes in NAD$^+$ and NADH concentration need to be considered. In rat liver, the [lactate]/[pyruvate] ratio increases from about 10 to 14 in the absence of ethanol to 20 to 50 in the presence of ethanol.[37,43,62] This represents a decrease in free cytosolic [NAD$^+$]/[NADH] from 700 to 1000 down to 200 to 500.[37] Since the initial free NAD$^+$ is about 0.5 m$M$ and the free NADH, therefore, about 0.5 $\mu M$, the ratio change probably represents an increase in NADH to 1.0 to 2.5 $\mu M$, assuming the total free nucleotide concentration is to remain approximately the same.[63,64] This is a small net change in NADH concentration and its significance for any particular enzyme will depend on the kinetic parameters of that enzyme for NADH. For example, it has been calculated that for cytosolic malate dehydrogenase from rat liver, this change in NADH concentration is sufficient to alter the overall direction of the enzyme reaction from NADH production to NADH oxidation,[63] which is necessary for NADH reoxidation via the malate-aspartate shuttle. The kinetic properties for the human enzyme have been found to be similar to those of the rat.[65]

The sensitivity of liver cytosolic malate dehydrogenase to changes in [NADH] suggests that the kinetic parameters of this enzyme are important in determining the extent of free cytosolic [NAD$^+$]/[NADH] ratio change during ethanol metabolism.[63,66] A preliminary study of red blood cell malate dehydrogenase isozymes in nondrinking alcoholic and nonalcoholic subjects showed no differences between the two, although the total enzyme activity was slightly higher in alcoholics.[67] Elevation of mitochondrial malate dehydrogenase activity has been observed in liver biopsy material from drinking alcoholics.[68] In view of these observations, further investigation of the role of malate dehydrogenase in the regulation of hepatic redox state during ethanol metabolism is needed.

The work on the kinetics of malate dehydrogenase also indicates that the hepatic redox state is regulated by the rate of ethanol metabolism, rather than the hepatic redox state regulating ethanol metabolism, as has been suggested.[69] For a further discussion of this interaction, Volume II, Chapter 1 should be consulted. The observation that the redox state does not regulate the rate of ethanol metabolism has been supported by the findings of Cronholm,[70] who estimated that, during ethanol metabolism, the change in the ratio of NAD$^+$ and NADH bound to ADH was about 345-fold. This is a large change when compared with the two- to fivefold change usually observed in [lactate]/[pyruvate] ratio and, hence, in free cytosolic [NAD$^+$]/[NADH] ratio, with ethanol. Cronholm[70] suggested that the free cytosolic [NAD$^+$]/[NADH] ratio is not in equilibrium with coenzyme bound to ADH and that, therefore, the redox state of the free cytosolic [NAD$^+$]/[NADH] system cannot affect the rate of ethanol metabolism. Some recent papers[71,72] have suggested that there may be compartmentation of NADH pools in the cytoplasm, but other workers[73] have claimed that this is not so, at least with respect to lactate dehydrogenase and ADH. Thus, the significance of multiple cytosolic coenzyme pools in relation to redox state changes during ethanol metabolism remains to be clarified. As previously noted,[36] functional heterogeneity of liver cells needs to be excluded as a possible cause of apparent intracellular compartmentation of NADH pools.

## B. Chronic Effects

It has been shown that chronic ethanol treatment attenuates the increase in [lactate]/[pyruvate], and, hence, the decrease in free cytosolic [NAD$^+$]/[NADH], produced in rat liver by an acute ethanol dose.[74] Similar findings have been reported in baboons[48,75] and in humans.[76,77] This could be due to increased rates of NADH reoxidation resulting from either a "hypermetabolic state" (see Section IV.B) or increased transhydrogenation to NADP$^+$ due to increased microsomal ethanol oxidation.[74] Development of alcoholic liver injury impairs circulation and decreases liver oxygenation, so offsetting potential benefits of attenuation in hepatic redox state.[48] Some studies in rats have now shown that chronic ethanol treatment can increase hepatic ADH activity (see Volume II, Chapters 1 and 3). This might be expected to accentuate rather than attenuate the decrease in free cytosolic [NAD$^+$]/

[NADH] in liver during ethanol metabolism, as has, indeed, been shown in male spontaneously hypertensive rats.[78]

More work is needed to confirm the effects of chronic ethanol intake on liver redox state, but the effects may well be different in different strains of animal and could differ widely between individuals in human studies.

## IV. OXYGEN UPTAKE AND OXYGENATION

### A. Acute Effects

*1. Effect of Moderate Ethanol Concentrations*

Ethanol added as a substrate to perfused liver or isolated hepatocytes has been shown to increase, decrease, or not change rates of oxygen uptake.[79] It appears that the effect of ethanol in such preparations depends on whether the liver is from fed or starved animals, what additional substrates are provided, and also, perhaps, on the particular experimental procedures that are used.[79,80] The changes observed in rates of oxygen uptake are linked to the effects of ethanol on other metabolic pathways, particularly those of glucose and fatty acid metabolism.[79-81] For example, in rats starved for 48 h, both gluconeogenesis and rates of oxygen uptake are inhibited by ethanol in the presence of lactate and stimulated by ethanol in the presence of pyruvate.[79] From these results, it can be predicted that the effect of a single acute dose of ethanol on oxygen uptake in human liver will depend on the metabolic state of the liver at the time.

It has recently been shown that the basal rate of oxygen consumption by the perfused rat liver can be decreased by free-radical scavengers without changing mitochondrial respiration or glycolysis.[82-84] The term "antioxidant-sensitive respiration" (ASR) has been introduced to describe the portion of basal oxygen uptake which is suppressed by antioxidants: basal ASR corresponds to 5 to 7% of total respiration.[83,84] Ethanol stimulates ASR,[82,83,85] as does acetaldehyde.[82] It appears that ASR could account for as much as 70% of the increase in oxygen uptake obtained by adding ethanol to perfused livers.[82] The magnitude of ASR depends on the substrates used for perfusion and is enhanced by fasting. Thus, the presence of ASR is a further factor which could cause variation in the effects of acute ethanol administration on hepatic oxygen uptake.

Hepatic oxygen uptake has been shown to be unchanged by acute ethanol administration in human subjects.[46] However, in intact animals, the end effect of acute ethanol treatment on liver oxygenation may be determined by blood flow as well as by rates of oxygen uptake.[16,48] An acute dose of ethanol increases blood flow through the liver, and this may actually lead to a net increase in hepatic oxygen tension.[16] In baboons, increased splanchnic oxygen uptake with ethanol has been shown to be balanced by increased blood flow, so that the hepatic venous oxygen tension was unchanged.[48] These results suggest that, even if moderate acute ethanol intake causes increased hepatic oxygen uptake, it is unlikely to produce anoxia in the liver. A recent study[86] has shown that the increase in portal blood flow with ethanol may be transient, lasting less than 60 min, so the time course of effects of acute ethanol on hepatic oxygenation may be important.

*2. Effect of a Single, Large Ethanol Dose*

It has been shown that pretreatment of rats with a single, very large (5.0 g/kg) ethanol dose will increase rates of oxygen uptake, as well as of ethanol metabolism, in perfused liver. The effect is known as the swift increase in alcohol metabolism (SIAM) and is discussed in detail in Volume II, Chapter 2. The effect on oxygen uptake, although not on ethanol oxidation, has been reproduced using isolated hepatocytes.[79] In hepatocytes the stimulation of oxygen uptake was less than in perfused liver (17 to 29%[79] compared with 38 to 80%[87]). It only occurred in cells from starved rats, whereas the effect in perfused liver occurred primarily in liver from fed rats. The mechanism of stimulation of oxygen uptake by a single, large ethanol dose is unclear. The perfusion studies led to the conclusion that the effect was caused by inhibition of glycolysis,[87] but the studies with hepatocytes[79] seem to suggest that the primary effect could be on mitochondrial respiration, with gluconeogenesis increased as a result. Further work is needed to clarify this point.

These studies do suggest, however, that even a single bout of very heavy drinking could have adverse consequences for the liver in terms of damage due to generation of anoxic zones by the increased oxygen uptake. It should be stressed that the increased oxygen uptake persists in the absence of ethanol, as it is apparent in hepatocyte preparations which were obtained from ethanol-pretreated animals, but not provided with ethanol as a substrate.[79] The time course of the increased oxygen uptake has not been determined in hepatocytes, but the effect in perfusion experiments was shown to be maximal at 2.5 h after ethanol treatment, and to persist for up to 16 h.[87] A selective increase in the pericentral oxygen gradient in isolated perfused liver has been observed in mice treated with a single, high ethanol dose administered in the vapor phase.[88]

Some studies have also suggested a genetic predisposition in some strains of rats and mice towards increased rates of ethanol oxidation, and, hence, perhaps of oxygen uptake, following a single, large ethanol dose (see Volume II, Chapter 2). Thus, some individuals could be particularly susceptible to anoxic liver damage following "binge-style" drinking.

## B. Chronic Effects

Chronic ethanol treatment has been shown to increase oxygen uptake in liver slices and perfused liver from rats, and this has led to the proposal that a "hypermetabolic state" develops in liver after prolonged ethanol intake (for a review, see Reference 11). It has been suggested that the increased oxygen uptake leads to hypoxia in the pericentral liver regions, hypoxic damage, and necrosis.[4,11] Hepatic vein oxygenation has been shown to be lower in alcoholics with hepatic necrosis than in control subjects.[89] The increased hepatic oxygen uptake has been attributed to development of a hyperthyroid state,[4,11,90,91] which can be reversed by the antithyroid drug, propylthiouracil.[4] Consequently, propylthiouracil has been tried in treatment of liver damage due to ethanol, but with limited success.[92] Some studies using rat liver preparations have failed to demonstrate increased hepatic oxygen uptake after chronic ethanol treatment[93-97] and the development of a hyperthyroid state has also been questioned.[1,2,96,97]

The hypermetabolic state and increased oxygen uptake have been claimed to lead to increased rates of ethanol metabolism, but it is now recognized that chronic alcohol intake can also cause an increase in ADH activity (see Volume II, Chapter 3). In some instances, both factors may contribute to development of metabolic tolerance.[91] The interactions between thyroid status, hepatic oxygen uptake, and rates of ethanol metabolism are complex (see also Volume II, Chapter 1) and clearly require further investigation. In view of the influence of liver metabolic state on acute ethanol-induced changes in oxygen uptake (see Reference 79 and Section IV.A.1), the changes due to chronic ethanol should be carefully reexamined, with due consideration for such confounding variables as age.[98]

A recent study[88] demonstrated that the pericentral oxygen gradient in perfused rat liver is selectively increased after chronic ethanol treatment. This led to the proposal that ethanol-induced inhibition of glycolysis (which occurs primarily in pericentral hepatocytes[99]) may lead to increased oxygen uptake and, hence, to oxygen depletion and liver damage in pericentral liver regions. Hypoxia per se has been shown to lead to hepatocellular injury in some experiments.[100] Sato et al.[16] suggested that decreased liver blood flow is an important determinant of hypoxia after chronic ethanol treatment of rats, although increased oxygen uptake was not excluded as a contributing factor. Decreased blood flow was found to be the major cause of decreased hepatic venous oxygen tension in baboons,[48] and in this study, chronic ethanol treatment did not increase hepatic oxygen uptake. The low oxygen tension was thought to aggravate the redox-linked toxicity of alcohol to the extent of inhibiting protein synthesis, this leading to cellular damage. This finding is particularly significant because the alcohol-treated baboons exhibited liver changes similar to the early stages of liver cirrhosis in humans.

In summary, there appear to be two theories regarding the role of anoxia in the pericentral regions of liver in the development of alcohol liver damage. One theory is that chronic ethanol administration leads to increased oxygen uptake, and that this, in turn, leads to anoxia of the pericentral regions, which already have the lowest oxygen tension. Cellular damage then results directly from the anoxia.[4] The alternative theory is that, even if oxygen uptake is increased in the presence of ethanol, this is compensated for by increased blood flow, at least in the early stages of chronic ethanol intake. The lower oxygen tensions normally prevailing in the pericentral region exacerbate the redox changes caused by ethanol, so that these are sufficient to interfere extensively with processes such as protein synthesis and lipid metabolism. This disruption of metabolic processes eventually leads to cellular damage, which, in turn, leads to decreased blood flow, development of anoxia, and a further cycle of damage.[1]

## V. EFFECTS OF ETHANOL ON PROTEIN SYNTHESIS AND SECRETION

Various aspects of the effects of ethanol on hepatic protein metabolism have been discussed in recent articles.[1-3,101,102] The following section provides an overall summary of the available information.

### A. Acute Effects
Acute ethanol administration has been shown to inhibit protein synthesis in numerous studies *in vitro*.[103-108] Results of studies *in vivo,* however, have been inconsistent, with some showing an inhibition[105,109,110] and others none.[102,111,112]

The enhanced effects seen in liver preparations *in vitro* are probably due to a relative depletion of metabolites and hormones in these systems,[61,102] pH shifts,[108] and enhanced reduction of the free cytosolic [NAD$^+$]/[NADH] ratio.[61] The inconsistent results with studies *in vivo* may reflect problems with techniques used to monitor protein synthesis. The studies which showed no effects were recent ones in which care was taken to ensure constant precursor-specific radioactivity[111] or in which radiolabeled puromycin was used to avoid the problem of changing precursor pool sizes.[112]

Thus, present evidence suggests that acute ethanol treatment does not significantly affect rates of protein synthesis *in vivo*. However, the enhanced redox shift in pericentral liver regions may mean that protein synthesis is inhibited in this region *in vivo*.[48] Some studies have shown that acute ethanol administration appears to inhibit protein export from the liver *in vivo,*[105,113,114] but, again, this has not been found in other studies *in vivo*[102,111] or *in vitro*.[115] The differing results may again reflect experimental design.[102]

### B. Chronic Effects
*1. Protein Synthesis*
The reported effects of chronic ethanol administration on protein synthesis have also varied between different studies, with some showing inhibition *in vivo* and *in vitro*,[102,116] while others have not.[112] There is even some indication that protein synthesis could be increased.[61,117] Differences between the studies probably reflect differences in methods, particularly in relation to handling of animals[102] and diet.[61] Thus, the effects of chronic alcohol intake on protein synthesis still require further definition.

*2. Protein Export*

In spite of the lack of consistency in reports of the effects of ethanol on protein synthesis, it appears to be widely accepted that chronic alcohol intake causes accumulation of protein in the liver,[3,61,101,112] although there has been some contradictory evidence.[102,111] With chronic alcohol intake, enlargement or "ballooning" of hepatocytes occurs, leading to hepatomegaly. The accumulation of export protein within the hepatocytes, together with concomitant retention of water, has been regarded as the major cause of this hepatomegaly, although lipid accumulation also contributes.[3] Some studies, however, have suggested that proteins play only a small part in the osmotic changes leading to increased intracellular water (which accounts for 50 to 60% of the increase in liver weight) and that the osmotic changes are due to accumulation of $K^+$ and other, probably low molecular weight, compounds.[4,118]

It appears that disruption of microtubules, essential to the secretory process, is a possible mechanism for hepatic intracellular protein accumulation, although some workers have failed to find evidence to support this (for discussion, see Reference 3). Most of the accumulated secretory protein appears to be located in the Golgi apparatus after acute ethanol treatment, although some is also found in the endoplasmic reticulum and the cytosol.[114] On the basis of these observations it has been suggested that ethanol interferes with movement of protein via vesicular structures between the Golgi and plasma membrane, and possibly also between the rough endoplasmic reticulum and the Golgi.[114]

Increased hepatocyte size may play an important role in the increased hepatic resistance to blood flow and, thus, contribute to the development of portal hypertension. This effect, with accompanying decreased oxygenation of the liver, could be an important factor in initiation of liver damage.[3,4] Thus, if decreased protein export is a major factor leading to increased hepatocyte size, further clarification of the mechanism is critical for understanding and treatment of the early stages of alcoholic liver damage. Lieber[61] has pointed out that the nature of the protein accumulating in the liver with chronic alcohol intake has yet to be fully defined, since export proteins such as albumin and transferrin account for only a fraction of the total accumulated protein. Mallory bodies, a characteristic inclusion of hepatocytes from patients with alcoholic hepatitis, have been shown to be composed of glycoprotein and may derive from the liver cell cytoskeleton.[24] Further clarification of the nature of the accumulated protein should help to determine the mechanism of accumulation.

During chronic ethanol exposure, much of the protein retained in the liver is proteolysed[61] with accumulation of some amino acids, including $\alpha$-amino-*n*-butyric acid which is a metabolite of methionine, serine, and threonine. The plasma level of this amino acid has been nominated as a marker for alcohol consumption in humans,[119] but has subsequently been shown to have low specificity for chronic alcohol intake and, therefore, to be of limited value.[8]

*3. Collagen Deposition*

In addition to the intracellular accumulation of protein that occurs with chronic alcohol intake, there is considerable deposition of extracellular collagen. Even in the early stages of fatty liver degeneration, myofibroblasts proliferate and the collagen they produce is deposited, particularly around the terminal hepatic venules.[120] Although products of ethanol metabolism have been suggested as causative agents,[121] the mechanisms responsible for increased collagen deposition have not yet been fully defined.[3] The deposition is associated with increases in proline availability,[122] peptidylproline hydroxylase,[123] and, in baboon liver, with increased levels of type I procollagen mRNA.[124] The measurement of serum collagen propeptides has been suggested as a diagnostic tool for assessing the extent of alcoholic liver damage and as an indicator of resumed drinking.[125] While these responses suggest increased collagen synthesis, decreased collagen breakdown may also contribute.[126]

Collagen precursor formation and collagen deposition also increase in the inflammatory reaction of alcoholic hepatitis which develops in some alcoholics.[125] Such inflammation-dependent fibrosis may depend on the triggering of an immune cytotoxic reaction.[24] At this stage it seems that research on the mechanism of collagen accumulation has produced a series of observations, with no clear overall picture being formed.

The reduced access of hepatocytes to blood supply[127] and the increased resistance to flow leading to portal hypertension[128] have been blamed on collagen deposition in sinusoids and the space of Disse in cirrhosis. However, a causal role for collagen accumulation in the development of portal hypertension has been questioned, with hepatocyte enlargement being viewed as playing the major role.[4]

## VI. SUMMARY

There now seems little doubt that liver damage in heavy drinkers results from the metabolism of ethanol per se, rather than from primary nutritional deficiency. There is a wealth of information about the appearance and functional impairment of the ethanol-damaged liver, as well as about the diverse metabolic changes produced by ethanol in the livers of experimental animals. Nevertheless, it still remains quite uncertain which of these metabolic effects are the most significant in the etiology of liver damage.

## ACKNOWLEDGMENT

We would like to thank Carole Partridge for typing this manuscript.

## REFERENCES

1. **Lieber, C. S.,** Alcohol and the liver: 1984 update, *Hepatology,* 4, 1243, 1984.
2. **Lieber, C. S.,** Metabolism and metabolic effects of alcohol, *Med. Clin. North Am.,* 68, 3, 1984.
3. **Leiber, C. S.,** Alcohol and the liver: metabolism of ethanol, metabolic effects and pathogenesis of injury, *Acta Med. Scand.,* Suppl. 703, 11, 1985.
4. **Orrego, H., Blake, J. E., Medline, A., and Israel, Y.,** Interrelation of the hypermetabolic state, necrosis, anemia and cell enlargement as determinants of severity in alcoholic liver disease, *Acta Med. Scand.,* Suppl. 703, 81, 1985.
5. **Lundquist, F.,** Production and utilization of free acetate in man, *Nature,* 193, 579, 1962.
6. **Nuutinen, H., Lindros, K., Hekali, P., and Salaspuro, M.,** Elevated blood acetate as an indicator of fast ethanol elimination in alcoholics, *Alcohol,* 2, 623, 1985.
7. **Korri, U.-M., Nuutinen, H., and Salaspuro, M.,** Increased blood acetate: a new laboratory marker of alcoholism and heavy drinking, *Alcoholism Clin. Exp. Res.,* 9, 468, 1985.
8. **Salaspuro, M.,** Conventional and coming laboratory markers of alcoholism and heavy drinking, *Alcoholism Clin. Exp. Res.,* 10, 5S, 1986.
9. **Christensen, E. L. and Higgins, J. J.,** Effect of acute and chronic administration of ethanol on the redox states of brain and liver, in *Biochemistry and Pharmacology of Ethanol,* Vol. 1, Majchrowicz, E. and Noble, E. P., Eds., Plenum Press, New York, 1979, chap. 10.
10. **Reitz, R. C.,** Effects of ethanol on the intermediary metabolism of liver and brain, in *Biochemistry and Pharmacology of Ethanol,* Vol. 1, Majchrowicz, E. and Noble, E. P., Eds., Plenum Press, New York, 1979, chap. 12.
11. **Israel, Y., Kalant, H., Orrego, H., Khanna, J. M., Phillips, M. J., and Stewart, D. J.,** Hypermetabolic state, oxygen availability, and alcohol-induced liver damage, in *Biochemistry and Pharmacology of Ethanol,* Vol. 1, Majchrowicz, E. and Noble, E. P., Eds., Plenum Press, New York, 1979, chap. 15.
12. **Collins, M. A., Ed.,** *Aldehyde Adducts in Alcoholism,* Alan R. Liss, New York, 1985.
13. **Hawkins, R. and Kalant, H.,** The metabolism of ethanol and its metabolic effects, *Pharmacol. Rev.,* 24, 67, 1972.
14. **Lieber, C. S. and Savolainen, M.,** Ethanol and lipids, *Alcoholism Clin. Exp. Res.,* 8, 409, 1984.
15. **Andersson, S., Cronholm, T., and Sjovall, J.,** Redox effects of ethanol on steroid metabolism, *Alcoholism Clin. Exp. Res.,* 10, 55S, 1986.

16. **Sato, N., Kamada, T., Kawano, S., Hayashi, N., Kishida, Y., Meren, H., Yoshihara, H., and Abe, H.,** Effect of acute and chronic ethanol consumption on hepatic tissue oxygen tension in rats, *Pharmacol. Biochem. Behav.,* 18 (Suppl. 1), 443, 1983.

17. **Watson, R. R., Mohs, M. E., Eskelson, C., Sampliner, R. E., and Hartman, B.,** Identification of alcohol abuse and alcoholism with biological parameters, *Alcoholism Clin. Exp. Res.,* 10, 364, 1986.

18. **Takase, S., Takada, A., Tsutsumi, M., and Matsuda, Y.,** Biochemical markers of chronic alcoholism, *Alcohol,* 2, 405, 1985.

19. **Thompson, A. D.,** Alcohol and nutrition, *Clin. Endocrinol. Metab.,* 7, 405, 1978.

20. **Koop, D. R. and Coon, M. J.,** Ethanol oxidation and toxicity: role of alcohol P-450 oxygenase, *Alcoholism Clin. Exp. Res.,* 10, 44S, 1986.

21. **Teschke, R. and Gellert, J.,** Hepatic microsomal ethanol-oxidising system (MEOS): metabolic aspects and clinical implications, *Alcoholism Clin. Exp. Res.,* 10, 20S, 1986.

22. **French, S. W.,** Role of mitochondrial damage in alcoholic liver disease, in *Biochemistry and Pharmacology of Ethanol,* Majchrowicz, E. and Noble, E. P., Eds., Plenum Press, New York, 1979, chap. 14.

23. **Lauterburg, B. H., Davies, S., and Mitchell, J. R.,** Ethanol suppresses hepatic glutathione synthesis in rats *in vivo, J. Pharmacol. Exp. Ther.,* 230, 7, 1984.

24. **Johnson, R. D. and Williams, R.,** Immune responses in alcoholic liver disease, *Alcoholism Clin. Exp. Res.,* 10, 471, 1986.

25. **Nanji, A. A. and French, S. W.,** Dietary factors and alcoholic cirrhosis, *Alcoholism Clin. Exp. Res.,* 10, 271, 1986.

26. **Davidson, C. S.,** Changing concepts in the pathogenesis of alcoholic liver disease, *Alcoholism Clin. Exp. Res.,* 10, 3S, 1986.

27. **Zimmerman, H. J.,** Effects of alcohol on other hepatotoxins, *Alcoholism Clin. Exp. Res.,* 10, 3, 1986.

28. **French, S. W., Miyamoto, K., and Tsukamoto, H.,** Ethanol-induced hepatic fibrosis in the rat: role of the amount of dietary fat, *Alcoholism Clin. Exp. Res.,* 10, 13S, 1986.

29. **Lindros, K. O., Stowell, L., Väänänen, H., Sipponen, P., Lamminsivu, U., Pikkarainen, P., and Salaspuro, M.,** Uninterrupted prolonged ethanol oxidation as a main pathogenetic factor of alcohol liver damage: evidence from a new liquid diet animal model, *Liver,* 3, 79, 1983.

30. **Roselle, G. A., Mendenhall, C. L., Muhleman, A. F., and Chedid, A.,** The ferret: a new model of oral ethanol injury involving the liver, bone marrow, and peripheral blood lymphocytes, *Alcoholism Clin. Exp. Res.,* 10, 279, 1986.

31. **Ratner, A. V., Carter, E. A., Pohost, G. M., and Wands, J. R.,** Nuclear magnetic resonance spectroscopy and imaging in the study of experimental liver diseases, *Alcoholism Clin. Exp. Res.,* 10, 241, 1986.

32. **Cunningham, C. C.,** Use of nuclear magnetic resonance spectroscopy to study the effects of ethanol consumption on liver metabolism and pathology, *Alcoholism Clin. Exp. Res.,* 10, 246, 1986.

33. **Slater, T. F., Sawyer, B. C., and Strauli, U. D.,** Changes in liver nucleotide concentrations in experimental liver injury. II. Acute ethanol poisoning, *Biochem. J.,* 93, 267, 1964.

34. **Cherrick, G. R. and Leevy, C. M.,** The effect of ethanol metabolism on levels of oxidised and reduced nicotinamide-adenine dinucleotide in liver, kidney and heart, *Biochim. Biophys. Acta,* 107, 29, 1965.

35. **Williamson, D. H., Lund, P., and Krebs, H. A.,** The redox state of free nicotinamide-adenine dinucleotide in the cytoplasm and mitochondria of rat liver, *Biochem. J.,* 103, 514, 1967.

36. **Sies, H.,** Nicotinamide nucleotide compartmentation, in *Metabolic Compartmentation,* Sies, H., Ed., Academic Press, New York, 1982, 205.

37. **Veech, R. L., Guynn, R., and Veloso, D.,** The time-course of the effects of ethanol on the redox and phosphorylation states of rat liver, *Biochem. J.,* 127, 387, 1972.

38. **Veech. R. L.,** The effects of ethanol on the free nucleotide systems and related metabolites in liver and brain, in *Alcohol and Aldehyde Metabolising Systems,* Thurman, R. G., Yonetani, T., Williamson, J. R., and Chance, B., Eds., Academic Press, New York, 1974, 383.

39. **Forsander, O. A.,** Influence of ethanol on the redox state of the liver, *Q. J. Stud. Alcohol,* 31, 550, 1970.

40. **Grunnet, N., Thieden, H. I. D., and Quistorff, B.,** Oxidation of reducing equivalents during ethanol metabolism. Studies with 1-$^3$H ethanol, in *Alcohol and Aldehyde Metabolising Systems,* Vol. 3, Thurman, R. G., Williamson, J. R., Drott, H. R., and Chance, B., Eds., Academic Press, New York, 1977, 195.

41. **Davis, E. J. and Lumeng, L.,** Thermodynamic aspects of hydrogen transfer via the malate-aspartate shuttle under conditions of steady-state phosphorylation and reduction potentials, in *Alcohol and Aldehyde Metabolising Systems,* Vol. 3, Thurman, R. G., Williamson, J. R., Drott, H. R., and Chance, B., Eds., Academic Press, New York, 1977, 243.

42. **Dawson, A. G.,** Oxidation of cytosolic NADH formed during aerobic metabolism in mammalian cells, *Trends Biochem. Sci.,* 4, 171, 1979.

43. **Lindros, K. O. and Stowell, A. R.,** Effects of ethanol-derived acetaldehyde on the phosphorylation potential and on the intramitochondrial redox state in intact rat liver, *Arch. Biochem. Biophys.,* 218, 429, 1982.

44. **Stubbs, M., Veech, R. L., and Krebs, H. A.,** Control of the redox state of the nicotinamide-adenine dinucleotide couple in rat liver cytoplasm, *Biochem. J.,* 126, 59, 1972.
45. **Wilson, D. F., Stubbs, M., Veech, R. L., Erecinska, M., and Krebs, H. A.,** Equilibrium relations between the oxidation-reduction reactions and the adenosine triphosphate synthesis in suspensions of isolated liver cells, *Biochem. J.,* 140, 57, 1974.
46. **Tygstrup, N., Ranek, L., Ramsøe, K., and Keiding, S.,** Effect of submaximal ethanol elimination on hepatic redox levels in man, in *Alcohol and Aldehyde Metabolising Systems,* Thurman, R. G., Yonetani, T., Williamson, J. R., and Chance, B., Eds., Academic Press, New York, 1974, 469.
47. **Nuutinen, H. U., Salaspuro, M. P., Valle, M., and Lindros, K. O.,** Blood acetaldehyde concentration gradient between hepatic and antecubital venous blood in ethanol-intoxicated alcoholics and controls, *Eur. J. Clin. Invest.,* 14, 306, 1984.
48. **Baraona, E., Jauhonen, P., Miyakawa, H., and Lieber, C. S.,** Zonal redox changes as a cause of selective perivenular hepatotoxicity of alcohol, *Pharmacol. Biochem. Behav.,* 18 (Suppl. 1), 449, 1983.
49. **Jauhonen, P., Baraona, E., Miyakawa, H., and Lieber, C. S.,** Mechanism for selective perivenular hepatotoxicity of ethanol, *Alcoholism Clin. Exp. Res.,* 6, 350, 1982.
50. **Greenberger, N. J., Cohen, R. B., and Isselbacher, K. J.,** The effect of chronic ethanol administration on liver alcohol dehydrogenase activity in the rat, *Lab. Invest.,* 14, 264, 1965.
51. **Morrison, G. and Brock, F. E.,** Quantitative measurement of alcohol dehydrogenase in the lobule of normal livers, *J. Lab. Clin. Med.,* 70, 116, 1967.
52. **Väänänen, H., Salaspuro, M., and Lindros, K. O.,** The effect of chronic ethanol ingestion on ethanol metabolising enzymes in isolated periportal and perivenous rat hepatocytes, *Hepatology,* 4, 862, 1984.
53. **Väänänen, H. and Lindros, K. O.,** Comparison of ethanol metabolism in isolated periportal or perivenous hepatocytes: effects of chronic ethanol treatment, *Alcoholism Clin. Exp. Res.,* 9, 315, 1985.
54. **Maly, I. P. and Sasse, D.,** The intra-acinar distribution patterns of alcohol dehydrogenase activity in the liver of juvenile, castrated and testosterone-treated rats, *Biol. Chem. Hoppe Seyler,* 368, 315, 1987.
55. **Quistorff, B. and Grunnet, N.,** Dual-digitonin-pulse perfusion, *Biochem. J.,* 243, 87, 1987.
56. **Lindros, K. O. and Penttilä, K. E.,** Digitonin-collagenase perfusion for efficient separation of periportal or perivenous hepatocytes, *Biochem. J.,* 228, 757, 1985.
57. **Quistorff, B.,** Gluconeogenesis in periportal and perivenous hepatocytes of rat liver, isolated by a new high-yield digitonin/collagenase perfusion technique, *Biochem. J.,* 229, 221, 1985.
58. **Quistorff, B. and Chance, B.,** Redox scanning in the study of metabolic zonation of liver, in *Regulation of Hepatic Metabolism,* Thurman, R. G., Kauffman, F., and Jungermann, K., Eds., Plenum Press, New York, 1986, 185.
59. **Quistorff, B., Chance, B., and Takeda, H.,** Two- and three-dimensional redox heterogeneity of rat liver. Effects of anoxia and alcohol on the lobular redox pattern, in *Frontiers of Biological Energetics,* Vol. 2, Dutton, P. L., Leigh, J. S., and Scarpa, A., Eds., Academic Press, New York, 1978, 1487.
60. **Jauhonen, V. P., Baraona, E., Lieber, C. S., and Hassinen, I. E.,** Dependence of ethanol-induced redox shift on hepatic oxygen tensions prevailing *in vivo, Alcohol,* 2, 163, 1985.
61. **Lieber, C. S.,** Alcohol, protein metabolism, and liver injury, *Gastroenterology,* 79, 373, 1980.
62. **Braggins, T. J. and Crow, K. E.,** The effects of high ethanol doses on rates of ethanol oxidation in rats, *Eur. J. Biochem.,* 119, 633, 1981.
63. **Crow, K. E., Braggins, T. J., Batt, R. D., and Hardman, M. J.,** Rat liver cytosolic malate dehydrogenase: purification, kinetic properties, role in control of free cytosolic NADH concentration, *J. Biol. Chem.,* 257, 14217, 1982.
64. **Crow, K. E.,** Ethanol metabolism by the liver, *Rev. Drug Metab. Drug Int.,* 5, 113, 1985.
65. **Crow, K. E., Braggins, T. J., and Hardman, M. J.,** Human liver cytosolic malate dehydrogenase: purification, kinetic properties, and role in ethanol metabolism, *Arch. Biochem. Biophys.,* 225, 621, 1983.
66. **Crow, K. E., Braggins, T. J., Batt, R. D., and Hardman, M. J.,** Kinetics of malate dehydrogenase and control of rates of ethanol metabolism in rats, *Pharmacol. Biochem. Behav.,* 18 (Suppl. 1), 233, 1983.
67. **Willis, J. A., Stowell, K. M., Crow, K. E., Batt, R. D., and Hardman, M. J.,** Increased malate dehydrogenase activity in blood from non-drinking alcoholics, *Alcohol Alcoholism,* 20, 293, 1985.
68. **Jenkins, W. J. and Peters, T. J.,** Mitochondrial enzyme activities in liver biopsies from patients with alcoholic liver disease, *Gut,* 19, 341, 1978.
69. **Khanna, J. M. and Israel, Y.,** Ethanol metabolism, in *Liver and Biliary Tract Physiology 1. International Review of Physiology,* Vol. 21, Javitt, N. B., Ed., University Park Press, Baltimore, 1980, 275.
70. **Cronholm, T.,** $NAD^+$-dependent ethanol oxidation: redox effects and rate limitation, *Pharmacol. Biochem. Behav.,* 18 (Suppl. 1), 229, 1983.
71. **Cronholm, T.,** Incorporation of the 1-pro-R and 1-pro-S hydrogen atoms of ethanol in the reduction of acids in the liver of intact rats and in isolated hepatocytes, *Biochem. J.,* 229, 323, 1985.
72. **Berry, M. N., Grivell, A. R., and Wallace, P. G.,** Energy-dependent regulation of the steady-state concentrations of the components of the lactate dehydrogenase reaction in liver, *FEBS Lett.,* 119, 317, 1980.

73. **Vind, C. and Grunnet, N.,** The reversibility of cytosolic dehydrogenase reactions in hepatocytes from starved and fed rats, *Biochem. J.,* 222, 437, 1984.

74. **Domschke, S., Domschke, W., and Lieber, C. S.,** Hepatic redox state: attenuation of the acute effects of ethanol induced by chronic ethanol consumption, *Life Sci.,* 15, 1327, 1975.

75. **Salaspuro, M. P., Shaw, S., Jayatilleke, E., Ross, W. A., and Lieber, C. S.,** Attenuation of the ethanol-induced hepatic redox change after chronic alcohol consumption in baboons: metabolic consequences *in vivo* and *in vitro, Hepatology,* 1, 33, 1981.

76. **Salaspuro, M. P. and Kesaniemi, Y. A.,** Intravenous galactose elimination tests with and without ethanol loading in various clinical conditions, *Scand. J. Gastroenterol.,* 8, 681, 1973.

77. **Salaspuro, M. P., Lindros, K. O., and Pikkarainen, P.,** Effect of 4-methylpyrazole on ethanol elimination rate and hepatic redox changes in alcoholics with adequate or inadequate nutrition and in nonalcoholic controls, *Metab. Clin. Exp.,* 6, 631, 1978.

78. **Wahid, S., Khanna, J. M., Carmichael, F. J., Lindros, K. O., Rachamin, G., and Israel, Y.,** Alcohol-induced redox changes in the liver of the spontaneously hypertensive rat, *Biochem. Pharmacol.,* 30, 1277, 1981.

79. **Stowell, K. M. and Crow, K. E.,** The effect of acute ethanol treatment on rates of oxygen uptake, ethanol oxidation and gluconeogenesis in isolated rat hepatocytes, *Biochem. J.,* 262, 595, 1985.

80. **Phillips, J. W., Berry, M. N., Grivell, A. R., and Wallace, P. G.,** Some unexplained features of hepatic ethanol oxidation, *Alcohol,* 2, 57, 1985.

81. **Berry, M. N., Clark, D. G., Grivell, A. R., and Wallace, P. G.,** The calorigenic nature of hepatic ketogenesis, *Eur. J. Biochem.,* 131, 205, 1983.

82. **Videla, L. A. and Villena, M. I.,** Effect of ethanol, acetaldehyde and acetate on the antioxidant-sensitive respiration in the perfused rat liver: influence of fasting and diethylmaleate treatment, *Alcohol,* 3, 163, 1986.

83. **Videla, L. A.,** Hepatic antioxidant-sensitive respiration. Effect of ethanol, iron and mitochondrial uncoupling, *Biochem. J.,* 223, 885, 1984.

84. **Videla, L. A., Villena, M. I., Donoso, G., Giulivi, C., and Boveris, A.,** Changes in oxygen consumption induced by t-butyl hydroperoxide in perfused rat liver. Effect of free radical scavengers, *Biochem. J.,* 223, 879, 1984.

85. **Videla, L. A.,** Chemically-induced antioxidant-sensitive respiration: relation to glutathione content and lipid peroxidation in the perfused rat liver, *FEBS Lett.,* 178, 119, 1984.

86. **Yoshihara, H., Sato, N., Sasaki, Y., Uchima, E., Inoue, A., Matsumura, T., Hayashi, N., Kawano, S., Kamada, T., and Abe, H.,** Effect of ethanol on portal venous blood flow in healthy volunteers: comparison between the subjects with and without AlDH 1 isozyme, *Alcohol,* 2, 463, 1985.

87. **Yuki, T. and Thurman, R. G.,** The swift increase in alcohol metabolism. Time course for the increase in hepatic oxygen uptake and the involvement of glycolysis, *Biochem. J.,* 186, 119, 1980.

88. **Ji, S., Lemasters, J. J., Christenson, V., and Thurman, R. G.,** Selective increase in pericentral oxygen gradient in perfused rat liver following ethanol treatment, *Pharmacol. Biochem. Behav.,* 18 (Suppl. 1), 439, 1983.

89. **Iturriaga, H., Ugarte, G., and Israel, Y.,** Hepatic vein oxygenation, liver blood flow and the rate of ethanol metabolism in recently abstinent alcoholic patients, *J. Clin. Invest.,* 10, 211, 1980.

90. **Israel, Y., Kalant, H., Orrego, H., Khanna, J. M., Videla, L., and Phillips, M. J.,** Experimental alcohol-induced hepatic necrosis: suppression by propylthiouracil, *Proc. Natl. Acad. Sci. U.S.A.,* 72, 1137, 1975.

91. **Rachamin, G., Okuno, F., and Israel, Y.,** Inhibitory effect of propylthiouracil on the development of metabolic tolerance to ethanol, *Biochem. Pharmacol.,* 34, 2377, 1985.

92. **Orrego, H., Kalant, H., Israel, Y., Blake, J., Medline, A., Rankin, J. G., Armstrong, A., and Kapur, B.,** Effect of short-term therapy with propylthiouracil in patients with alcoholic liver disease, *Gastroenterology,* 76, 105, 1979.

93. **Gordon, E. R.,** ATP metabolism in the ethanol-induced fatty liver, *Alcoholism Clin. Exp. Res.,* 1, 21, 1977.

94. **Cederbaum, A. I., Dicker, E., Lieber, C. S., and Rubin, E.,** Factors contributing to the adaptive increase in ethanol metabolism due to chronic consumption of ethanol, *Alcoholism Clin. Exp. Res.,* 1, 27, 1977.

95. **Kondrup, J., Lundquist, F., and Damgaard, S. E.,** Metabolism of palmitate in perfused rat liver. Effect of ethanol in livers from rats fed on a high-fat diet with or without ethanol, *Biochem. J.,* 184, 89, 1979.

96. **Schaffer, W. T., Denckla, W. D., and Veech, R. L.,** The effect of chronic ethanol consumption on the rate of whole animal and perfused liver oxygen consumption, in *Alcohol and Aldehyde Metabolising Systems,* Vol. 4, Thurman, R. G., Ed., Plenum Press, New York, 1980, 587.

97. **Schaffer, W. T., Denckla, W. D., and Veech, R. L.,** Effects of chronic ethanol administration on $O_2$ consumption in whole body and perfused liver of the rat, *Alcoholism Clin. Exp. Res.,* 5, 192, 1981.

98. **Britton, R. S., Videla, L. A., Rachamin, G. S., Okuno, F., and Israel, Y.,** Effect of age on metabolic tolerance and hepatomegaly following chronic ethanol administration, *Alcoholism Clin. Exp. Res.,* 8, 528, 1984.

99. **Jungermann, K. and Katz, N.,** Functional hepatocellular heterogeneity, *Hepatology,* 2, 385, 1982.

100. **Lemasters, J. J., Ji, S., Stemkowski, C. J., and Thurman, R. G.,** Hypoxic hepatocellular injury, *Pharmacol. Biochem. Behav.,* 18 (Suppl. 1), 455, 1983.

101. **Baraona, E. and Lieber, C. S.,** Effects of alcohol in hepatic transport of proteins, *Annu. Rev. Med.,* 33, 281, 1982.

102. **Morland, J., Bessessen, A., Smith-Kielland, A., and Wallin, B.,** Ethanol and protein metabolism in the liver, *Pharmacol. Biochem. Behav.,* 18 (Suppl. 1), 251, 1983.

103. **Tuma, D. J., Jennett, R. B., and Sorrell, M. F.,** Effect of ethanol on the synthesis and secretion of hepatic secretory glycoproteins and albumin, *Hepatology,* 1, 590, 1981.

104. **Morland, J. and Bessessen, A.,** Inhibition of protein synthesis by ethanol in isolated rat liver parenchymal cells, *Biochim. Biophys. Acta,* 474, 312, 1977.

105. **Baraona, E., Pikkarainen, P., Salaspuro, M., Finkelman, F., and Lieber, C. S.,** Acute effects of ethanol on hepatic protein synthesis and secretion in the rat, *Gastroenterology,* 79, 104, 1980.

106. **Dich, J. and Tønnesen, I. C.,** Effects of ethanol nutritional status, and composition of the incubation medium on protein synthesis in isolated rat liver parenchymal cell, *Arch. Biochem. Biophys.,* 204, 640, 1980.

107. **Lakshmanan, M. R., Felver, M. E., and Veech, R. L.,** Alcohol and very low density lipoprotein synthesis and secretion by isolated hepatocytes, *Alcoholism Clin. Exp. Res.,* 4, 361, 1980.

108. **Wallin, B., Morland, J., and Fikke, A. M.,** Combined effects of ethanol and pH-change on protein synthesis in isolated rat hepatocytes, *Acta Pharmacol. Toxicol.,* 49, 134, 1981.

109. **Jeejeebhoy, K. N., Phillips, M. J., Bruce-Robertson, A., Ho, J., and Sodtke, U.,** The acute effect of ethanol on albumin, fibrinogen and transferrin synthesis in the rat, *Biochem. J.,* 126, 1111, 1972.

110. **Murty, C. N., Verney, E., and Sidransky, H.,** Acute effect of ethanol on membranes of the endoplasmic reticulum and on protein synthesis in rat liver, *Alcoholism Clin. Exp. Res.,* 4, 93, 1980.

111. **Wallin, B., Bessessen, A., Fikke, A.-M., Aarbakke, J., and Morland, J.,** No effect of acute ethanol administration on hepatic protein synthesis and export in the rat *in vivo, Alcoholism Clin. Exp. Res.,* 8, 191, 1984.

112. **Donohue, T. M., Sorrell, M. F., and Tuma, D. J.,** Hepatic protein synthesis activity *in vivo* after ethanol administration, *Alcoholism Clin. Exp. Res.,* 11, 80, 1987.

113. **Sorrell, M. F. and Tuma, D. J.,** Selective impairment of glycoprotein metabolism by ethanol and acetaldehyde in rat liver slices, *Gastroenterology,* 75, 200, 1978.

114. **Volentine, G. D., Tuma, D. J., and Sorrell, M. F.,** Subcellular location of secretory proteins retained in the liver during the ethanol-induced inhibition of hepatic protein secretion in the rat, *Gastroenterology,* 90, 158, 1986.

115. **Morland, J., Rothschild, M. A., Oratz, M., Mongelli, J., Donor, D., and Schreiber, S. S.,** Protein secretion in suspensions of isolated rat hepatocytes: no influence of acute ethanol administration, *Gastroenterology,* 80, 159, 1981.

116. **Smith-Kielland, A., Svendsen, L., Bessessen, A., and Morland, J.,** Effect of chronic alcohol consumption on *in vivo* protein synthesis in livers of female and male rats fed two different dietary regimens, *Alcohol Alcoholism,* 18, 285, 1983.

117. **Baraona, E., Leo, M. A., Borowsky, S. A., and Lieber, C. S.,** Pathogenesis of alcohol-induced accumulation of protein in the liver, *J. Clin. Invest.,* 60, 546, 1977.

118. **Israel, Y. and Orrego, H.,** On the characteristics of alcohol-induced liver enlargement and its possible hemodynamic consequences, *Pharmacol. Biochem. Behav.,* 18 (Suppl. 1), 433, 1983.

119. **Shaw, S. and Lieber, C. S.,** Plasma amino acids in the alcoholic: nutritional aspects, *Alcoholism Clin. Exp. Res.,* 7, 22, 1983.

120. **Nakano, M., Worner, T. M., and Lieber, C. S.,** Perivenular fibrosis in alcoholic liver injury; ultrastructure and histological progression, *Gastroenterology,* 83, 777, 1982.

121. **Savolainen, E. R., Leo, M. A., Timpl, R., and Lieber, C. S.,** Acetaldehyde and lactate stimulate collagen synthesis of cultured baboon liver myofibroblasts, *Gastroenterology,* 87, 777, 1984.

122. **Hakkinen, H.-M. and Kulonen, E.,** Effects of ethanol on the metabolism of alanine, glutamic acid and proline in rat liver, *Biochem. Pharmacol.,* 24, 199, 1975.

123. **Feinman, L. and Lieber, C. S.,** Hepatic collagen metabolism: effects of alcohol consumption in rats and baboons, *Science,* 176, 795, 1972.

124. **Zern, M. A., Leo, M., Giambrone, M.-A., and Lieber, C. S.,** Increased type-1 procollagen mRNA levels and *in vitro* protein synthesis in the baboon model of chronic alcoholic liver disease, *Gastroenterology,* 89, 1123, 1985.

125. **Savolainen, E.-R., Goldberg, B., Leo, M., Velez, M., and Lieber, C. S.,** Diagnostic value of serum procollagen peptide measurements in alcoholic liver disease, *Alcoholism Clin. Exp. Res.,* 8, 384, 1984.

126. **Henley, K. S., Laughrey, E. G., Appleman, H. D., and Flecker, K.,** Effect of ethanol on collagen formation in dietary cirrhosis in the rat, *Gastroenterology,* 72, 502, 1977.
127. **Mak, K. M. and Lieber, C. S.,** Alterations in endothelial fenestrations in liver sinusoids of baboons fed alcohol: a scanning electron-microscopic study, *Hepatology,* 4, 386, 1984.
128. **Miyakawa, H., Iida, S., Leo, M. A., Greenstein, R., Zimmon, D., and Lieber, C. S.,** Pathogenesis of precirrhotic portal hypertension in alcohol-fed baboons, *Gastroenterology,* 88, 143, 1985.

Chapter 2

# STRUCTURAL CHANGES IN LIVER CAUSED BY ETHANOL

**Robin Fraser**

## TABLE OF CONTENTS

# I. INTRODUCTION

Alcoholic liver disease is common in Western society.[1] Its least dangerous and reversible manifestation, fatty liver, inevitably follows even one session of heavy drinking.[2-4] In its most dangerous form, alcoholic cirrhosis, it is ultimately fatal. Although this condition occurs in only a small proportion of long-term heavy drinkers, it is one of the leading causes of death in pregeriatric adults.[1-5]

A wide range of structural changes in the liver results from ethanol abuse. These morphological patterns have been grouped into separate entities such as fatty liver, alcoholic hepatitis, and cirrhosis. There has been speculation on the causal relationships between these entities and on the tendency for the less severe states to revert to normal morphology or to lead to the permanently disordered hepatic architecture of cirrhosis and its consequent abnormal function.[2-6]

The normal liver has a histological pattern made up of a multitude of small anatomical units known as lobules[7] or acini.[8] Columns of hepatocytes are separated by a honeycomb of blood sinusoids through which pass the white and red blood corpuscles surrounded by the sinusoidal endothelial cells. A better understanding of the interactions between these different cells and their organelles will help to elucidate the role of alcohol in abnormal liver morphology and function.

The dimensions of the liver lobules are of optimal proportions to ensure the functional efficiency of the hepatocytes or liver parenchymal cells. Portal blood from the stomach, intestines, and spleen containing the products of digestion toxins and bilirubin, as well as oxygen-rich blood from the hepatic artery, enters the lobules from the terminal vessels in the portal tracts. The blood then percolates through the honeycomb of sinusoids between the hepatocytes and leaves the lobule by the postsinusoidal or central venule. The hepatocytes process the materials transported from the sinusoidal blood through the fenestrated sinusoidal endothelium and subjacent plasma-filled space of Disse. Substances synthesized by the hepatocytes are excreted into the space of Disse and than pass back through the fenestrae into the sinusoidal blood. Those substances to be excreted in the bile, such as bilirubin, cholesterol, and bile acids, enter the bile cannaliculi running between the hepatocytes. The cannaliculi carry bile towards the portal tracts where they then join the bile ducts.

Abnormal liver architecture leads to disordered hepatic function as well as causing anoxia, degeneration, and necrosis of the hepatocytes and sinusoidal cells. The latter include the fenestrated endothelial cells, the phagocytic Kupffer cells, and the Ito cells, which are potential fibroblasts.[9,10]

The study of structural changes related to the abuse of ethanol has involved varying techniques, including meticulous studies of liver biopsies,[6] electron microscopy of the various liver cells,[9-12] and separation of their component organelles by ultracentrifugation.[13] The use of probes such as antibodies for structural analysis,[14-17] of cellular separation in the fluorescent activated cell sorter,[18] and of suitable experimental animals[19-23] will all be necessary to unravel the pathogenesis of alcoholic liver disease.

# II. THE MORPHOLOGICAL COMPONENTS OF ALCOHOLIC LIVER DISEASE

## A. Fatty Change

The terms fatty change, steatosis, or fatty liver are now preferred to fatty degeneration and fatty infiltration, which have as yet unproven etiological implications. Alcohol abuse is only one of a multitude of insults leading to fatty change. Other factors are poisoning by substances such as carbon tetrachloride and phosphorus; sensitivity to tetracyclines, steroids, methotrexate, and other drugs; diseases such as diabetes, obesity, malnutrition, venous congestion from heart failure, some apolipoprotein abnormalities, some viral diseases, Reye's syndrome in children, and an occasional complication of pregnancy.[3,6]

Following the ingestion of ethanol, fatty change in the liver is common in both man and animals.[24,25] Macroscopically, the organ is enlarged and, although it is less dense, it is abnormally heavy with its cut surface exhibiting a pale yellowish color and a greasy feel. On microscopic examination, some or all of the hepatocytes are distended by fat droplets. These may be small and numerous, but are often large, with a single droplet taking up most of the cell cytoplasm and pushing the nucleus to one side. Ultrastructurally, the droplets are membrane bound or microvesicular at first, but tend to lose their membranes as they expand or coalesce.[3,6]

The fat-laden hepatocytes are usually arranged in groups which are separated by normal or enlarged nonfatty hepatocytes throughout the liver lobule[3,6,26] (Figure 1). They are present in greatest numbers in the zone around the postsinusoidal or central venule.[3,4,6] However, in severe cases, all of the hepatocytes are distended by fat. This fatty change may vary from mild to severe, thus, ranking grades have been devised.[2,4]

Morphologically, the pathologist cannot be dogmatic about the underlying cause of fatty change, and the clinical history and findings must be taken into account. For example, neither the microvesicles of fat present in hepatocytes in cases of tetracycline sensitivity and in the fatty change associated with pregnancy, or the larger vesicles associated with alcohol abuse, are confined exclusively to these conditions.[3]

It is very rare for the alcohol abuser to die from liver failure following fatty change.[4] The function of the liver remains essentially normal and the morphology reverts to normal after a few days of abstinence.[2,4,6,26] Very occasionally, cholestasis or portal hypertension may occur in the liver following acute alcohol abuse, probably because of enlarged hepatocytes narrowing the bile cannaliculi or hepatic sinusoids.[3,4] The evolution of and recovery from fatty change are usually studied by liver biopsy or animal experimentation.[4,19,21,24,25] However, fatty change of the liver due to alcohol abuse is seen at the coroner's autopsy because of the increased incidence of accidental death associated with the abuse of alcohol.

The pathogenesis of the morphological changes in alcoholic steatosis is not well understood. The concurrent enlargement and "cloudy swelling" of some hepatocytes, as well as the early membrane-bound lipid-containing microvesicles, which may represent swollen smooth endoplasmic reticulum, suggest that alcohol hinders the export of newly synthesized triglycerides.[2,6] A known example of fatty liver produced by the inhibition of excretion of newly synthesized triglyceride in the form of very low density lipoproteins is seen in the orotic acid-deficient rat.[27]

In the chronic alcohol abuser, drinking heavily for weeks at a time, the composition of fat in the hepatocytes resembles recently ingested triglyceride rather than that stored in adipose tissue.[24,25] This suggests that hepatocytes are swamped by dietary lipid in the form of the relatively large lipoproteins of dietary origin, the chylomicrons or their remnants. An increased cholesterol content of chylomicrons and their remnants also leads to a greater uptake of triglyceride by the liver.[28] These large circulating lipoproteins are usually separated from the hepatocytes by the barrier of an intact, but fenestrated, sinusoidal endothelium. Increased endothelial porosity occurs with continued drinking of alcohol due to dilation of the fenestrae.[19,21] This hypothesis will be expanded in a later section discussing the effect of alcohol on the sinusoidal cells.

The important question of the transition from the relatively innocuous fatty change of acute alcohol abuse to the more serious abnormalities of irreversible alcoholic liver disease following prolonged abuse will now be examined. This question is important clinically as it will help to pinpoint patients at risk of permanent liver damage from continued drinking. No adequate biochemical test has yet been found to make this differentiation, hence, the tell-tale morphological signs seen in liver biopsy are still the best indicators for assessing future risk.[4,6]

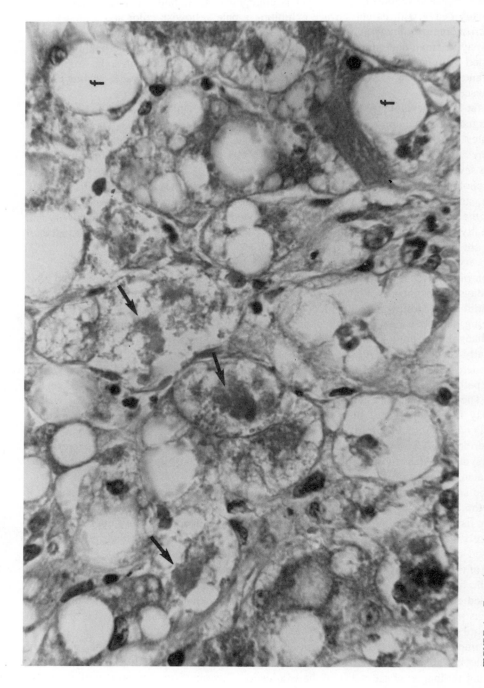

FIGURE 1.   Fatty change of some hepatocytes (f) and, in addition, degenerate swollen hepatocytes containing Mallory bodies (arrows) indicative of human alcoholic hepatitis. (Original magnification × 140.)

## B. Lipogranulomas

After cell death, fatty droplets coalesce to form localized areas of chronic inflammation and fibrosis, but these lipogranulomas do not lead to the widespread fibrosis and disordered architecture of cirrhosis.[4,6] Lipogranulomas have often been observed in second-generation rats fed alcohol throughout life (Figure 2). True cirrhosis, however, has never been noted in these animals.[72]

## C. Central or Hepatic Vein Sclerosis

Until recently, alcohol hepatitis, with widespread foci of necrosis and inflammation, was considered a necessary step between fatty liver and cirrhosis.[3,4,6] However, meticulous studies of liver biopsies from baboons fed alcohol over many years have failed to confirm this. Instead, they have shown a creeping sclerosis or fibrosis around the central postsinusoidal venules of the lobules, often associated with venous thrombosis and organization, and eventually leading to a bridging fibrosis across and between the lobules.[23,29] Rabbits in a recent pilot experiment conducted for 3 years in the author's laboratory developed cirrhosis and occasional venous thrombi. During this period 10% ethanol was added to their drinking water, and cholesterol to their food over two separate months during their first year of life.[20] Several studies on liver biopsies from human alcohol abusers show central lobular or acinar zone 3 venous sclerosis and thrombosis. These histological findings may be important in the prediction of irreversible alcoholic cirrhosis.[3,6,30-35]

## D. Alcoholic Hepatitis

Although it is potentially reversible, this inflammatory condition is thought by many authorities to be the essential prerequisite for the formation of irreversible cirrhosis in humans.[3,6] Morphologically, the liver affected by alcoholic hepatitis undergoes a series of changes which include degeneration and necrosis of hepatocytes associated with acute inflammation, as evidenced by surrounding neutrophils. Intracellular or alcoholic hyaline inclusions within the degenerate hepatocytes can be seen by light microscopy as eosinophilic glassy or hyaline bodies within the cytoplasm (Figure 1). Other associated changes may include the formation of lipogranulomas, noninflammatory acidophilic necrosis of hepatocytes or apoptosis, cholestasis, lymphocyte infiltration, Kupffer cell proliferation, and fibrosis.[3,4,6]

## E. Mallory Bodies

Alcoholic hyalin was originally described by Mallory in 1911 as an irregular, coarse, hyaline meshwork, which stains deeply with eosin and which is present in the cytoplasm of liver cells of alcoholic patients.[11] These relatively large, irregular bodies can sometimes be distinguished by light microscopy from smaller, round hyaline bodies which electron microscopy has revealed as giant mitochondria. The latter have been subdivided into several types, depending on their ultrastructure and enzyme histochemistry. Although rarely present in the same cells as Mallory bodies, they are frequently seen in neighboring hepatocytes.[11,12]

Mallory bodies with their tangled fibrillary ultrastructure (Figure 3) are not peculiar to hepatocytes damaged by alcohol. They are also seen where degeneration is due to other causes including primary biliary cirrhosis, Wilson's disease, Indian childhood cirrhosis, intestinal bypass, diabetes mellitus, long-standing bile duct obstruction, perhexiline maleate ingestion, and abetalipoproteinemia.[15,17,36,37] In experiments with mice in which Mallory bodies have been produced by microtubular poisons such as colchicine or griseofulvin, it is uncertain whether they represent tangles of microtubules or intermediate filaments.[14,15,17,38,39] Various subtypes of alcoholic hyalin have been described, depending on the thickness of their filaments. It has been suggested that Mallory bodies represent a pathological form of keratinization in hepatocytes, brought about by vitamin A deficiency associated with alcoholism.[6]

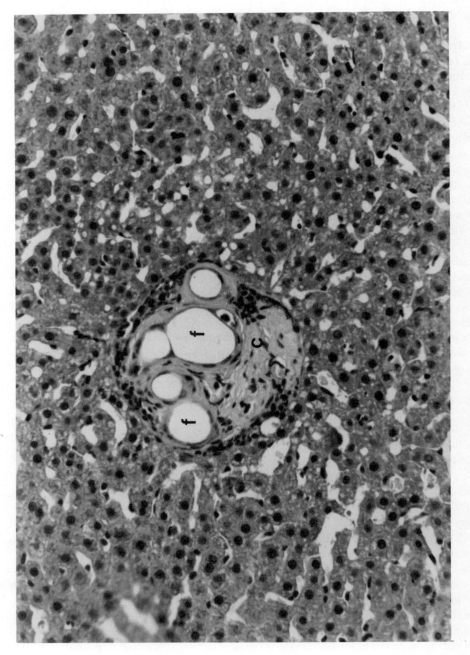

FIGURE 2. Localized lipogranuloma within a rat liver showing large fatty droplets (f), surrounded by collagen (c). (Original magnification × 40.)

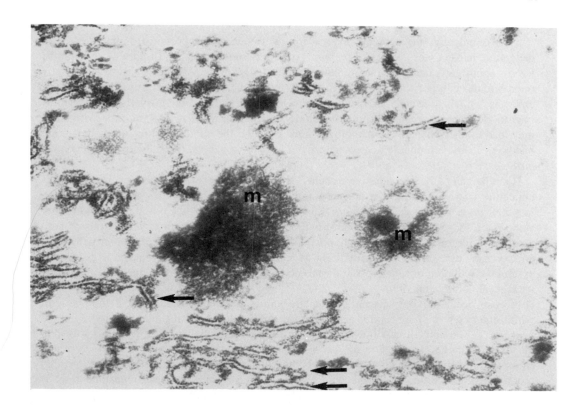

FIGURE 3.    Transmission electron micrograph of fibrillary tangles of Mallory bodies (m) and granular endoplasmic reticulum (arrows) within a human hepatocyte from the same tissue as depicted in Figure 1. (Original magnification × 16,750.)

### F. Alcoholic Cirrhosis

Cirrhosis is an irreversible disease of the liver in which, as defined by the World Health Organization, the architecture throughout the entire organ is disordered by parenchymal or hepatocytic necrosis, regeneration and repair, or fibrosis.[40] Necrosis is not necessarily a continual process and, due to the reserve capacity of the liver, no biochemical abnormality or dysfunction may be apparent at any given time. Currently, biopsy is the only method of diagnosis.[4,6,41]

### 1. Occurrence

In Western societies such as North America, where the sale and consumption of alcohol are an integral part of social and economical life, about two thirds of the cases of cirrhosis are alcohol related. In regions of Asia, however, where alcohol consumption is low, only about one case in ten is related to alcohol abuse, and the majority follow viral hepatitis.[42] Cirrhosis only develops in a proportion of heavy drinkers, affecting up to one third of those who abuse alcohol over a long period.[2,3] The daily limit of alcohol consumption above which cirrhosis can occur is being lowered. Although some authorities believe that cirrhosis will occur in 20% of alcohol abusers drinking daily the equivalent of 120 to 180 g of pure ethanol for 15 years (15 whisky-pint years),[4] others have found that a threshold of as little as 50 g/d for 10 years is sufficient for 15% of abusers to develop cirrhosis.[43] These figures depend, of course, on the abuser's weight and sex, since it has been shown recently that women are particularly susceptible to hepatic damage from alcohol.[6,41]

## 2. Structural Changes

Macroscopically, the cirrhotic liver is irregular, due to the presence of nodules of re-generative hepatocytes, separated by bands of scar tissue or collagen (Figure 4). At first the liver is abnormally heavy and large, but may eventually shrink. Alcoholic cirrhosis is usually micronodular in type, with nodules less than 5 mm in diameter, in contrast with the postviral cirrhosis, which is usually macronodular.[40] However, there is much overlap, and in time micronodular cirrhosis may become macronodular. As in fatty liver, a multitude of insults may lead to cirrhosis and the pathologist must consider the clinical history, as well as other histological and immunological evidence, before pronouncing the diagnosis of alcoholic cirrhosis.[40,43]

Histologically, the regenerative nodules of cirrhosis have lost the perfect, repetitive ar-chitecture characteristic of the normal lobules (Figure 5). The nodules are large and have lost their three-dimensional symmetry, and the distance between the portal tracts and central veins is no longer standard. The differing lengths of sinusoids with their columns of hep-atocytes result, on the one hand, in hypoxia, degeneration, and necrosis of some hepatocytes, and, on the other, in the short circuiting of hepatocytes and the exit from the liver into the systemic circulation of unprocessed portal blood. In this way, abnormal hepatic architecture leads to abnormal function.

Fibrosis occurs not only around the hepatic venule, but under the sinusoidal endothelium, within the space of Disse, and it also bridges from one lobule or regenerative nodule to the next.[30,31,33] The portal tracts thicken, with an increase in collagen and proliferation of bile ducts. This marked increase in hepatic collagen further alters the hepatic architecture and sinusoidal dimensions, leading to increased vascular resistance within the liver and to high blood pressure within the portal vein.[44,45] Hemorrhage from the gastrointestinal tract is common in portal hypertension and the resultant hypotension and release of ammonia from bacterial breakdown of blood in the intestine leads to hepatic failure in the already com-promised cirrhotic liver.

## 3. Associated Risk Factors

Other compounding factors associated with alcoholic cirrhosis are under investigation, since only some heavy drinkers develop cirrhosis, irrespective of the alcoholic beverage consumed.[46] These factors include concurrent hepatic insults such as viral hepatitis,[47,48] effects of autoimmunity,[16,49-52] genetic abnormalities such as hemochromatosis and $\alpha_1$ an-titrypsin deficiency,[3] and the lack of vitamin A and other nutritional substances,[53] and the combination of alcohol with other toxins such as carbon tetrachloride.[54] These studies involve not only clinicians and pathologists, but epidemiologists and experimental pathologists. So far, in studies with experimental animals, only the baboon and, more recently, the cholesterol-fed rabbit have developed alcoholic cirrhosis.[20,21]

If the patient with alcoholic cirrhosis abstains from alcohol, survival may continue for some years with a compromised but compensated liver. However, continuing alcohol abuse will lead to early death from hemorrhage, hepatic failure, ascites, septicemia, or endotox-emia. Unfortunately, those who do abstain and who live for over 5 years have an increased risk of developing primary liver cancer, especially if associated with a past infection of hepatitis B.[55-57]

## III. SINUSOIDAL CELLS AND ETHANOL

The study of the ultrastructure of sinusoidal cells has gained impetus since Wisse introduced perfusion fixation of liver specimens.[58] The three types of sinusoidal cells known to be involved in alcoholic liver disease are the fenestrated endothelial cells, the intrasinusoidal phagocytic Kupffer cells, and the perisinusoidal fat-storing cells of Ito, which are facultative

FIGURE 4. Macroscopic appearance of human liver with micronodular cirrhosis. (Original magnification × 2; scale 1 cm².)

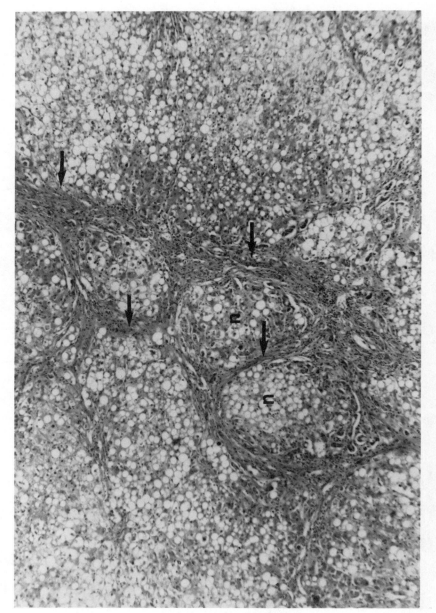

FIGURE 5.    Alcoholic cirrhosis with the disordered architecture resulting from regenerative nodules (n), separated by bands of fibrosis (arrows). Note, also, hyperplasia of bile cannaliculi within the fibrous bands and fatty change of hepatocytes within the nodules. (Original magnification × 35.)

fibroblasts (Figure 6). Interactions between these cells, the hepatocytes, and the circulation play a role in pathological changes produced by alcohol abuse with the liver and other organs.[10]

The "liver sieve" formed by the sinusoidal fenestrated endothelium shields hepatocytes from excess dietary fat, which is transported in the circulation as large chylomicrons until these are depleted of enough triglyceride to enable them to pass through the fenestrae.[9,10,59] It has been shown that alcohol dilates the endothelial fenestrae in rats[19] and baboons,[21] as well as in tissue culture,[60] and this effect is postulated as one factor leading to the accumulation of dietary fat in animals fed alcohol.[19]

It has also been shown that, although the fenestrae dilate, their total number decreases. Long sections of sinusoidal endothelium with a basement membrane are present, but these lack fenestrae.[21,60-63] The exchange of lipoproteins between the circulation and some hepatocytes is therefore hindered and may account for the patchy nature of steatosis often seen in cirrhosis.[10]

Decreased phagocytosis by the Kupffer cells is a direct effect of alcohol abuse; it is also caused when these cells are bypassed as in the disordered architecture of cirrhosis.[64] The resultant decrease in phagocytosis of endotoxins from the intestine, for example, may play a role in the pathogenesis of alcoholic hepatitis and cirrhosis. Those endotoxins entering the systemic circulation may lead to disseminated intravascular coagulation and shock.[10,65]

The Ito cell is currently receiving much attention in the study of alcoholic liver disease.[6,10] Ito cells can be examined and quantified by light microscopy,[66,67] while electron microscopy shows transitions between these cells and fibroblasts.[68-71] In sclerosis around the central venule, Ito cells increase in number, leading to an increase in basement membrane and collagen under the sinusoidal endothelium and within the space of Disse.[33,68] This stimulation of Ito cells may be triggered by a decrease in their vitamin A content, since chylomicron remnants which transport dietary retinol to the liver are blocked by alcoholic defenestration of the sinusoidal endothelium.[63,72]

## IV. CONCLUSION

Although the early and inevitable changes in liver morphology caused by alcohol abuse, such as fatty liver, are of little clinical consequence and are reversible, the later changes and their consequences are less common, but more severe. In the case of alcoholic cirrhosis, they are irreversible. Finally, death is inevitable, as a result of portal hypertension and gastrointestinal hemorrhage, overwhelming hepatic failure, or hepatocellular carcinoma.

The detection of altered histology by liver biopsy is the only certain way, at present, to discern the extent of alcoholic disease. Meticulous studies of biopsy material will allow us to predict which heavy drinkers will progress to the potentially fatal malady of cirrhosis. The correlation of changes in hepatic morphology, as seen at the gross, histological, or ultrastructural level, with changes in biochemistry and function, will improve our understanding of the pathogenesis of alcoholic liver disease.

## ACKNOWLEDGMENTS

I wish to thank the Medical Research Council and National Heart Foundation of New Zealand for supporting my research interest in hepatic morphology, Fredrika Murray for her scientific and secretarial skills, Sue Townsend for critical appraisal of the literary style, and Roger Davis, Tony Day, Vivienne Heslop, and Alastair McGill for preparing the photographs.

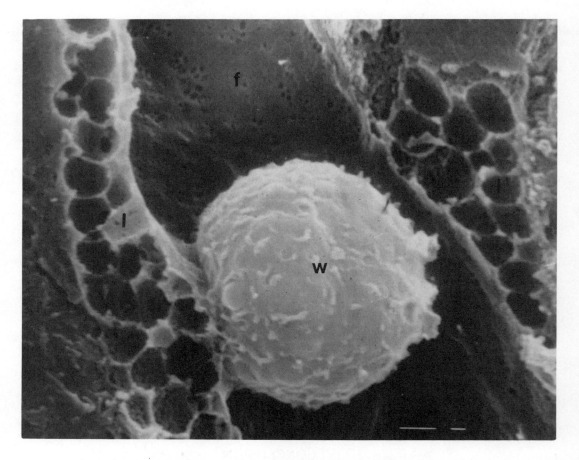

FIGURE 6.    Scanning electron micrograph of a liver sinusoid lined by fenestrated endothelium (f), with perisinusoidal Ito cells (I) in the space of Disse and a white blood cell (W) in the sinusoidal lumen. (Original magnification × 5000.)

# REFERENCES

1. **Saunders, J. B.,** Alcoholic liver disease in the 1980s, *Br. Med. J.,* 287, 1819, 1983.
2. **Rubin, E. and Lieber, C. S.,** Alcohol-induced hepatic injury in nonalcoholic volunteers, *N. Engl. J. Med.,* 278, 869, 1968.
3. **Fleming, K. A. and McGee, J. O'D.,** Review article. Alcohol induced liver disease, *J. Clin. Pathol.,* 37, 721, 1984.
4. **Pimstone, N. R. and French, S. W.,** Alcoholic liver disease, *Med. Clin. North Am.,* 68, 39, 1984.
5. **Orholm, M., Sorenson, T. I. A., Bentsen, K., Hoybye, G., Eghoje, K., and Christoffersen, P.,** Mortality of alcohol abusing men prospectively assessed in relation to history of abuse and degree of liver injury, *Liver,* 5, 253, 1985.
6. **Hall, P.,** Pathology and pathogenesis of alcoholic liver disease, in *Alcoholic Liver Disease,* Hall, P., Ed., Edward Arnold, London, 1985, chap. 2.
7. **Elias, H. and Petty, D.,** Gross anatomy of the blood vessels and bile ducts within the human liver, *Am. J. Anat.,* 90, 59, 1952.
8. **Rappaport, A. M., Boroway, Z. J., Lougheed, W. M., and Lotto, W. N.,** Subdivision of hexagonal liver lobules into a structural and functional unit, *Anat. Rec.,* 119, 11, 1954.
9. **Wisse, E., De Zanger, R. B., Charels, K., Van Der Smissen, P., and McCuskey, R. S.,** The liver sieve: considerations concerning the structure and function of endothelial fenestrae, the sinusoidal wall and the space of Disse, *Hepatology,* 5, 683, 1985.
10. **Fraser, R., Day, W. A., and Fernando, N. S.,** A review: the liver sinusoidal cells. Their role in disorders of the liver, lipoprotein metabolism and atherogenesis, *Pathology,* 18, 5, 1986.

11. **Iseri, O. A. and Gottlieb, L. S.,** Alcoholic hyalin and megamitochondria as separate and distinct entities in liver disease associated with alcoholism, *Gastroenterology,* 60, 1027, 1971.
12. **Uchida, T., Kronborg, I., and Peters, R. L.,** Giant mitochondria in the alcoholic liver diseases — their identification frequency and pathologic significance, *Liver,* 4, 29, 1984.
13. **Lieber, C. S.,** Alcohol metabolism, in *Alcoholic Liver Disease,* Hall, P., Ed., Edward Arnold, London, 1985, chap. 1.
14. **Denk, H., Krepler, R., Lackinger, E., Artlieb, U., and Franke, W. W.,** Immunological and biochemical characterisation of the keratin-related component of Mallory bodies: a pathological pattern of hepatocytic cytokeratins, *Liver,* 2, 165, 1982.
15. **Kimoff, R. J. and Huang, S.-N.,** Immunocytochemical and immunoelectron microscopic studies on Mallory bodies, *Lab. Invest.,* 45, 491, 1981.
16. **MacSween, R. N. M. and Anthony, R. S.,** Immune mechanisms in alcoholic liver disease, in *Alcoholic Liver Disease,* Hall, P., Ed., Edward Arnold, London, 1985, chap. 3.
17. **Peters, M., Tinberg, H. M., and Govindarajan, S.,** Immunocytochemical identity of hepatocellular hyalin in alcoholic and non-alcoholic liver diseases, *Liver,* 2, 361, 1982.
18. **Tanaka, Y., Hirata, R., Minato, Y., Hasumura, Y., and Takeuchi, J.,** Isolation and higher purification of fat-storing cells from rat liver with flow cytometry, in *Cells of the Hepatic Sinusoid,* Kirn, A., Knook, D. L., and Wisse, E., Eds., The Kupffer Cell Foundation, Rijswijk, the Netherlands, 1986, 473.
19. **Fraser, R., Bowler, L. M., and Day, W. A.,** Damage of rat liver sinusoidal endothelium by ethanol, *Pathology,* 12, 371, 1980.
20. **Fraser, R., Fernando, N., Heslop, V., and Joyce, S. L.,** Cirrhosis in the alcoholic rabbit: a new model (abstract), *N.Z. Med. J.,* 99, 247, 1986.
21. **Mak, K. M. and Lieber, C. S.,** Alterations in endothelial fenestrations in liver sinusoids of baboons fed alcohol: a scanning electron microscopic study, *Hepatology,* 4, 386, 1984.
22. **Lindros, K. O., Stowell, L., Vaananen, H., Sipponen, P., Lamminsivu, U., Pikkarainen, P., and Salaspuro, M.,** Uninterrupted prolonged ethanol oxidation as a main pathogenetic factor of alcholic liver damage: evidence from a new liquid diet animal model, *Liver,* 3, 79, 1983.
23. **Popper, H. and Lieber, C. S.,** Histogenesis of alcoholic fibrosis and cirrhosis in the baboon, *Am. J. Pathol.,* 98, 695, 1980.
24. **Lieber, C. S. and Spritz, N.,** Effects of prolonged ethanol intake in man: role of dietary, adipose, and endogenously synthesized fatty acids in the pathogenesis of the alcoholic fatty liver, *J. Clin. Invest.,* 45, 1400, 1966.
25. **Lieber, C. S., Spritz, N., and DeCarli, L. M.,** Role of dietary, adipose, and endogenously synthesized fatty acids in the pathogenesis of the alcoholic fatty liver, *J. Clin. Invest.,* 45, 51, 1966.
26. **Israel, Y., Britton, R. S., and Orrego, H.,** Liver cell enlargement induced by chronic alcohol consumption: studies on its causes and consequences, *Clin. Biochem.,* 15, 189, 1982.
27. **Pottenger, L. A. and Getz, G. S.,** Serum lipoprotein accumulation in the livers of orotic acid-fed rats, *J. Lipid Res.,* 12, 450, 1971.
28. **Vessby, B. and Ontko, J. A.,** Cholesterol content as a possible determinant in the removal of chylomicron triglyceride by the liver, *Atherosclerosis,* 47, 111, 1983.
29. **Nakano, M. and Lieber, C. S.,** Ultrastructure of initial stages of perivenular fibrosis in alcohol-fed baboons, *Am. J. Pathol.,* 106, 145, 1982.
30. **Brunelli, E., Macarri, G., Jezequel, A. M., and Orlandi, F.,** Diagnostic value of the fibrosis of the terminal hepatic venule in fatty liver and chronic hepatitis due to ethanol or other aetiology, *Liver,* 5, 261, 1985.
31. **Burt, A. D. and MacSween, R. N. M.,** Hepatic vein lesions in alcoholic liver disease: retrospective biopsy and necropsy study, *J. Clin. Pathol.,* 39, 63, 1986.
32. **Edmondson, H. A., Peters, R. L., Reynolds T. B., and Kuzma, O. T.,** Sclerosing hyaline necrosis of the liver in the chronic alcoholic, *Ann. Intern. Med.,* 59, 646, 1963.
33. **Horn, T., Junge, J., and Christoffersen, P.,** Early alcoholic liver injury: changes of the Disse space in acinar zone 3, *Liver,* 5, 301, 1985.
34. **Nakano, M., Worner, T. M., and Lieber, C. S.,** Perivenular fibrosis in alcoholic liver injury: ultrastructure and histologic progression, *Gastroenterology,* 83, 777, 1982.
35. **Nasrallah, S. M., Nassar, V. H., and Galambos, J. T.,** Importance of terminal hepatic venule thickening, *Arch. Pathol. Lab. Med.,* 104, 84, 1980.
36. **Hay, D. R. and Gwynne, J. F.,** Cirrhosis of the liver following therapy with perhexiline maleate, *N.Z. Med. J.,* 96, 204, 1983.
37. **Lewis, D., Wainwright, H. C., Kew, M. C., Zwi, S., and Isaacson, C.,** Liver damage associated with perhexiline maleate, *Gut,* 20, 186, 1979.
38. **Irie, T., Benson, N. C., and French, S. W.,** Relationship of Mallory bodies to the cytoskeleton of hepatocytes in griseofulvin-treated mice, *Lab. Invest.,* 47, 336, 1982.

39. **Uchida, T., Kronborg, I., and Peters, R. L.,** Alcoholic hyalin-containing hepatocytes — a characteristic morphologic appearance, *Liver,* 4, 233, 1984.
40. **Anthony, P. P., Ishak, K. G., Nayak, N. C., Poulsen, H. E., Scheuer, P. J., and Sobin, L. H.,** The morphology of cirrhosis, *J. Clin. Pathol.,* 31, 395, 1978.
41. Editorial, Who gets alcoholic cirrhosis?, *Lancet,* 2, 263, 1984.
42. **Lelbach, W. K.,** Cirrhosis in the alcoholic and its relation to the volume of alcohol abuse, *Ann. N.Y. Acad. Sci.,* 252, 85, 1975.
43. **Sorensen, T. I. A., Orholme, M., Bentsen, K. D., Hoybye, G., Eghoje, K., and Christoffersen, P.,** Propsective evaluation of alcohol abuse and alcoholic liver injury in men as predictors of development of cirrhosis, *Lancet,* 2, 241, 1984.
44. **Christensen, U., Sorensen, T. I. A., Jensen, L. I., Aagaard, J. and Burcharth, F.,** The free portal pressure in awake patients with and without cirrhosis of the liver, *Liver,* 3, 147, 1983.
45. **Krogsgaard, K., Gluud, C., Henriksen, J. H., and Christoffersen, P.,** Correlation between liver morphology and haemodynamics in alcoholic liver disease, *Liver,* 5, 173, 1985.
46. **Tuyns, A. J., Esteve, J., and Pequignot, G.,** Ethanol is cirrhogenic, whatever the beverage, *Br. J. Addict.,* 79, 389, 1984.
47. **Feller, A., Uchida, T., and Rakela, J.,** Acute viral hepatitis superimposed on alcoholic liver cirrhosis: clinical and histopathologic features, *Liver,* 5, 239, 1985.
48. **Inoue, K., Kojima, T., Koyata, H., Matsui, S., Aoyama, K., Konda, T., Ichida, T., and Sasaki, H.,** Hepatitis B virus antigen and antibodies in alcoholics, *Liver,* 5, 247, 1985.
49. **Actis, G., Mieli-Vergani, G., Portmann, B., Eddleston, A. L. W. F., Davis, M., and Williams, R.,** Lymphocyte cytotoxicity to autologous hepatocytes in alcoholic liver disease, *Liver,* 3, 8, 1983.
50. **Crapper, R. M., Bhathaland, P. S., and Mackay, I. R.,** Chronic active hepatitis in alcoholic patients, *Liver,* 3, 327, 1983.
51. **Kurki, P., Gripenberg, M., Teppo, A.-M., and Salaspuro, M.,** Profiles of antinuclear antibodies in chronic active hepatitis, primary biliary cirrhosis and alcoholic liver disease, *Liver,* 4, 134, 1984.
52. **Poralla, T., Hutteroth, T. H., and Buschenfelde, K.-H. M. Z.,** Cellular cytotoxicity against autologous hepatocytes in alcoholic liver disease, *Liver,* 4, 117, 1984.
53. **Leo, M. A. and Lieber, C. S.,** Interaction of ethanol with vitamin A, *Alcoholism Clin. Exp. Res.,* 7, 15, 1983.
54. **Zimmerman, H. J.,** Effects of alcohol on other hepatotoxins, *Alcoholism Clin. Exp. Res.,* 10, 3, 1986.
55. **Bassendine, M. F., Seta, L. D., Salmeron, J., Thomas, H. C., and Sherlock, S.,** Incidence of hepatitis B virus infection in alcoholic liver disease, HBsAg negative chronic active liver disease and primary liver cell cancer in Britain, *Liver,* 3, 65, 1983.
56. **Ferenci, P., Dragosics, B., Marosi, L., and Kiss, F.,** Relative incidence of primary liver cancer in cirrhosis in Austria. Etiological considerations, *Liver,* 4, 7, 1984.
57. **Nakanuma, Y. and Ohta, G.,** Morphology of cirrhosis and occurrence of hepatocellular carcinoma in alcoholics with and without HBsAg and in non-alcoholic HBsAg-positive patients. A comparitive autopsy study, *Liver,* 3, 231, 1983.
58. **Wisse, E.,** An electron microscopic study of the fenestrated endothelial lining of rat liver sinusoids, *J. Ultrastruct. Res.,* 31, 125, 1970.
59. **Fraser, R., Bosanquet, A. G., and Day, W. A.,** Filtration of chylomicrons by the liver may influence cholesterol metabolism and atherosclerosis, *Atherosclerosis,* 29, 113, 1978.
60. **Charels, K., De Zanger, R. B., Van Bossuyt, H., Van der Smissen, P., and Wisse, E.,** Influence of acute alcohol administration on endothelial fenestrae of rat livers, in *Cells of the Hepatic Sinusoid,* Kirn, A., Knook, D. L., and Wisse, E., Eds., The Kupffer Cell Foundation, Rijswijk, the Netherlands, 1986, 497.
61. **Fraser, R., Day, W. A., and Wright, P. L.,** Fatty liver and the sinusoidal cells, in *Festschrift for F.C. Courtice,* Garlick, D., Ed., University of New South Wales, Sydney, 1981, 139.
62. **Henriksen, J. H., Horn, T., and Christoffersen, P.,** The blood-lymph barrier in the liver. A review based on morphological and functional concepts of normal and cirrhotic liver, *Liver,* 4, 221, 1984.
63. **Horn, T., Christoffersen, P., and Henricksen, J. H.,** Alcoholic liver injury: defenestration in non cirrhotic livers — a scanning electron miroscopic study, *Hepatology* 7, 77, 1987.
64. **Liehr, H. and Grun, M.,** Clinical aspects of Kupffer cell failure in liver diseases, in *Kupffer Cells and Other Liver Sinusoidal Cells,* Wisse, E. and Knook, D. L., Eds., Elsevier/North-Holland, Amsterdam, 1977, 427.
65. **Nolan, J. P.,** The role of endotoxin in liver injury, *Gastroenterology,* 69, 1346, 1975.
66. **Hall, P., Smith, R. D., and Gormley, B. M.,** Routine stains on osmicated resin embedded hepatic tissue, *Pathology,* 14, 73, 1982.
67. **Bronfenmajer, S., Schaffner, F., and Popper, H.,** Fat-storing cells (lipocytes) in human liver, *Arch. Pathol.,* 82, 447, 1966.

68. **Ballardini, G., Esposti, S. D., Bianchi, F. B., DeGiorgi, L. B., Faccani, A., Biolchini, L., Bushachi, C. A., and Pisi, E.,** Correlation between Ito cells and fibrogenesis in an experimental model of hepatic fibrosis. A sequential stereological study, *Liver,* 3, 58, 1983.

69. **Friedman, S. L., Roll, F. J., Boyles, J., and Bissell, D. M.,** Hepatic lipocytes: the principal collagen-producing cells of normal rat liver, *Proc. Natl. Acad. Sci. U.S.A.,* 82, 8681, 1985.

70. **Mak, K. M., Leo, M. A., and Lieber, C. S.,** Alcoholic liver injury in baboons: transformation of lipocytes to transitional cells, *Gastroenterology,* 87, 188, 1984.

71. **James, O'D., McGee, M. B., and Patrick, R. S.,** The role of perisinusoidal cells in hepatic fibrogenesis, *Lab. Invest.,* 26, 429, 1972.

72. **Fraser, R.,** unpublished observations.

Chapter 3

THE EFFECTS OF ETHANOL AND ITS METABOLISM ON THE BRAIN

**Joseph N. Santamaria**

TABLE OF CONTENTS

# I. INTRODUCTION

Twenty years ago, alcohol-related brain damage seemed to be a clearly defined entity with a variety of clinical syndromes, neatly categorized according to the major impact of some desctructive process. Wernicke's encephalopathy and Korsakoff's psychosis were described in the late 19th century.[1,2] To these were added cerebellar ataxia, central pontine myelinolysis, Marchiafava-Bignami syndrome, amblyopia (alcohol and tobacco), dementia, and various confusional states.[3]

Efforts were made to determine pathogenesis. It was postulated that neural damage could be caused by the direct action of alcohol (ethyl alcohol or ethanol), by its major metabolite, acetaldehyde, by nutritional deficiency, head trauma, severe infections, bouts of anoxia, genetic abnormalities, and by alteration to the passage of blood in smaller blood vessels. Some changes were thought to be due to metabolic disturbances such as those associated with severe withdrawal or due to advanced liver disease.[4]

In more recent years, the whole subject has been extensively reviewed[5] and many studies are now being conducted in various parts of the world. The literature is dispersed throughout many journals of various disciplines — neurology, neuropsychology, neuropathology, radiology, and related sciences, using newer techniques of investigations. The information is bewildering and its interpretation is fraught with danger.

There are major changes occurring in the investigative sciences with the introduction of computerized axial tomography (CAT), auditory-, visual-, and somatosensory-evoked responses, nuclear magnetic resonance, positron emission tomography, cognitive evoked potentials, and stimulus-specific evoked potentials. To these must be added the various batteries of psychometric tests. With many of these measuring devices there are problems of determining normal values as well as the interpretation of the results. Moreover, studies need to be of a longitudinal nature to determine the effect of such variables as continued drinking, abstinence, age, and the reversibility of effects. Given these developments, the earlier descriptions of brain damage are undergoing a major reappraisal while the pathogenesis and prognosis are still being debated. The spectrum of brain damage in heavy drinkers has changed so that there is a greater heterogeneity in clinical syndromes, and the classical presentations of the Wernicke's and Korsakoff's syndromes are less frequently seen.

Alcohol-related brain damage is a term deeply entrenched in the medical literature. Rarely is it defined, but it is usually meant to describe those neurological and psychobehavioral disorders seen in chronic heavy drinkers. Because of recent studies which have detected cerebral abnormalities in the absence of clinical findings, a new term has been introduced, that of cerebral dysfunction. This has the advantage of covering those disorders which seem to recover over time, even those which take longer to recover than one would expect if the effects were simply pharmacological. Moreover, the neuropsychological and CAT scan abnormalities often found in so-called "intact alcoholics"[6] have raised new questions about the effects of drinking on cerebral function and on neural tissue. This chapter will briefly survey the acute effects of alcohol on the brain, but will not cover the phenomena of tolerance and dependence. Then will follow a short presentation of the classical clinical syndromes which result from chronic alcohol intake. The remainder of the chapter will be devoted to a review of more recent knowledge, particularly in relation to modern investigative techniques and the various theories about pathogenesis and prognosis.

# II. ACUTE EFFECTS

Ethyl alcohol or ethanol is a depressant of the central nervous system, being closely related to the volatile anesthetics. In higher doses it possesses analgesic properties, but the dose required for surgical anesthesia carries the real danger of failure of the respiratory center.[7]

Ethanol is mainly used as a recreational drug with significant mood-altering properties. In lower doses the main effects on the central nervous system are those on mood and behavior. It is often difficult to determine how alcohol will act in these lower doses as environmental factors and premorbid states play a significant role in mood changes. Whether a person becomes morose, vivacious, or drowsy is often determined by the social situation, the reasons for drinking, and the expectations of the drinker.[8]

As the blood alcohol level begins to rise a process of behavioral disinhibition becomes manifest. In social company, the person becomes less self-conscious and more talkative with a tendency to lose emotional control and discretion. Between 0.03 and 0.08 g%, judgment becomes impaired and there is a failure to appreciate the significance of events. Vision is affected so that visual acuity diminishes and tunnel vision develops as the field of vision narrows.[9]

Cognitive functions begin to deteriorate so that thinking becomes difficult, concentration is impaired, together with learning and remembering. Judgment becomes faulty and there is a lack of self-awareness so that a false self-confidence emerges. Between 0.05 and 0.08 g%, muscle control becomes affected as reaction time is slowed and a mild degree of incoordination can be detected.[7] These relatively early signs affecting cognition, judgment, concentration, coordination, and vision result in impairment of driving and other skills and the earliest manifestations appear at quite low blood alcohol levels.[10]

As the blood alcohol concentration (BAC) rises, these effects are exaggerated and new features emerge. Nystagmus becomes apparent, usually about 0.1 g%, and incoordination of movements becomes more prominent. There is difficulty with articulation with evidence of dysphoria in performing fine movements and skilled operations. The features of ataxia emerge. Behavior can be highly variable, depending on the social circumstances, personality, and the nature of the external stimuli. It can vary widely from vivacity to moroseness, from maudlin conduct to violent actions, from increasing apathy to deepening somnolence.[7] As intoxication progresses, drowsiness gives way to sleep and then to stupor. Deepening coma may lead to respiratory paralysis and death can result. These latter features usually develop when the blood alcohol level exceeds 0.3 g% and the danger of death from respiratory failure is likely to occur when the BAC approaches or exceeds 0.5 g%.[7]

A state of "pathological" intoxication has also been described. This is thought to be idiosyncratic and short lived and may be induced by relatively small doses of alcohol. Behavior tends to be violent and there may be delusions and hallucinations. The person is likely to fall into a deep sleep and on awakening has no recollection of the event. In such cases, there may be a history of previous head injury or familial epilepsy. There is, however, conflicting evidence about its true existence and its nature.[11]

## III. CLASSICAL CLINICAL SYNDROMES

These syndromes arise in a relatively small percentage of chronic alcoholics (there may be genetic factors influencing their occurrence — see Volume I, Chapter 10) and are indicative of advanced stages of brain damage.

### A. Korsakoff's Syndrome

The main psychological features of Korsakoff's syndrome (or Korsakoff's psychosis) are

1.  There is poor memory for recent events. The patient has great difficulty in acquiring any new information irrespective of its nature and irrespective of how it may be presented (e.g., auditory, visual, or tactile modalities).
2.  The immediate recall is relatively intact so that new information is immediately registered. The patient will be able to repeat a series of six or seven digits without any

difficulty, but is unable to retain the material beyond the immediate span of time and, consequently, the information is rapidly lost even after a delay of only a few seconds.

3. There is evidence of retrograde amnesia and the patient will have great difficulty in narrating in a coherent fashion details and events from the past. This is associated with a loss of the chronology of events so that past events may be recalled, but without appreciation of their proper time frame.

4. The patient is often unaware of the memory disorder.

5. There may be confabulation. However, confabulation is not an invariable component of the syndrome and may be secondary to memory difficulties, with actual past events being projected into a recent setting and often mixed with confabulated events to give a touch of plausibility to the patient's story.[12]

This form of amnesia seems to be associated with the bilateral destruction of certain structures close to the central axis of the brain, so that the condition is sometimes spoken of as "axial amnesia".[13] The cerebral structures involved include the hippocampus, the fornix, the mammillary bodies, and certain thalamic nuclei which form part of what is known as the Circuit of Papez. Barbizet referred to this system as constituting a midline or axial functional structure intervening in the memory process.[14] The amnesic disorder dominates the syndrome. There is usually well-preserved general intellectual ability as measured by the standard tests of the Wechsler Adult Intelligence Scale (WAIS). The scores are often towards the lower range of intellectual functioning which is a feature of axial amnesia, irrespective of its casuality.

Very often other abnormalities are detected in patients with this syndrome, especially abnormalities of frontal lobe activity. There is frequently evidence of frontal lobe atrophy, especially in the dorsolateral areas.[15]

The Korsakoff syndrome often emerges from the characteristic acute syndrome of Wernicke's encephalopathy. However, many patients with the typical amnesic disorder described above have no clear history of a previous Wernicke's encephalopathy, although the pathological findings at post-mortem show lesions in the same areas of the brain as are found in patients dying from Wernicke's encephalopathy. Some workers have described a gradual evolution of the amnesic syndrome. Cutting described differences between what he called acute and chronic groups. The most striking difference between them was that the verbal and performance IQ was well preserved in the patients with the acute form of the disorder.[16]

### B. Wernicke's Encephalopathy

This syndrome was first described in 1881 by Wernicke in three patients, two of whom were alcoholics.[1] Originally, a clinical triad was recognized, consisting of disturbed consciousness and confusion, abnormalities of ocular movements, and ataxia. There is a global confusion and patients are unable to concentrate and are usually apathetic. Drowsiness is usually present, but patients are nearly always rousable. The abnormalities of ocular movements are variable, with bilateral paralysis of the sixth cranial nerves, nystagmus, and gaze palsy being common. Occasionally, there may be a complete ophthalmoplegia. Finally, there is a staggering gait often associated with peripheral neuropathy.

In recent years, recorded cases have tended not to be so clinically characteristic and a heterogeneity of features has been described.[17]

### C. Cerebellar Atrophy

Clinical evidence of damage to the cerebellum is occasionally seen in heavy drinkers, manifesting itself as a cerebellar ataxia. It is associated with degenerative changes in the anterior vermis and the adjacent paramedian areas of the cerebellum. The clinical disorder may develop rapidly, but often there is a gradual deterioration over years.[3] There is often evidence of other cerebral defects, both on the CAT scan and on psychometric testing.

### D. Central Pontine Myelinolysis

This condition is associated with the widespread demyelination of nerve fibers about the ventral part of the pons. The disorder usually appears acutely with progressive involvement of the medulla. It is nearly always fatal, although cases of recovery have been reported. It is thought to be caused by thiamin deficiency and may be precipitated by the unwise rapid correction of an associated hyponatremia.[3]

### E. Marchiafava Bignami Syndrome

The Marchiafava Bignami syndrome is thought to be due to widespread demyelination of the corpus callosum with an extension of the lesion into adjacent areas. It was originally described in heavy consumers of red wine, but it may occur with other types of alcohol. It is associated with coma and fits and usually has a fatal outcome. Very occasionally a progressive frontal dementia with a generalized spasticity dominates the clinical picture.[3]

## IV. METHODS OF CLINICAL ASSESSMENT AND RESULTS

### A. Neuropathology

Evidence of brain damage has been demonstrated by the post-mortem examination of the brains of deceased alcoholics. In 1955 Courville described cortical atrophy with diffuse neuronal loss in chronic alcoholics and drew attention to the damage found in the dorsolateral convolutions of the frontal lobes. These findings were confined to the frontal lobes.[15]

In their classical study of the Wernicke-Korsakoff syndrome, Victor et al.[18] described the vascular congestion and hemorrhages around the third and fourth ventricles, with foci of damage in the thalamus, the mammillary bodies, and the periventricular area around the brain stem. In the acute phase there was little evidence of neuronal loss, but this was a prominent feature in chronic disease. These authors also noted that cortical atrophy was present in 25% of their cases.[18] More recently, Harper reported on a study at the Royal Perth Hospital in Western Australia.[19,20] He examined 4677 brains at post-mortem and the typical features of Wernicke-Korsakoff syndrome were found in 131 cases (2.8%). Of these, 90% were associated with alcohol abuse, but only 20% had a clinical diagnosis of the syndrome(s) before death. Other studies have been conducted, particularly in rodents. Irle and Markowitsch demonstrated cerebral lesions in rats given a prolonged intake of alcohol which were similar to those found in chronic alcoholics.[21]

### B. Pneumoencephalography (PEG)

PEG studies in chronic alcoholics have been widely reported since the early 1950s. Tumarin et al. studied seven young alcoholics and found cortical atrophy and ventricular enlargement in four cases.[22] The work of Lafon et al.[23] Lereboullet et al.[24] Riboldi and Garavaglia,[25] Ferrer et al.,[26] Brewer and Perrett,[27] and many others confirmed Tumarin's findings and revealed a high incidence of cerebral damage. Ohara and Homma reported on the close correlation between the aggregate dose of alcohol and the degree of cerebral damage.[28] Brewer and Perrett found a very high prevalence of brain damage in their series (over 90%) and commented that even heavy social drinkers were at risk.[27]

### C. Computerized Axial Tomography (CAT Scans)

The CAT scan has largely superseded PEG and electroencephalography in the investigation of cerebral damage in heavy drinkers. As an investigative tool it has added greatly to our knowledge by revealing the frequent reversibility of apparent damage and the dissociation that may occur between the CAT scan features and the clinical findings. There are several problems that need to be resolved, especially when attempts are made to compare the results of different research centers. There are no universal standards of methodology of measuring changes and employing criteria of definitive change.[29] However, it remains a very valuable noninvasive method of investigating an organ which is otherwise inaccessible in the human subject, and it allows for correlations to be made with other measures (psychometric tests) and other variables such as age, consumption of alcohol, head injury, and sex.

Fox et al. presented some early findings which revealed a significant incidence of ventricular enlargement in chronic alcoholics.[30] This was confirmed by Bergman et al.[31] who also described widened frontal and parietal sulci in over 60% of cases. Bergman et al. observed that cortical atrophy was more common than ventricular enlargement and they seemed to be independent of each other.[31] Carlen et al., in a series of studies (1976, 1978, and 1980), reported similar findings, but their work and that of others raised the problem of the confounding variable of age.[32-34] This has drawn attention to the urgent need to develop normative CAT scan data for various age groups, as well as the determination of when to perform the test after abstinence has been achieved. Lishman et al. studied 100 consecutive male admissions to an alcoholism treatment unit.[35] The scans were performed after at least 2 weeks of abstinence and the mean age was 43.5 years. The radiological findings in the alcoholic group were age related, but the alleged dose consumed did not correlate significantly with the severity of the CAT scan abnormalities. Lishman et al. also commented that in a follow-up study of 23 patients, 9 remained abstinent while 14 continued to drink. In four of the abstinent group, there was evidence of improvement in the CAT scan, but no improvement was observed in those who continued to drink. Carlen et al.[34] reported similar findings as did Cala et al.[36] Carlen's group studied 52 recently abstinent chronic alcoholics with CAT scans (and psychometric tests) and repeated the scans on 20.[37] They observed that reversible changes may be manifest soon after the cessation of drinking and that the degree of reversibility was greater in those who remained abstinent between the scans. This relationship between reversibility of damage and abstinence is now widely reported.

The significance of minor degrees of cerebral atrophy is unclear and it may relate to changes in interstitial fluid volume rather than cellular components. Effective regeneration of neurons in the central nervous system is not known to occur in any disease. Therefore, recovery from "atrophy" is unlikely to represent neuronal regeneration.

Of considerable interest are the reports of abnormal CAT scans in young alcoholics. Lee et al. found a higher than expected incidence of cerebral atrophy in young males aged 21 to 35.[38] Similar findings were reported by Bergman et al.[39] They found significant CAT scan changes in 46% of young alcoholics aged 20 to 29 and in 62% of those aged 30 to 39.

Most CAT studies in humans over the last 8 to 10 years have combined neuropsychological measurements and occasionally nutritional assessments. The CAT scan findings, especially in younger people, have raised such issues as premorbid risk factors and the possibility that other organic disorders, such as liver disease, may affect cerebral function. Before discussing further significance of the changes observed on CAT scans, particularly the question of confounding variables, it is pertinent to review the present state of knowledge derived from psychometric tests.

## D. Neuropsychometric Tests

Over the last 20 years, neuropsychologists have become deeply interested in alcohol-related cerebral dysfunction. Much of the earlier work was concerned with the phenomenon of amnesia, especially in Korsakoff's syndrome, but greater attention is now being given to measures of cognitive and intellectual functioning. There is a plethora of material being published in this field, often being correlated with other tests of cerebral damage. Although some early studies were reported in the 1950s and 1960s,[39,40] the work in this field accelerated in the 1970s and a great volume of information has appeared in the present decade. Brewer and Perrett[27] correlated neuropsychological assessment with other indices of brain damage, while Horvath[41] reported on a similar study on 1100 consecutive alcoholics presenting to an alcoholism clinic in Melbourne, Australia. In Horvath's paper, it was reported that the "demented" patients had longer drinking histories and higher daily alcohol consumption than nondemented alcoholics, and he also found a sex disparity of three females to one male. Horvath's special study of 100 patients with brain damage revealed that a broad range of changes may occur, including cortical lesions involving the frontal, parietal, and temporal lobes, the cerebellum, as well as the subcortical axial structures. The neuropsychologists have devised "batteries" of tests to measure intellectual capacity, cognitive abilities, perceptual and adaptive responses, as well as the components of memory and motor function. The most frequently used psychometric tests are known as the Halstead-Reitan Neuropsychological Battery, including the extended measurements.[42] Also widely used are the WAIS, the Wechsler Memory Scale, and objective measurements of personality, for example, the Minnesota Multiphasic Personality Inventory (MMPI).

### 1. Intellectual Capacity — Cognitive Skills

In general, alcoholics have not shown a global impairment of intellect and intelligence, although they often tend to be at the lower limits of the normal range.[43] However, they have been shown to have performance deficits on several cognitive tasks. Verbal abstracting difficulties have been regularly observed.[40,44,45] Visuospatial function is also impaired as are some verbal abstracting tasks.[46]

In 1977, Parker and Noble observed that decrements of higher mental functions could be demonstrated and there was a significant relationship between such cerebral dysfunction and the amount of alcohol consumed per drinking occasion.[47] Of particular importance was the "decrement in adaptive abilities, concept formation, and the capacity to shift from one idea to another." This decrement in the ability to adapt to change means that the alcoholic may function well if all that he is required to do is to use long-established skills and call upon consolidated stores of old memories. However, he is likely to flounder when called upon to acquire new skills or perform in new situations.

### 2. Psychomotor Performance

Simple motor skills tend to remain intact, but impairment of response becomes evident when the motor tasks are more complex and require greater mental involvement.[48] Walsh reported that with an increase in drinking there is a continuing decrease in the ability to perform tasks requiring speed — perceptuomotor retardation.

### 3. Memory

A great deal of study has attempted to unravel the nature of memory deficients in chronic alcoholics. Molloy (1976) demonstrated that in patients with Korsakoff's syndrome there was a marked reduction in the memory quotient.[12] For the chronic alcoholic without the severe amnesic features of the Korsakoff's syndrome, there was a small lowering of memory relative to intelligence (quoted by Walsh[13]). Molloy[12] noted that the Korsakoff syndrome could be categorized by several outstanding features:

1. Relatively preserved general intelligence ability
2. Relatively intact immediate memory
3. Anterograde amnesia, leading to poor scores on the Wechsler Memory Scale
4. Retrograde amnesia with a loss of chronological ordering of events

However, other incapacities were also noted on visuoperceptual tasks, complex problem solving, abstraction, and concept formation.

## V. DEGREE AND PROGRESSION OF CEREBRAL DYSFUNCTION

It is apparent that alcoholics and even heavier "social" drinkers reveal levels of cerebral dysfunction that range from such gross disorders as Korsakoff's syndrome or advanced dementia to the normal functioning of young nonalcoholics. Such findings have led to two hypotheses:

1.    A continuum of impairment or the progression model
2.    A process of accelerated aging

The continuum hypothesis was elaborated by Grant et al. in an article entitled: "The Natural History of Alcohol and Drug Related Brain Disorder".[49] From the time of starting to use the drug, there occurs a latency period of indefinite duration, in part determined by the frequency of ingestion of the drug and the actual doses consumed. During this period, even sensitive psychometric tests may not reveal any significant impairment. As the drinking progresses over time, the tests will detect progressive impairment of cognitive ability to a point of severe disablement.

Grant et al. studied a group of alcoholics who on average began heavy drinking in their mid 20s, and they found evidence of cerebral impairment around the age of 40.[49] When these changes appeared, the disorder was described as preclinical. Later, when the degree of impairment was more marked and persistent the clinical stage had been reached. Tarter reported that the severity of the cognitive defects was closely related to the drinking history.[50] Alcoholics who had been drinking excessively for 10 years or more showed greater cognitive incapacity than did alcoholics with a shorter history of alcohol abuse.

While this theory has heuristic value, it suffers from certain problems. Evidence is accumulating that many cerebral deficits are reversible, whether by abstinence or improved nutrition or both. Cala et al. have shown that there is no easy separation between acute and chronic effects, and between reversible and irreversible damage.[36] More recently, Adams and Grant have drawn attention to the importance of premorbid risk factors on the neuropsychological performance of alcoholics.[51] They found that recently detoxified alcoholics (sober 1 month) with a positive premorbid risk history had worse neuropsychological performance than did those without antecedent risk events.

The second hypothesis of accelerated or premature aging is based on the striking similarities that exist between the memory deficits of older people and those found in much younger alcoholics. The description of "senile dementia" by Adams and Victor lends further credence to this hypothesis.[52] However, again there is the problem of the reversibility of the CAT scan findings and of neuropsychological deficits in many abstinent alcoholics. Moreover, a recent study by Brown et al. showed that the Randt Memory Test differentiated the effects of aging from the pattern of memory disturbances seen in Korsakoff's syndrome.[53] Bergman et al. concluded that learning and memory deficits are due to central subcortical brain damage, while cortical changes contribute to the impairments detected in the Halstead-Reitan Battery and the Halstead Impairment Index.[39]

The assessment of cerebral function is currently a watershed. The CAT scan and neuropsychological tests have added greatly to our knowledge about the association between heavy drinking and cerebral dysfunction. The CAT scans have revealed how extensive the changes may be, with cortical atrophy involving particularly the frontal and parietal regions as well as central subcortical structures. Often these objective changes are reversible with prolonged abstinence, but even when present they may be disproportionate to the clinical picture. Molloy has drawn attention to the need to develop new perspectives in the study of cognitive defects in alcoholics.[12] She alludes to the faculties of attention and concentration which are critically important in the performance of many neuropsychological tests. These faculties can be affected by mood, training, educational achievements, and extraneous factors such as previous head injury and the use of addictive-type drugs. These confounding variables may be partly overcome by the recent introduction of the microcomputer into the field of psychometric studies.

The newer techniques of investigation are very largely experimental tools and it is presently not possible to comment on their future role in this field of research.

## VI. REVERSIBILITY OF BRAIN DAMAGE/DYSFUNCTION

It has long been known that the administration of thiamin alone will bring about a rapid improvement in several of the clinical features of Wernicke's encephalopathy. There is a rapid resolution of the ophthalmoplegia, while consciousness and confusion are improved. However, there are frequently residual deficits in neuropsychological tests, especially for memory and complex cognitive tasks.[54]

The introduction of the CAT scan revolutionized our knowledge of the gross morphological changes in chronic alcoholics and so-called "heavy social" drinkers, but has led to considerable speculation as to what happens in patients who maintain abstinence at least between scans. It was Carlen and his co-workers in 1978 who reported on the improvement in CAT scans which occurred in four out of the eight abstinent patients in their study.[33] However, none of these showed complete recovery within the period of the examinations.

Since 1978 several prospective studies have been reported in the English literature.[31,36,55-58] These all reveal that a majority of chronic alcoholics (about 70%) manifest cerebral atrophy on the CAT scan, but partial reversibility occurs in many with continued abstinence. The process that occurs to account for these changes is, however, not known, although there is considerable speculation. Discussion is compromised by the fact that the number of alcoholics who have been reexamined still remains few and duration of abstinence between scans has varied greatly. Carlen et al. observed that the earlier the first CAT scan was done, the greater was the improvement at reexamination.[33] Moreover, the degree of recovery was greater in younger than in older abstinent alcoholics.

It is possible that some of the early changes may result from relatively simple metabolic processes, such as shifts in water and electrolytes. Carlen and Wilkinson also suggested that increased protein synthesis in the central nervous system may contribute to the changes.[34]

However, the evidence of continued improvement extending over several months has led to the theory that there may be a regrowth of neural tissues, especially of dendritic and axonal processes, with reestablishment of synaptic connections and possibly also of neurotransmitter processes. However, effective regeneration of central axonal processes is not known to occur in any human disease. The poor responses in older alcoholics may be the result of longer exposure to neurotoxicity and the irrecoverability of the damaged nerve cells.

## VII. NEUROPATHOLOGICAL MECHANISMS OF BRAIN DAMAGE

The work of Victor et al. still remains the classical study of the Wernicke-Korsakoff syndrome.[18] In these conditions, they described the lesions as being located in periventricular gray matter surrounding the third and fourth ventricles. The lesions were symmetrical and more severe around this central axial area, but also extended into the medial dorsal nuclei of the thalamus (88%) and the pulvinar (85%). In 50%, the fornix and cerebellum were affected. Not all the lesions were subcortical, as in 27% there was evidence of cortical atrophy, particularly of the frontal lobes, but also in the temporal and parietal regions.

In the acute phase of Wernicke's encephalopathy there is vascular congestion about the third and fourth ventricles and the aqueduct. Occasionally, macroscopic hemorrhages are seen. Microscopically, foci of damage are seen in the thalamus, mammillary bodies, and central areas of the brain stem, but there is little loss of neurons or myelin sheaths. In the chronic stages, while the distribution of lesions is much the same, there was greater loss of neurons and more marked degeneration of the mammillary bodies. These syndromes accounted for 3.1% of all alcoholic disorders seen at the Boston City Hospital over a 2-year period. The highest incidence rates were in the fifth and sixth decades and women were overrepresented. Harper found similar lesions in 1.7% of a large series of brains examined at post-mortem over the years 1973—1976.[19] On retrospective analysis, he was able to determine that 90% of such patients were alcoholics.

Wernicke's encephalopathy has long been regarded as due to thiamin deficiency. Studies before the 1939—1945 war had already shown that vitamin B preparations, administered early in acute Wernicke's encephalopathy, dramatically brought about improvement in the ocular signs and state of consciousness. During the war, prisoners who were nutritionally deprived developed a similar syndrome which responded to vitamin B administration. Victor et al. reported that in Wernicke's encephalopathy many of the neurological features responded to thiamin alone.[18] Wood et al. studied 32 cases of carefully documented Wernicke's encephalopathy and found that 60% were deficient in thiamin while another 16% had borderline values.[17] Those found to have an abnormal thiamin status responded rapidly to the administration of thiamin, especially with regard to the ophthalmoplegia and the mental state, while other signs such as nystagmus and ataxia cleared more slowly.

Korsakoff's psychosis is not so clearly attributable to thiamin deficiency, especially as the memory disorder shows little, if any, response to thiamin. The evidence is by association. The pathological changes in Korsakoff's psychosis show the same distribution as those seen in Wernicke's encephalopathy and frequently the classical amnesic features develop in patients who previously were treated for Wernicke's encephalopathy. Harper's[19] observation that only 20 % of patients found at post-mortem with the lesions of the Wernicke-Korsakoff syndrome were diagnosed before death raises several speculative issues. The end stage of the pathological features may be the outcome of recurrent or slowly progressive damage and the causes may be multifactorial. Moreover, the classical alcohol-related cerebral disorders are less commonly seen than in the past. The clinical features tend to be scrambled or partial so that the etiology is often controversial.

The only consistent association with these disorders is the heavy consumption of alcohol, although heavy tobacco consumption is also recorded in a majority of heavy drinkers.[59] The specific role of alcohol in brain damage remains uncertain. Phillips et al. were unable to demonstrate any direct toxic action on the cerebral cortex of a group of rats given a massive dose to a small site in the parietal cortex;[60] nor were they able to demonstrate any damage when acetaldehyde, ammonia, or bilirubin was used. However, the artificial route of exposure to alcohol makes these results difficult to interpret. In a further experiment to test the effect of chronic heavy consumption of alcohol on the brains of C57 black mice, Phillips and Cragg administered high daily doses of alcohol over a period of 4 months.[61]

One group of mice was then sacrificed and studied, while a second group was withdrawn from alcohol and fed an alcohol-free diet for another 4 months before sacrifice. They then closely studied the hippocampal regions of the brain. They found that in the alcohol-withdrawal group the cell counts were significantly lower than in the simple alcohol group and in age-matched controls. These results were reproducible in different strains of mice.[62] These workers also studied the neurophils in the stratum radiatum of the hippocampus. They found a reduction of 10 to 15% in synaptic connections in the alcohol-withdrawal group. Riley and Walker had previously reported a 50% reduction of such synaptic connections in a similar group of alcohol-withdrawal mice.[63] Once the neuron dies, it cannot be replaced, so that destruction of neurons leads to a permanent deficit. However, Sumner and Watson have suggested that neurophils can be repaired and interrupted synaptic connections can be reformed. This suggestion is doubted by many neurologists.[64] The proposition, therefore, is that alcohol and its metabolites are not directly toxic to cerebral tissue, but that withdrawal from the heavy chronic consumption of alcohol can result in neuronal death and damage to neurophils. How this occurs is not known.

It is apparent that much work still remains to be done to clarify the mechanisms of cerebral damage due to alcohol. Such experiments cannot ethically be performed in man so that animal models will need to be used. It remains to be seen what the new methods of cerebral investigation can contribute to this sphere of knowledge. Will the promise of progress suggested by the further development of new noninvasive techniques[65] be fulfilled?

# REFERENCES

1. **Wernicke, C.,** *Lehrbuch der Gerhirnkeiten,* Part 2, Kassel, Berlin, 1881.
2. **Victor, M and Yakoulev, P.,** Translation of Russian article by S. Korsakoff, 1889, *Neurology,* 5, 394, 1955.
3. **Byrne, E.,** Neurological evaluation of the alcoholic patient, in *Proc. Autumn School Stud. Alcohol and Drugs,* Santamaria, J., Ed., Department of Community Medicine, St. Vincent's Hospital, Fitzroy, Australia, 1984, 101.
4. **Santamaria, J.,** The epidemiology of alcohol related brain damage, in *Proc. Autumn School Stud. Alcohol and Drugs,* Santamaria, J., Ed., Department of Community Medicine, St. Vincent's Hospital, Fitzroy, Australia, 1984, 93.
5. **Sherlock, S.,** Alcohol and disease, *Br. Med. Bull.,* 38, 87, 1982.
6. **Grant, I., Adams, K., and Reed, R.,** Aging, abstinence and medical risk factors in the prediction of neuropsychological deficit among long term alcoholics, *Arch. Gen. Psychiatry,* 41, 710, 1984.
7. **Gilman, A., Goodman, L., and Gilman, A.,** *The Pharmacological Basis of Therapeutics,* Macmillan, New York, 1980, chap. 18.
8. **Shaw, G.,** Alcohol and the nervous system, in *Clinics in Endocrinology and Metabolism,* Marks, V., Wright, J., and Inglis, D., Eds., W. B. Saunders, East Sussex, U.K., 1978, 385.
9. **Bjerver, K. and Goldberg, L.,** Effect of alcohol ingestion on driving ability, *Q. J. Stud. Alcohol,* 11, 1, 1950.
10. **Flanagan, N., Strike, P., Rigby, C., and Lochridge, G.,** The effect of low doses of alcohol on driving performance, *Med. Sci. Law,* 23, 203, 1983.
11. **Maletsky, B.,** The diagnosis of pathological intoxication, *Q. J. Stud. Alcohol,* 37, 1215, 1976.
12. **Molloy, M.,** The psychometric assessment of alcohol related brain damage, in *Proc. Autumn School Stud. Alcohol and Drugs,* Santamaria, J., Ed., Department of Community Medicine, St. Vincent's Hospital, Fitzroy, Australia, 1984, 11.
13. **Walsh, K.,** Neuropsychological assessment in alcohol related brain damage, in *Proc. Natl. Workshop — Prevention of Alcohol Related Brain Damage,* Drew, L., Ed., Commonwealth Department of Health, Canberra, Australia, 1979, 31.
14. **Barbizet, J.,** *Human Memory and Its Pathology,* W. H. Freeman, San Francisco, 1970.
15. **Courville, C.,** *The Effects of Alcohol on the Nervous System of Man,* San Lucas Press, Los Angeles, 1955.
16. **Cutting, J.,** The relationship between Korsakoff's syndrome and alcoholic dementia, *Br. J. Psychiatry,* 132, 240, 1978.
17. **Wood, B., Currie, J., and Breen, K.,** Wernicke's encephalopathy in a metropolitan hospital, *Med. J. Aust.,* 144, 12, 1986.
18. **Victor, M., Adams, R., and Collins, G.,** *The Wernicke-Korsakoff Syndrome,* F. A. Davis, Philadelphia, 1971.
19. **Harper, C.,** Wernicke's encephalopathy. A more common disease than realized. A neuropathological study of 51 cases, *J. Neurol. Neurosurg. Psychiatry,* 42, 226, 1979.
20. **Harper, C.,** The incidence of Wernicke's encephalopathy in Australia — a neuropathological study of 131 cases, *J., Neurol. Neurosurg. Psychiatry,* 46, 593, 1983.
21. **Irle, E. and Markowitsch, H.,** Widespread neuroanatomical damage and learning deficits following chronic alcohol consumption or vitamin B (thiamine) deficiency in rats, *Behav. Brain Res.,* 9, 277, 1983.
22. **Tumarin, B., Wilson, J., and Snyder, G.,** Cerebral atrophy due to alcoholism in young adults, *U.S. Armed Forces Med. J.,* 6, 67, 1955.

23. **Lafon, R., Pages, P., Passouant, P., Labange, R., Mimvielle, J., and Cadilhac, J.,** Les donnes de la pneumoencephalographic et de l'ectroencephalogramme au course de l'alcoolisme chronique, *Rev. Neurol.,* 94, 611, 1956.

24. **Lereboullet, J., Pluvinage, R., and Amstutz, A.,** Aspects cliniques et electroencephalographiques des atrophies cerebrales alcooliques, *Rev. Neurol.,* 94, 674, 1956.

25. **Riboldi, A. and Garavaglia, G.,** Sulle turbe minisiche in un grupo di seggetti olcoolïsti gronici, *G. Psichiatr. Neuropatol.,* 3, 775, 1966.

26. **Ferrer, S., Santibanez, I., Castro, M., Krauskopf, D., and Saint Jean, H.,** Permanent neurological complications of alcoholism, in *Alcohol and Alcoholism,* Popham, R., Ed., Addiction Research Foundation, Toronto, 1970, 265.

27. **Brewer, C. and Perrett, L.,** Brain damage due to alcohol consumption: an air encephalographic, psychometric and electroencephalographic study, *Br. J. Addict.,* 66, 170, 1971.

28. **Ohara, K. and Homma, O.,** Ethanol and the central nervous system, *Int. J. Neurol.,* 9, 168, 1974.

29. **Tomlinson, B.,** Computerized tomography, in *Proc. Autumn School Stud. Alcohol and Drugs,* Santamaria, J., Ed., Department of Community Medicine, St. Vincent's Hospital, Fitzroy, Australia, 1984, 107.

30. **Fox, J., Ramsey, R., Huckman, H., and Proske, A.,** Cerebral ventricular enlargement: chronic alcoholics examined by computerized tomography, *J.A.M.A.,* 236, 365, 1976.

31. **Bergman, H., Borg, S., Hindmarsh, T., Idestrom, C., and Mutzell, S.,** Computed tomography of the brain and neuro-psychological assessment of alcoholic patients, in *Addiction and Brain Damage,* Richter, D., Ed., Croon Helm, London, 1980, 201.

32. **Carlen, P., Wilkinson, D., and Kiraly, L.,** Dementia in alcoholics: a longitudinal study including some reversible aspects, *Neurology,* 26, 355, 1976.

33. **Carlen, P., Wortzman, G., Holgate, R., Wilkinson, D., and Rankin, J.,** Reversible cerebral atrophy in recently abstinent alcoholics measured by computed tomography scans, *Science,* 200, 1076, 1978.

34. **Carlen, P. and Wilkinson, D.,** Alcoholic brain damage and reversible deficits, *Acta Psychiatr. Scand.,* 62 (Suppl. 286), 103, 1980.

35. **Lishman, W., Ron, M., and Acker, W.,** Computed tomography of the brain and psychometric assessment of alcoholic patients — a British study, in *Addiction and Brain Damage,* Richter, D., Ed., Croon Helm, London, 1980, 215.

36. **Cala, L., Jones, B., Burns, P., Davis, R., Stenhousen, N., and Mastaglia, F.,** The results of computerized tomography, psychometric testing and dietary studies in social drinking, with emphasis of reversibility after abstinence,. *Med. J. Aust.,* 2, 264, 1983.

37. **Carlen, P., Wilkinson, D., Wortzman, G., and Holgate, R.,** Partially reversible cerebral atrophy and functional improvement in recently abstinent alcoholics, *Can. J. Neurol. Sci.,* 11, 441, 1984.

38. **Lee, K., Moller, L., and Harot, F.,** Alcohol induced brain damage and liver damage in young males, *Lancet,* 2, 761, 1979.

39. **Bergman, H., Borg, S., Hindmarsh, T., Idestrom, C., and Mutzell, S.,** Computerized tomography of the brain and psychometric assessment of male alcoholic patients and a random sample of the general male population, *Acta Psychiatr. Scand.,* 62 (Suppl. 286), 47, 1980.

40. **Fitzhugh, L., Fitzhugh, K., and Reitan, R.,** Adaptive abilities and intellectual functioning in hospitalized alcoholics, *Q. J. Stud. Alcohol,* 21, 414, 1960.

41. **Horvath, T.,** Clinical spectrum and epidemiological features of alcohol dementia, in *Alcohol, Drugs and Brain Damage,* Rankin, J., Ed., House of Lind, Toronto, 1975, 1.

42. **Reitan, R. and Davison, L.,** *Clinical Neuropsychology: Current Status and Application,* John Wiley & Sons, New York, 1974.

43. **Walsh, K.,** Neuropsychological assessment in alcohol related brain damage, in *Prevention of Alcohol Related Brain Damage,* Drew, L., Ed., Commonwealth Department of Health, Canberra, Australia, 1974, 31.

44. **Jones, B. and Parsons, O.,** Impaired abstracting ability in chronic alcoholics, *Arch. Gen. Psychiatry,* 24, 71, 1971.

45. **Jones, B. and Parsons, O.,** Specific versus generalized deficits of abstracting ability in chronic alcoholics, *Arch. Gen. Psychiatry,* 26, 380, 1972.

46. **Tarter, R.,** Brain damage in chronic alcoholics, in *Addiction and Brain Damage,* Richter, D., Ed., Croon Helm, London, 1980, 267.

47. **Parker, E. and Noble, E.,** Alcohol consumption and cognitive functioning in social drinkers, *Q. J. Stud. Alcohol,* 38, 1224, 1977.

48. **Loberg, T.,** Neuropsychological findings in the early and middle phases of alcoholism, in *Neuropsychological Assessment of Neuropsychiatric Disorders,* Grant, I. and Adams, K., Eds., Oxford University Press, New York, 1986, chap. 18.

49. **Grant, I., Reed, R., and Adams, K.,** Natural history of alcohol and drug related brain disorder: implications for neuropsychological research, *J. Clin. Neuropsychol.,* 2, 321, 1980.

50. **Tarter, R.,** Analysis of cognitive deficits in chronic alcoholics, *J. Nerv. Ment. Dis.,* 157, 138, 1973.

51. **Adams, K. and Grant, I.,** Influence of premorbid risk factors in neuropsychological performance in alcoholics, *J. Clin. Exp. Neuropsychol.,* 8, 362, 1986.
52. **Adams, R. and Victor, M.,** *Harrison's Principles of Internal Medicine,* McGraw-Hill, New York, 1983, chap. 22.
53. **Brown, E., Randt, C., and Osborne, D.,** Assessment of memory disturbances in aging, in *The Aging Brain and Ergot Alkaloids,* Agnoli, A., Crepoldi, G., Spano, P., and Trabucci, M., Eds., Raven Press, New York, 1983.
54. **Thompson, A.,** Alcohol related structural brain changes, *Br. Med. Bull.,* 38, 87, 1982.
55. **Haubek, A. and Lee, K.,** Computer tomography in alcoholic cerebellar atrophy, *Neuroradiology,* 18, 77, 1979.
56. **Wilkinson, D. and Carlen, P.,** Relationship of neuropsychological test performance to brain morphology in amnesic and new amnesic chronic alcoholics, *Acta Psychiatr. Scand.,* 62 (Suppl. 286), 86, 1980.
57. **Ron, M., Acker, W., and Lishman, W.,** Morphological abnormalities in the brains of chronic alcoholics, a clinical, psychological and computerized trial tomographic study, *Acta Psychiatr. Scand.,* 62 (Suppl. 286), 41, 1980.
58. **Porjesz, B. and Begleiter, H.,** Brain dysfunction and alcohol, in *The Pathogenesis of Alcoholism: Biological Factors,* Kissin, B. and Begleiter, H., Eds., Plenum Press, New York, 1983, chap. 11.
59. **Wilkinson, R., Kornaczewski, A., Rankin, J., and Santamaria, J.,** Physical disease in alcoholism: initial survey of 1000 patients, *Med. J. Aust.,* 1, 1217, 1971.
60. **Phillips, S., Cragg, B., and Singh, S.,** The short term toxicity of ethanol to neurones in rat. Cerebral cortex treated by topical applications in vivo, *J. Neurol. Sci.,* 49, 353, 1981.
61. **Phillips, S. and Cragg, B.,** Chronic consumption of alcohol by adult mice. Effect on hippocampal cells and synapses, *Exp. Neurol.,* 80, 218, 1983.
62. **Phillips, S. and Cragg, B.,** Withdrawal from alcohol causes a loss of cerebellar purkinje cells in mice, *J. Stud. Alcohol,* 45, 475, 1984.
63. **Riley, J. and Walker, D.,** Morphological alterations in the hippocampus after long term alcohol consumption in mice, *Science,* 201, 646, 1978.
64. **Sumner, B. and Watson, W.,** Retraction and expansion of the dendritic tree of motor neurones of adult rats produced in vivo, *Nature,* 233, 273, 1971.
65. Symposium on Imaging Research on Alcoholism (multiple authors), *Alcholism Clin. Exp. Res.,* 10, 223, 1986.

Chapter 4

# EFFECTS OF ETHANOL AND ITS METABOLITES ON THE HEART

**Markku Kupari and Antti Suokas**

## TABLE OF CONTENTS

# I. INTRODUCTION

Chronic excessive alcohol ingestion was first mentioned as a causative factor in heart failure as early as over a century ago,[1] and in 1873 Walshe[2] described a "form of patchy cirrhosis in the myocardium" in a chronic alcoholic patient. Subsequently, cardiac dilation, hypertrophy, and congestive heart failure were reported in chronic users of alcohol, but these conditions were attributed to nutritional deficiencies rather than to alcohol itself. More recently, heart disease in alcoholics was associated with thiamin deficiency[3] or with cobalt[4] added to beer. These reports also failed to recognize the causative role of alcohol per se in heart disease.

The establishment of inherent cardiotoxicity from alcohol in the late 1950s[5] led to an upsurge in experimental and clinical research into its acute and chronic effects on the heart and circulation. These studies have shown that alcohol produces a number of biochemical, electrophysiological, and morphological alterations in the myocardial cell and its suborganelles. These changes seem to result in an acute negative inotropic effect and they probably underlie the potential of chronic alcohol abuse to produce a more persistent heart muscle disease (alcoholic cardiomyopathy). Some of the human studies have given discrepant results, however, which have directed research to the extramyocardial circulatory actions of alcohol as well. Recently, the role of the metabolites of alcohol — acetaldehyde and acetate — has aroused interest and it seems that under certain circumstances both can contribute to the acute effects of drinking. Concomitant with biomedical research, epidemiological studies have generated much data suggesting associations between alcohol consumption, coronary heart disease, and hypertension.[6,7] In this review, however, we will concentrate on the clinical and experimental studies dealing with the effects of ethanol and its metabolites on the heart muscle and its function.

# II. ACUTE EFFECTS OF ALCOHOL INGESTION

## A. Effects of Ethanol
### 1. Morphological, Metabolic, and Electrophysiologic Changes

Single doses of ethanol do not detectably alter myocardial structure as seen by light microscopy.[8] Electron microscopy, however, has shown acute morphological disarray in mitochondria and sarcoplasmic reticulum in humans,[9] as well as mitochondrial changes, increased amounts of lipid and ribosomes, and interstitial edema in rats.[8] These myopathic changes are reflected in increases of the serum enzymes released from the myocardium.[10] In mice, the acute changes in cardiac ultrastructure can be partially prevented with tocopherol pretreatment,[11] suggesting that lipid peroxidation contributes to the myocardial toxicity of ethanol. Although the mammalian heart muscle lacks alcohol dehydrogenase and the microsomal ethanol-oxidizing system, it is relatively rich in catalase.[12] This enzyme regulates the levels of free oxygen radicals and oxidizes ethanol in a reaction consuming hydrogen peroxide and generating acetaldehyde. In chronic experiments,[13] catalase has been shown to protect the myocardium from the injurious actions of ethanol, but there are no data about an acute protective effect.

Acute ethanol exposure changes myocardial cell metabolism. It decreases the activity of some plasma membrane-bound enzymes[14] and inhibits mitochondrial and sarcoplasmic AT-Pase,[15] as well as the oxidative enzymes of mitochondria.[16] Utilization of fatty acids is impaired, while their esterification to triglycerides increases.[17] Calcium uptake and binding in sarcoplasmic reticulum-enriched microsomes are also impaired,[18] but only with concentrations of ethanol that are clearly in the lethal range. In skeletal muscle ethanol renders the heavy sarcoplasmic reticulum more permeable to calcium ions, thereby decreasing the amount of storable calcium and potentially disturbing muscle contractions.[19] Whether this holds also for the striated heart muscle is unknown at present.

Ethanol shortens the duration of the action potential in the myocardium.[20] This effect is concentration dependent and probably reflects physical alterations in the cell membranes occurring when ethanol interacts with the hydrophobic regions of the lipid bilayer of the sarcolemma. Ethanol increases the fluidity of biomembranes and so, gains the potential to affect membrane-bound enzymes and transmembrane ionic fluxes.[21]

### 2. Myocardial Contractility and Cardiac Pump Function

Ethanol exerts an acute negative inotropic effect on isolated strips of atrial or ventricular myocardium.[22] The depression of contractility is linearly dependent on ethanol concentration and appears first when this exceeds about 100 mg/dl. The inotropic influence of different alcohols parallels the length of their carbon side chain, i.e., their lipophilic activity.[23] Animal experiments *in vivo* using either open-[24] or closed-chest anesthetized animals[25,26] or conscious chronically instrumented animals[27,28] have also shown myocardial depression after ethanol administration. In some studies, ethanol impaired cardiac performance even at concentrations as low as 110 to 120 mg/dl,[25,27] but the majority found this to occur only at blood ethanol concentrations approaching or exceeding 200 mg/dl.

The results of the human studies are more conflicting. Studies using cardiac and peripheral vascular catheterization in healthy subjects have shown that modest ethanol intake either does not affect or improves left ventricular pump function.[29,30] Noninvasive imaging in similar subjects has revealed either impairment[31,32] or no changes[33,34] in left ventricular ejection performance. In patients with stable coronary, valvular, or pericardial heart disease, early invasive data[29,35] suggested an acute myocardial depression from rather tiny amounts of ethanol. Measurements of the systolic time intervals[36,37] have consistently demonstrated postethanol increases in the preejection period and its ratio to left ventricular ejection time both in normals and in cardiac patients. This pattern of change has been interpreted as indicating ethanol-induced depression of myocardial contractility. Recent research[34] strongly suggests, however, that these changes reflect decreased preload rather than impaired myocardial performance. It is important to recognize that ethanol is also an arterial and venous dilator[38] and a diuretic.[39] By these effects it can change cardiac loading conditions and pump function even without affecting myocardial performance. The importance of the peripheral effects is nicely demonstrated in a recent study[40] which showed that in patients with severe heart failure, alcohol can evoke an acute relief of pulmonary congestion in association with a tendency of improvement in cardiac pump function.

Many of the human studies have thus failed to show any myocardial depression after alcohol intake, some even during the stress of isotonic[30] or isometric[41] exercise. At first this may seem irreconcilable with the *in vitro* data about negative inotropic capacity of ethanol. It seems very likely, however, that in an intact organism the vasodilating and indirect sympathomimetic effects of ethanol outweigh its myocardial depressant capacity at low blood concentrations. The reductions in peripheral arterial resistance and left ventricular size[34] unload the heart, while the increased sympathetic activity and secretion of adrenaline simultaneously counteract the ethanol-induced impairment of myocardial contractility. When the contribution of these effects is excluded — by autonomic blockade and controlling the loading conditions — the negative inotropic influence of ethanol can be demonstrated also in man.[32] A useful analogy can be drawn between ethanol and calcium channel blockers. Nifedipine, for instance, is a negative inotropic agent *in vitro*, but improves cardiac pump function *in vivo* through its vasodilating capacity.

### B. Effects of Acetaldehyde

Acetaldehyde is the first intermediary metabolite of ethanol. It is a very potent compound and has indirect sympathomimetic and vasodilating effects even in micromolar concentrations. At higher concentrations it also has direct effects on the myocardial cells, but these are usually totally overshadowed by its strong indirect influence on the heart and circulation. Acetaldehyde could certainly contribute to the cardiovascular effects of alcohol were it not that only negligible and frequently undetectable concentrations of acetaldehyde are normally released into the systemic circulation after its hepatic oxidation.[42] (See also Volume II, Chapter 13.) Acetaldehyde may, however, accumulate in blood if the activity of liver aldehyde dehydrogenase is inhibited with drugs[43] or if its isoenzyme pattern is genetically defective as in Orientals.[44] Elevated blood acetaldehyde levels during ethanol oxidation have been detected also in chronic alcoholics.[45]

### 1. Metabolic, Morphological, and Electrophysiological Changes

Acetaldehyde can produce acute adverse biochemical effects on the myocardium. In the isolated guinea pig heart, acetaldehyde decreases myocardial protein synthesis.[46] It also decreases oxygen consumption in isolated heart mitochondria,[47] depresses the activity of Na/K-activated ATPase of isolated cardiac plasma membranes,[14] and inhibits calcium-activated myofibrillar ATPase.[47] In rats, brief exposure to massive doses of acetaldehyde (peak serum level 1.55 m$M$) caused structural changes in myocardium including contraction bands, mitochondrial swelling, and cristal disarray concomitant with myofibrillar changes.[48] The electrophysiological effects of acetaldehyde are similar to the actions of noradrenaline: it prolongs the duration of the myocardial action potential and increases its area.[20]

### 2. Myocardial Contractility and Cardiac Pump Function

Acetaldehyde has a biphasic direct effect on the contractility of isolated myocardium: positive inotropy at low concentration (0.3 to 3.0 m$M$) is followed by depression of contractility at concentrations exceeding 10 m$M$.[49] Animal studies *in vivo* have shown that intravenously infused acetaldehyde increases heart rate, cardiac output, blood pressure, and myocardial contractile force and evokes peripheral vasodilation.[24,28,50] The positive chrono- and inotropic effects can be totally suppressed by β-adrenergic blockade,[24] whereas vasodilation persists even during combined α- and β-blockade.[50]

In man, the effects of acetaldehyde can be studied either during the antialcohol drug-ethanol interaction (see Volume II, Chapters 9 and 15) or during the spontaneous alcohol flush in Orientals (see Volume II, Chapter 16). Most of the changes recorded in these human experiments have been similar to those induced by exogenous acetaldehyde in animals. During the calcium carbimide- and nitrefazole-ethanol interactions,[43,51] accumulation of up to 0.15 m$M$ acetaldehyde was associated with increases in heart rate (53 to 70%), cardiac output (78 to 107%), and ejection fraction (25%) and decreases in diastolic blood pressure (19 to 30%) and peripheral vascular resistance (46 to 54%). The changes correlated with the blood acetaldehyde[43,51] and catecholamine[51] levels and were relieved by lowering blood acetaldehyde with an alcohol-dehydrogenase inhibitor (4-methylpyrazole) or by β-adrenergic blockade with propranolol.[43] In ethanol-sensitive Japanese, an ingestion of 0.5 g/kg of ethanol was associated with elevated blood acetaldehyde (0.02 to 0.08 m$M$) and peripheral vasodilation, as well as intense enhancement of left ventricular pump function.[44] Cardiovascular responses correlated with blood acetaldehyde levels and differed significantly from those in Caucasians and Orientals with normal alcohol metabolism. Although acetaldehyde generally is an indirect sympathomimetic agent, high concentrations may occasionally produce centrally mediated vagal reflexes resulting in sudden hypotension and sinus bradycardia.[44,51] A number of fatalities have been associated with alcohol-aversive drug reactions.[52]

## C. Effects of Acetate

The end product of hepatic ethanol metabolism is acetate, and shortly after ethanol ingestion its concentration in arterial blood increases from below 0.2 m$M$ to about 0.8 to 0.9 m$M$[53] Acetate is rapidly oxidized in peripheral tissues, especially in skeletal and cardiac muscle.

## 1. Metabolic Changes in Myocardium

Acetate very efficiently monopolizes myocardial energy metabolism. At 5 m$M$ acetate, 80% of the oxygen consumption in a perfused rat heart can be accounted for by acetate oxidation,[54] and in humans the myocardial utilization of acetate increases in parallel to its concentration in arterial blood.[53] Acetate also inhibits the oxidation of oleate and increases its incorporation into triacylglycerols.[55] Therefore, it may be partly responsible for the accumulation of lipids in myocardium after ethanol ingestion. The rapid intramitochondrial activation of acetate to acetyl-CoA causes an accumulation of AMP[54] and a secondary increase in the formation of adenosine. In concentrations encountered during ethanol metabolism acetate has no significant electrophysiological effects on cardiac tissue.[20]

## 2. Myocardial Contractility and Cardiac Pump Function

Both animal and human studies[56-58] have suggested that acetate is an arterial vasodilator. In a recent study[58] intravenously infused acetate increased cardiac output and decreased peripheral vascular resistance and both diastolic and mean blood pressure in concentrations encountered during ethanol metabolism. It is questionable whether acetate has any direct effect on myocardial contractility. The available data suggest, however, that at very high concentrations *in vitro* and after bolus injection *in vivo* acetate transiently depresses contractility,[57] whereas it may improve contractility both in animals and in man when given intravenously in lower concentrations.[56,58,59] The mechanism by which acetate dilates blood vessels and augments cardiac pump function is probably not a direct one, but reflects the increased production of adenosine,[56] which is a very potent vasodilator both in the systemic and coronary circulation. Acetate also seems to have diuretic capacity[58] and overall it can contribute to the cardiac and circulatory actions of alcohol in humans.

## III. CHRONIC ALCOHOL CONSUMPTION

Knowledge about the cardiac effects of chronic alcohol use is derived in the first place from a number of long-term exposure studies in animals, and, second, from work done to describe the structure and function of the heart in alcoholic persons. All available data focus on ethanol and there is no information about the chronic effects of its metabolites. This is unfortunate since the metabolism of ethanol changes with chronic use and alcoholics tend to have abnormally high concentrations of both acetaldehyde[45] and acetate[60] in peripheral blood.

### A. Studies in Animals
### 1. Morphological, Biochemical, and Electrophysiological Changes

Light and electron microscopic studies in animals given chronic alcoholic treatment have shown myocardial cell abnormalities, including dilatation and swelling of sarcoplasmic reticulum[61,62] and intercalated disks,[63] the presence of glycoproteinaceous material between cardiac muscle fibers,[62] alterations in mitochondria,[63] and triglyceride deposits within the myocardial cell.[64] A modest hypertrophy may develop in the alcoholic rat heart,[65] but the larger animals do not show such a change.[62,64] The biochemical changes with chronic use include increased left ventricular collagen,[61] diminished β-oxidation of fatty acids,[66] depressed mitochondrial respiration,[62,63,65] decreased intracytoplasmic calcium transport,[63,67] depressed myofibrillar ATPase activity,[67] impaired protein synthesis,[68] and a relative preponderance of glycolysis over glycogenolysis.[66] Chronic alcohol ingestion also alters the constituents of biomembranes: it increases both phospholipid methylation[69] and synthesis of phospholipids rich in saturated fatty acids.[70]

Recently, the production and accumulation of fatty acid ethyl esters have been demonstrated in the ethanol-exposed heart.[71] These compounds derive from esterification of free fatty acids and may induce mitrochondrial dysfunction by being hydrolyzed to a toxic free fatty acid. Another new finding is that myocardium of several mammalian species contains catalase and that prolonged ethanol consumption may increase its activity.[72] Furthermore, when ethanol was given to rats with inhibited catalase activity, extensive structural abnormalities were observed in the myocardium.[13] This suggests that myocardial catalase may protect the heart from the cardiotoxicity of ethanol.

The density of cardiac beta-adrenergic receptors is decreased in chronic alcoholic rats but increased during the early withdrawal period.[73] The decrease of cardiac β-adrenergic receptors suggests augmented catecholaminergic stimulation during chronic alcohol use. Interestingly, the hypertrophic response seen in alcoholic rat heart correlates significantly with increases in urinary excretion of adrenaline and noradrenaline.[74]

### 2. Myocardial Contractility and Cardiac Pump Function

Cardiac contractile behavior has generally been altered after long-term consumption of ethanol, but differences between animal species exist. In dogs fed ethanol for 18 or 52 months, preload challenges elicited significantly larger increases in end-diastolic pressure than in control animals, indicating impaired myocardial reserve.[61] Increased left ventricular collagen was the apparent basis for the compliance abnormality. Contractility was, however, reduced only in the long-term group (52 months).[61] In another study,[63] dogs were fed ethanol for 29 months: *in vitro* studies of glycerinated cardiac muscle fibers demonstrated reduced contractility. However, *in vivo* functional abnormalities were not present.

Hearts from rats fed ethanol for periods varying from 12 weeks to 12 months have been studied either as isolated preparations[65,75,76] or *in situ*.[77] No abnormalities in basal cardiac function have been shown. However, perfused hearts have shown decreased left ventricular response to challenges with dobutamine,[75] increased afterload,[65] electrical pacing,[76] and isoproterenol.[76] After ethanol was withdrawn for 4 to 6 months,[75] cardiac response to dobutamine was no more significantly different from controls. The exact biochemical mechanisms responsible for the impaired contractile function of the chronic alcoholic heart have not been identified. A number of changes that could contribute have been discussed above. One additional possible mechanism is the influence of alcohol on the rate of phosphorylation and dephosphorylation of the contractile and regulatory proteins, myosin, troponin, and phospholamban. Some results[78] indicate major alterations in calcium- and phospholipid-sensitive protein kinases in hearts obtained from chronically alcohol-treated rats.

It is important to recognize that none of the chronic ethanol experiments using adult animals has been able to produce a model of congestive dilated alcoholic cardiomyopathy. In a recent study, newborn domestic turkeys were fed ethanol for 8 weeks,[67] resulting in enlargement and hypokinesis of the left ventricle compatible with incipient dilated cardiomyopathy. The study, however, failed to provide data to indicate a congestive state.

### B. Human Alcoholic Cardiomyopathy

There is much circumstantial evidence in humans for the development of progressive cardiac dysfunction and ultimately cardiomyopathy with long-standing excessive ethanol consumption. First, chronic heavy use of alcohol is uncovered in a significant proportion of patients with unexplained myocardial disease.[79] Second, necropsy studies show that cardiomyopathy is much more common among alcoholics than among nonalcoholics.[80] Third, alcoholics even without symptoms or signs of cardiac disease commonly have abnormalities in left ventricular function, suggesting myocardial disease. The precise quantity and duration of alcohol exposure required to produce cardiomyopathy are difficult to assess with any confidence. Nevertheless, it has been suggested[80,81] that at least 10 years of daily intake exceeding 100 to 150 g of ethanol (2/5 to 3/5 of a bottle of spirits) is required to produce clinically overt congestive heart failure. As likely as not, a considerably less intense consumption is sufficient to produce a milder, but yet significant, degree of myocardial injury.

## 1. Preclinical Heart Muscle Disease

The most prominent structural abnormality in human alcohol-related heart disease is myocardial hypertrophy. Echocardiography has revealed ventricular septal and/or posterior wall thickness above the normal range in 27 to 47% of asymptomatic alcoholics.[80,82,83] The earliest functional abnormality is diminished diastolic compliance. As a consequence the left ventricular end-diastolic pressure is frequently increased even at rest, although the cavity size is normal or only slightly increased.[84] With excerise or increased afterload, end-diastolic pressure rises further, while ejection fraction and stroke volume remain unchanged or fall.[85,86] This shows that the left ventricular reserve capacity is diminished.

Sensitive invasive methods have shown abnormalities of left ventricular function in up to 80% of chronic alcoholics.[85-87] These patients are usually free from symptoms and signs of heart disease, but may experience chest pain, dyspnea, and palpitations, especially during heavy drinking bouts and early abstinence. The heart size is either normal or slightly increased on chest radiography. The electrocardiogram is typically normal, but absent septal Q waves, negative T waves, QRS changes of left ventricular hypertrophy, and especially prolonged QT interval have been observed.[84,87] Disturbances in cardiac rhythm constitute a rather conspicuous feature of early alcoholic heart muscle disease. The most commonly seen arrhythmia is atrial fibrillation, but other atrial as well as ventricular arrhythmias can be encountered.[87] An immediate heart muscle effect of ethanol,[88] in association with conduction delays due to patchy myocardial injury,[87] constitutes the background for reentrant arrhythmias. Indirect effects through catecholamine release and electrolyte disturbances may certainly also contribute. A relatively high incidence of sudden death has been observed in heavy drinkers.[89] Ventricular arrhythmias could have been involved, but heavy alcohol use may rarely also trigger off coronary spasm[90] and an acute myocardial infarction, despite normal epicardial coronary arteries.[91]

## 2. Dilated Alcoholic Cardiomyopathy

In 1 to 2% of chronic alcoholics, cardiac dysfunction progresses to congestive dilated cardiomyopathy. Symptoms and clinical findings reflect low-output failure with pulmonary and frequently also systemic venous congestion. Ectopic beats as well as sustained arrhythmias of atrial or ventricular origin are common. The clinical picture cannot be distinguished from primary or secondary forms of dilated cardiomyopathy. This similarity also holds for echocardiographic features[83] and findings at cardiac catheterization,[84] but careful study of an endomyocardial biopsy[92,93] — including measurement of myocardial enzyme activities — may help to differentiate between alcoholic and idiopathic cardiomyopathies. At necropsy,[94] the heart is enlarged and mildly to moderately hypertrophied. Mural thrombi are very common, but coronary arteries are usually completely patent. Microscopically, there is patchy fibrosis, interstitial edema, myocytolysis, and small areas of necrosis. Ultrastructurally, abnormalities of mitochondria and endoplasmic reticulum and accumulation of lipid droplets and lysosomes are frequently found.[95] These findings are identical to those in nonalcoholic dilated cardiomyopathy.

The natural course of alcoholic cardiomyopathy is relatively poor; over 40% of such patients died within an average time of 3 years in one follow-up study.[96] The therapeutic role of abstinence needs emphasis because 61 to 73% of abstaining patients were clinically improved in contrast with only 10 to 13% of those who continued drinking.[96,97] The potential reversibility of human alcoholic heart muscle disease has been shown also in carefully documented case histories and in studies using repeated endomyocardial biopsies after cessation of drinking.[92] A less benign prognosis has been associated with longstanding disease,[97] suggesting that at a certain stage of the disease, the pathogenetic mechanisms may continue despite abstinence from alcohol.

### 3. Dilated Cardiomyopathy in Alcoholics — Is Alcohol the Primary Cause or Just One Contributing Factor?

Since cardiomyopathy in alcoholics is so indistinguishable from the idiopathic variety, some have questioned the importance of alcohol and rather regard it as one of the many factors contributing to the genesis of idiopathic cardiomyopathy. This view gains support from the fact that congestive heart failure is surprisingly rare among alcoholics compared with the frequency of subclinical heart muscle disease. Furthermore, the lack of animal models also makes one suspicious as to whether alcohol as such can produce severe dilated cardiomyopathy. These arguments make a good case for reasoning that multiple factors may be involved in the genesis of cardiomyopathy in alcoholics. Malnutrition is one important element to be considered. In contrast to liver disease, however, clinically evident malnutrition is rare in the cardiac patient, and preclinical cardiac abnormalities in alcoholics have been shown to be independent of the nutritional status.[98] Deficiencies of vitamins have been suspected but not definitely related to cardiac damage, and deficiencies of essential metals[99] are no more likely to be involved. An immunological basis has also been considered. However, only a minority of patients with alcoholic cardiomyopathy have circulating muscle-specific antimyolemmal antibodies that are commonly found in patients with postmyocarditic cardiomyopathy.[100] Instead, alcoholics preferentially demonstrate low titer anti-interfibrillary and antifibrillary antibodies, a pattern which is typical for primary dilated cardiomyopathy. A genetic predisposition has been evaluated, but examination of two HLA loci (A and B) revealed no significant differences compared to controls.[101] Smoking can contribute to the genesis of heart muscle disease [102] and chronic cigarette use is very common among alcoholics. However, in animal studies, smoking did not potentiate the myocardial toxicity of ethanol,[103] and preclinical left ventricular abnormalities in alcoholics could not be explained by differences in cigarette smoking.[98] Data concerning the possible contribution of alcohol-induced arterial hypertension to the development of heart failure are controversial and further studies are clearly needed.[104]

## REFERENCES

1. **Wood, G. B.**, *A Treatise on the Practice of Medicine,* Vol. 2, Lippincott, Grambo, Philadelphia, 1855, 168.
2. **Walshe, W. H.**, *Disease of the Heart and Great Vessels,* Smith, Elder, London, 1873.
3. **Blankenhorn, M. A., Vilter, C. F., and Scheinker, I. M.**, Occidental beriberi heart disease, *J.A.M.A.* 131, 717, 1946.
4. **Alexander, C. S.**, Cobalt-beer cardiomyopathy. A clinical and pathological study of twenty-eight cases, *Am. J. Med.,* 53, 395, 1972.
5. **Eliaser, M. and Giansiracusa, F. J.**, The heart and alcohol, *Calif. Med.,* 84, 234, 1956.
6. **Klatsky, A. L., Friedman, G. D., and Siegelaub, A. B.**, Alcohol use and cardiovascular disease: the Kaiser-Permanente experience, *Circulation,* 64 (Suppl. 3), 32, 1981.
7. **Klatsky, A. L., Friedman, G. D., Siegelaub, A. B., and Gerard, M. J.**, Alcohol consumption and blood pressure, *N. Engl. J. Med.,* 296, 1194, 1977.
8. **von Lacerda, P. R., Knieriem, H. J., and Bozner, A.**, Die ultrastruktur des Herzmuskels der Ratte nach akuter Alkoholintoxikation, *Z. Kreislaufforsch.,* 58, 97, 1969.
9. **Klein, H. and Harmjanz, D.**, Effect of ethanol infusion on the ultrastructure of human myocardium, *Postgrad. Med. J.,* 51, 325, 1975.
10. **Lott, J. A. and Stang, J. M.**, Serum enzymes and isoenzymes in the diagnosis and differential diagnosis of myocardial ischemia and necrosis, *Clin. Chem.,* 26, 1241, 1980.
11. **Redetzki, J. E., Griswold, K. E., Nopajaroonsri, C., and Redetzki, H. M.**, Amelioration of cardiotoxic effects of alcohol by vitamin E, *J. Toxicol. Clin. Toxicol.,* 20, 319, 1983.
12. **Herzog, V. and Fahimi, H. D.**, Identification of peroxisomes (microbodies) in mouse myocardium, *J. Mol. Cell. Cardiol.,* 8, 271, 1976.

13. **Kino, M.,** Chronic effects of ethanol under partial inhibition of catalase activity in the rat heart: light and electron microscopic observations, *J. Mol. Cell. Cardiol.,* 13, 5, 1981.
14. **Williams, J. W., Tada, M., Katz, A. M., and Rubin E.,** Effect of ethanol and acetaldehyde on the (Na + K)-activated adenosine triphosphatase activity of cardiac plasma membranes, *Biochem. Pharmacol.,* 24, 27, 1975.
15. **Gvozdjak, A., Kruty, F., Bada, V., Niederland, T. E., and Gvozdjak, J.,** The effects of ethanol on myocardial mithochondrial and sarcoplasmic ATPase in rats, *Cor Vasa,* 19, 237, 1977.
16. **Bing, R. J.,** Cardiac metabolism: its contributions to alcoholic heart disease and myocardial failure, *Circulation,* 58, 965, 1978.
17. **Kako, K. J., Liu, M. S., and Thornton, M. J.,** Changes in fatty acid composition of myocardial triglyceride following a single administration of ethanol to rabbits, *J. Mol. Cell. Cardiol.,* 5, 473, 1973.
18. **Swartz, M., Repke, D. I., Katz, A. M. and Rubin, E.,** Effects of ethanol on calcium binding and uptake by cardiac microsomes, *Biochem. Pharmacol.,* 23, 2369, 1974.
19. **Ohnishi, S. T., Flick, J. L., and Rubin E.,** Ethanol increases calcium permeability to heavy sarcoplasmic reticulum of skeletal muscle, *Arch. Biochem. Biophys.,* 233, 588, 1984.
20. **Williams, E. S., Mirro, M. J., and Bailey, J. C.,** Electrophysiological effects of ethanol, acetaldehyde and acetate on cardiac tissues from dog and guinea pig, *Circ. Res.,* 47, 473, 1980.
21. **Rubin, E.,** Alcohol and the heart: theoretical considerations, *Fed. Proc.,* 41, 2460, 1982.
22. **Mason, D. T., Spann, J. F., Miller, R. R., Lee, G., Arbogast, R., and Segel, L. D.,** Effects of acute ethanol exposure on the contractile state of normal and failing cat papillary muscles, *Eur. J. Cardiol.* 7, 311, 1978.
23. **Nakano, J. and Moore, S. E.,** Effect of different alcohols on the contractile force of the isolated guinea-pig myocardium, *Eur. J. Pharmacol.,* 20, 266, 1972.
24. **Nakano, J. and Prancan, A. V.,** Effects of adrenergic blockade on cardiovascular responses to ethanol and acetaldehyde, *Arch. Int. Pharmacodyn. Ther.,* 196, 259, 1971.
25. **Regan, T. J., Koroxenidis, G., Moschos, C. B., Oldewurtel, H. A., Lehan, P. H., and Hellems, H. K.,** The acute hemodynamic and metabolic responses of the left ventricle to ethanol, *J. Clin. Invest.,* 45, 270, 1966.
26. **Goodkind, M. J., Gerber, N. H., Mellen, J. R., and Kostis, J. B.,** Altered intracardiac conduction after actue administration of ethanol in the dog, *J. Pharmacol. Exp. Ther.,* 194, 633, 1975.
27. **Horwitz, L. D., and Atkins, J. M.,** Acute effects of ethanol on left ventricular performance, *Circulation* 49, 124, 1974.
28. **Stratton, R., Dormer, I. C. J., and Zeiner, A. R.,** The cardiovascular effects of ethanol and acetaldehyde in exercising dogs, *Alcoholism Clin. Exp. Res.,* 5, 56, 1981.
29. **Gould, L., Zahir, M., DeMartino, A., and Gombrecht, R. F.,** Cardiac effects of a cocktail, *J. A. M. A.,* 218, 1799, 1971.
30. **Riff, D. P., Jain, A. C., and Doyle, J. T.,** Acute hemodynamic effects of ethanol on normal human volunteers, *Am. Heart J.,* 78, 592, 1969.
31. **Delgado, C. E., Fortuin, N., and Ross, R. S.,** Acute effects of low doses of alcohol on left ventricular function by echocardiography, *Circulation* 59, 535, 1975.
32. **Lang, R. M., Borow, K. M., Neumann, A., and Feldman, T.,** Adverse cardiac effects of acute alcohol ingestion in young adults, *Ann. Intern. Med.,* 102, 742, 1985.
33. **Child, J. S., Kovick, R. B., Levisman, J. A., and Pearce, M. L.,** Cardiac effects of acute ethanol ingestion unmasked by autonomic blockade, *Circulation* 59, 120, 1979.
34. **Kupari, M.,** Acute cardiovascular effects of ethanol. A controlled noninvasive study, *Br. Heart J.,* 49, 174, 1983.
35. **Gould, L., Zahir, M., DeMartino, A., Gomprecht, R. F., and Jaynal, F.,** Hemodynamic effects of ethanol in patients with cardiac disease, *Q. J. Stud. Alcohol.* 33, 714, 1972.
36. **Ahmed, S. S., Levinson, G. E., and Regan, T. J.,** Depression of myocardial contractility with low doses of ethanol in normal man, *Circulation,* 48, 378, 1973.
37. **Kuhn, H., Hust, M. H., Breidhart, G., and Wiebringhaus, E.,** Zur Frage der Kardiodepressiven Wirkung geringen Mengen von Alkohol bei Normalpersonen und Patienten mit koronarer Herzerkrankungen, *Z. Kardiol.,* 65, 1071, 1976.
38. **Altura, B. M., and Altura, B. T.,** Microvascular and vascular smooth muscle actions of ethanol, acetaldehyde, and acetate, *Fed. Proc.,* 41, 2447, 1982.
39. **Nicholson, W. M. and Taylor, H. M.,** The effect of alcohol on the water and electrolyte balance in man, *J. Clin. Invest.,* 17, 279, 1938.
40. **Greenberg, B. H., Schutz, R., Grunkemeier, G. L., and Griswold, H.,** Acute effects of alcohol in patients with congestive heart failure, *Ann. Intern. Med.,* 97, 171, 1982.
41. **Kupari, M., Heikkilä, J., and Ylikahri, R.,** Acute effects of alcohol on left ventricular dynamics during isometric exercise in normal subjects, *Clin. Cardiol.,* 6, 103, 1983.

42. **Lindros, K.,** Human blood acetaldehyde levels: with improved methods, a clearer picture emerges, *Alcoholism Clin. Exp. Res.,* 7, 70, 1983.
43. **Kupari, M., Lindros, K., Hillbom, M., Heikkilä, J., and Ylikahri, R.,** Cardiovascular effects of acetaldehyde accumulation after ethanol ingestion: their modification by beta-adrenergic blockade and alcohol dehydrogenase inhibition, *Alcoholism Clin. Exp. Res.,* 7, 283, 1983.
44. **Kupari, M., Eriksson, C. J. P., Heikkilä, J., and Ylikahri, R.,** Alcohol and the heart. Intense hemodynamic changes associated with alcohol flush in Orientals, *Acta Med. Scand.,* 213, 91, 1983.
45. **Nuutinen, H., Lindros, K., and Salaspuro, M.,** Determinants of blood acetaldehyde level during ethanol oxidation in chronic alcoholics, *Alcoholism Clin. Exp. Res.,* 7, 163, 1983.
46. **Schreiber, S. S., Briden, K., Oratz, M., and Rothschild, M. A.,** Ethanol acetaldehyde, and myocardial protein synthesis, *J. Clin. Invest.,* 51, 2820, 1972.
47. **Nayler, W. G., and Fassold, E.,** Effect of acetaldehyde on functioning of cardiac muscle at the ultrastructural level, *Rec. Adv. Cardiac Struct. Metab.,* 11, 489, 1978.
48. **Tomaru, A., Mizorogi, F., Fujita, K., Nishiyama, N., Miura, Y., Matsuda, F., Tanaka, T., and Horiguchi, M.,** Alcoholic cardiomyopathy — acetaldehyde poisoning rat: myocardial and serum changes in acute exposure, *Jpn. Circ. J.,* 47, 649, 1983.
49. **Truitt, E. B., Jr. and Walsh, M. J.,** The role of acetaldehyde in the actions of ethanol, in *The Biology of Alcoholism: Biochemistry,* Vol. 1, Kissin, B. and Begleiter, H., Eds., Plenum Press, New York, 1971, 161.
50. **McCloy, R. B., Prancan, A. V., and Nakano, J.,** Effects of acetaldehyde on the systemic, pulmonary, and regional circulations, *Cardiovasc. Res.,* 8, 216, 1974.
51. **Suokas, A., Kupari, M., Pettersson, J., and Lindros, K.,** Nitrefazole-ethanol interaction in man: cardiovascular responses and the accumulation of acetaldehyde and catecholamines, *Alcoholism Clin. Exp. Res.,* 9, 221, 1985.
52. **Sellers, E. M., Naranjo, C. A., and Peachey, J. E.,** Drugs to decrease alcohol consumption, *N. Engl. J. Med.,* 305, 1255, 1981.
53. **Lindeneg, O., Mellemgaard, K., Fabricius, J., and Lunquist, F.,** Myocardial utilization of acetate, lactate and free fatty acids after ingestion of ethanol, *Clin. Sci.,* 27, 427, 1964.
54. **Randle, P. J., England, J., and Denton, R. M.,** Control of the tricarboxylic acid cycle and its interactions with glycolysis during acetate utilization in rat heart, *Biochem. J.,* 117, 677, 1970.
55. **Kiviluoma, K. and Hassinen, I. E.,** Role of acetaldehyde and acetate in the development of ethanol-induced cardiac lipidosis, studied in isolated perfused rat hearts, *Alcoholism Clin. Exp. Res.,* 7, 169, 1983.
56. **Liang, C. S., and Lowenstein, J. M.,** Metabolic control of the circulation. Effects of acetate and pyruvate, *J. Clin. Invest.,* 62, 1029, 1978.
57. **Kirkendol, P. L., Pearson, J. E., Bower, J. D., and Holbert, R. D.,** Myocardial depressant effects of sodium acetate, *Cardiovasc. Res.,* 12, 127, 1978.
58. **Suokas, A., Kupari, M., Heikkilä, J., Lindros, K., and Ylikahri, R.,** Acute cardiovascular and metabolic effects of acetate in man, *Alcoholism Clin. Exp. Res.,* 12, 52, 1988.
59. **Nitenberg, A., Huyghebaert, M. F., Blanchet, F., and Amiel, C.,** Analysis of increased myocardial contractility during sodium acetate infusion in humans, *Kidney Int.,* 26, 744, 1984.
60. **Nuutinen, H., Lindros, K., Hekali, P., and Salaspuro, M.,** Elevated blood acetate as indicator of fast ethanol elimination in chronic alcoholics, *Alcohol,* 2, 623, 1985.
61. **Thomas, G., Haider, B., Oldewurtel, H. A., Lyons, M. M., Yeh, C. K., and Regan, T. J.,** Progression of myocardial abnormalities in experimental alcoholism, *Am. J. Cardiol.,* 46, 233, 1980.
62. **Regan, T. J., Ettinger, P., and Oldewurtel, H.,** Heart cell responses to ethanol, *Ann. N.Y. Acad. Sci.,* 252, 250, 1975.
63. **Sarma, J. S. M., Ikeda, S., Fischer, R., Maruyama, Y., Weishaar, R., and Bing, R. J.,** Biochemical and contractile properties of heart muscle after prolonged alcohol administration, *J. Mol. Cell. Cardiol.,* 8, 951, 1976.
64. **Regan, T. J., Khan, M. I., Ettinger, P. O., Haider, P., Lyons, M. M. and Oldewurtel, H. A.,** Myocardial function and lipid metabolism in the chronic alcoholic animal, *J. Clin. Invest.,* 54, 740, 1974.
65. **Whitman, V., Schuler, H. G., and Musselman, J.,** Effects of chronic ethanol consumption on the myocardial hypertrophic response to a pressure overload in the rat, *J. Mol. Cell. Cardiol.,* 12, 519, 1980.
66. **Edes, I., Ando, A., Csadany, M., Mazarean, H., and Guba, F.,** Enzyme activity changes in rat heart after chronic alcohol ingestion, *Cardiovasc. Res.,* 17, 691, 1983.
67. **Noren, G. R., Staley, N. A., Einzig, S., Mikell, F. L., and Asinger, R. W.,** Alcohol-induced congestive cardiomyopathy: an animal model, *Cardiovasc. Res.,* 17, 81, 1983.
68. **Schreiber, S. S., Evans, C. D., Reff, F., Rothschild, M. A., and Oratz, M.,** Prolonged feeding of ethanol to the young growing guinea pig. I. The effect on protein synthesis in the afterloaded right ventricle measured in vitro, *Alcoholism Clin. Exp. Res.,* 6, 384, 1982.
69. **Prasad, C., and Edwards, R. M.,** Increased phospholipid methylation in the myocardium of alcoholic rats, *Biochem. Biophys. Res. Commun.,* 111, 710, 1983.

70. **Littleton, J. M., Grieve, S. J., Griffiths, P. J., and John, G. R.,** Ethanol-induced alteration in membrane phospholipid composition: possible relationship to development of cellular tolerance to ethanol, *Adv. Exp. Med. Biol.,* 126, 9, 1980.

71. **Lange, L. G. and Sobel, B. E.,** Mitochondrial dysfunction induced by fatty acid ethyl esters, myocardial metabolites of ethanol, *J. Clin. Invest.,* 72, 724, 1983.

72. **Fahimi, H. D., Kino, M., Hicks, L., Throp, K. A., and Abelmann, W. H.,** Increased myocardial catalase in rats fed ethanol, *Am. J. Pathol.,* 96, 373, 1980.

73. **Banerjee, S. P., Sharma, V. K., and Khanna, J. M.,** Alterations of beta-adrenergic receptor binding during ethanol withdrawal, *Nature,* 276, 407, 1978.

74. **Adams, M. A., and Hirst, M.,** Ethanol-induced cardiac hypertrophy: correlation between development and the excretion of adrenal catecholamines, *Pharmacol. Biochem. Behav.,* 24, 33, 1986.

75. **Segel, L. D., Rendig, S. V., and Mason, D. T.,** Alcohol induced cardiac hemodynamic and $Ca^{2+}$ flux dysfunctions are reversible, *J. Mol. Cell. Cardiol.,* 13 , 443, 1981.

76. **Chan, T. C. K., and Sutter, M. C.,** The effects of chronic ethanol consumption on cardiac function in rats, *Can. J. Physiol. Pharmacol.,* 60, 777, 1982.

77. **Hepp, A., Rudolph, T., and Kochsiek, K.,** Is the rat a suitable model for studying alcoholic cardiomyopathy? Hemodynamic studies at various stages of chronic alcohol ingestion, *Basic Res. Cardiol.,* 79, 230, 1984.

78. **van Thiel, D. H., Gavaler, J. S., and Lehotay, D. C.,** Biochemical mechanisms responsible for alcohol-associated myocardiopathy, in *Recent Developments in Alcoholism,* Vol. 3, Galanter, M., Ed., Plenum Press, New York, 1985, chap. 12.

79. **Alexander, C. S.,** Idiopathic heart disease, *Am. J. Med.,* 41, 213, 1966.

80. **Schenk, E. A. and Cohen, J.,** The heart in chronic alcoholism. Clinical and pathologic findings, *Pathol. Microbiol.,* 35, 96, 1970.

81. **Goodwin, J. F.,** Prospects and predictions for the cardiomyopathies, *Circulation,* 50, 210, 1974.

82. **Askanas, A., Udoshi, M., and Sadhaji, S. A.,** The heart in chronic alcoholism: a noninvasive study, *Am. Heart J.,* 99, 9, 1980.

83. **Mathews, E. C., Jr., Gardin, J. M., Henry, W. L., DelNegro, A. A., Fletcher, R. D., Snow, J. A., and Epstein, S. E.,** Echocardiographic abnormalities in chronic alcoholics with and without overt congestive heart failure, *Am. J. Cardiol.,* 47, 570, 1981.

84. **Ahmed, S. S., Levinson, G. E., Fiore, J. J., and Regan, T. J.,** Spectrum of heart muscle abnormalities related to alcoholism, *Clin Cardiol.,* 3, 335, 1980.

85. **Regan, T. J., Levinson, G. E., Oldewurtel, H. A., Frank, M. J., Weisse, A. B., and Moschos, C. B.,** Ventricular function in noncardiacs with alcoholic fatty liver: role of ethanol in the production of cardiomyopathy, *J. Clin. Invest.,* 48, 397, 1969.

86. **Gould, L., Shariff, M., Zahir, M., and Di Lieto, M.,** Cardiac hemodynamics in alcoholic patients with chronic liver disease and a presystolic gallop, *J. Clin. Invest.,* 48, 860, 1969.

87. **Ettinger, P. O., Wu, C. F., De La Cruz, C., Jr., Weisse, A. B., Ahmed, S. S., and Regan, T. J.,** Arrhythmias and the "holiday heart": alcohol-associated cardiac rhythm disorders, *Am. Heart J.,* 95, 555, 1978.

88. **Gould, L., Reddy, C. V., Becker, W., Oh, K. C., and Kim, S. G.,** Electrophysiological properties of alcohol in man, *J. Electrocardiol.,* 11, 219, 1978.

89. **Kramer, K., Kuller, L., and Fisher, R.,** The increasing mortality attributed to cirrhosis and fatty liver in Baltimore (1957—1966), *Ann. Intern. Med.,* 69, 273, 1968.

90. **Takizawa, A., Yasue, H., Omote, S., Nagao, M., Hyon, H., Nishida, S., and Horie, M.,** Variant angina induced by alcohol ingestion, *Am. Heart J.,* 107, 25, 1984.

91. **Regan, T. J., Wu, C. F., and Weisse, A. B.,** Acute myocardial infarction in toxic cardiomyopathy without coronary obstruction, *Circulation* 51, 453, 1975.

92. **Ferriere, M., Rouy, S., Baissus, C., and Latour, H.,** Aspects histologiques de la cardiomyopathie congestive d'origine ethylique, *Ann. Cardiol. Angeiol.,* 32, 225, 1983.

93. **Richardson, P. J., Atkinson, L., and Wodak, A.,** The measurement of enzyme activities in endomyocardial biopsies from patients with congestive (dilated) cardiomyopathy and specific heart muscle disease, *Z. Kardiol.,* 71, 522, 1982.

94. **Steinberg, J. D., and Hayden, M. T.,** Prevalence of clinically occult cardiomyopathy in chronic alcoholism, *Am. Heart J.,* 101, 461, 1981.

95. **Edmonson, H. A.,** Pathology of alcoholism, *Am. J. Clin. Pathol.,* 74, 725, 1980.

96. **Demakis, J. G., Proskey, A., and Rahimtoola, S. H.,** The natural course of alcoholic cardiomyopathy, *Ann. Intern. Med.,* 80, 293, 1974.

97. **Gunnar, R. M., Demakis, J., and Rahimtoola, S. H.,** Clinical signs and natural history of alcoholic heart disease, *Ann. N.Y. Acad. Sci.,* 252, 264, 1975.

98. **Dancy, M., Bland, J. M., Leech, G., Gaitonde, M. K., and Maxwell, J. D.,** Preclinical left ventricular abnormalities in alcoholics are independent of nutritional status, cirrhosis, and cigarette smoking, *Lancet,* 1, 1122, 1985.

99. **Bogden, J. D. and Al-Rabiai, S.** Effect of chronic ethanol ingestion on the metabolism of copper, iron, manganese, selenium, and zinc in an animal model of alcoholic cardiomyopathy, *J. Toxicol. Environ. Health,* 14, 407, 1984.

100. **Maisch, B., Deeg, P., Liebau, G., and Kochsiek, K.,** Diagnostic relevance of humoral and cytotoxic immune reactions in primary and secondary dilated cardiomyopathy, *Am. J. Cardiol.,* 52, 1072, 1983.

101. **Kachru, R. B., Proskey, A. J., and Telischi, M.,** Histocompatibility antigens and alcoholic cardiomyopathy, *Tissue Antigens,* 15, 398, 1980.

102. **Hartz, A. J., Anderson, A. J., Brooks, H. L., Manley, J. C., Parent, G. T., and Barboriak, J. J.,** The association of smoking with cardiomyopathy, *N. Engl. J. Med.,* 311, 1201, 1984.

103. **Ahmed, S. S., Torres, R., and Regan, T. J.,** Interaction of chronic cigarette and ethanol use on myocardium, *Clin. Cardiol.,* 8, 129, 1985.

104. **Friedman, H. S., Vasavada, B. C., Malec, A. M., Hassan, K. K., Shah, A., and Siddiqui, S.,** Cardiac function in alcohol-associated systemic hypertension, *Am. J. Cardiol.,* 57, 227, 1986.

Chapter 5

ETHANOL AND THE FLOW PROPERTIES OF BLOOD

**Leslie O. Simpson and Robin J. Olds**

TABLE OF CONTENTS

# I. INTRODUCTION

Much of the subject matter in this volume is based upon biochemical considerations of the consequences of ingesting ethanol. In such studies it is assumed generally that the metabolic processes being studied will take place *in vivo* where the supply of oxygen and substrates will not be interrupted and metabolic wastes will continue to be removed. In this chapter we wish to draw attention to the possibility that the consequences of the ingestion of ethanol may affect the flow properties of blood adversely. Should this occur, then the flow of blood in the smallest capillaries and postcapillary venules may be impaired to the extent that blood flow ceases; the metabolic processes reliant upon the supply of oxygen and substrates will be compromised and the survival of the capillaries and the tissues they supply will be placed in jeopardy.

The subject matter will be presented in the following order. First, the relationship between red blood cell (RBC) and capillary dimensions will be discussed to underline the fundamental importance of normal capillary function for the maintenance of normal physiological processes. Second, the significance of the flow properties of the blood (blood rheology) as a factor in capillary function will be emphasized. Third, the effects of acutely raised levels of blood alcohol on blood rheology will be examined, with the direct effects of alcohol per se being considered separately from the possible effects of ethanol metabolites. Fourth, the effects of chronically raised levels of blood alcohol will be studied from the viewpoint that the persistent presence of ethanol metabolites may alter blood rheology sufficiently to play major roles in the pathogenesis of several of the complications of chronic alcoholism.

## A. Normal Capillary Function

In mammals, blood flows in a closed system in which substances absorbed from the small intestine, oxygen absorbed in the lungs, and hormones elaborated in the secreting glands are distributed throughout the body. By virtue of their simplicity of structure capillaries deliver the metabolic requirements of tissues. As many capillaries (average diameter 3 to 4 μm) are smaller than red cells (average diameter 7 to 8 μm) it is necessary for RBCs to change shape so that they can pass through the smaller capillaries. Thus, the deformability or resistance to change in curvature of RBCs is an important factor in determining the rate of capillary blood flow. Because of the importance of the microcirculation in the maintenance of normal physiological activities, it follows that impaired capillary blood flow will have pathophysiological significance in direct relationship to the extent that capillary blood flow is impaired. Before considering the possible effects and consequences of ethanol on the flow properties of blood, it is necessary to understand the relationship between capillary structure and the determinants of blood rheology. Capillaries are tubes of a jelly-like basement membrane, lined with endothelial cells and invested in a non-stretching tunic of collagen. Basement membranes exhibit pressure-dependent permeability as demonstrated by Landis.[1] It has been proposed that their permeability is due to their biological thixotropy.[2,3] As liquids and small solutes can pass freely through the internal lattice of the basement membrane gel, any increase in intracapillary pressure is associated with an increase in the passage of water and small solutes through the capillary wall into the surrounding tissues. Should capillary pressure continue to rise, plasma proteins of increasing size are able to penetrate the basement membrane and pass out through the capillary wall. When extremely high pressure prevails even RBCs may be extravasated.[4] In accordance with the degree of change in intracapillary pressure, the pressure-dependent permeability of the basement membrane determines the size of molecules passing through the capillary wall. However, the amount of material extravasated is modulated by endothelial cells varying the dimensions of the intercellular gaps, thus, changing the area of exposed basement membrane. The most permeable capillaries are lined with fenestrated endothelium (for example, in kidneys and secreting glands), while the least permeable capillaries have very close intercellular gaps (for example, those in the brain). This concept gains support from observations of increased capillary permeability when endothelial cells are lost, as in burns, or contract to a "rounded up" form as a response to an insufficiency of oxygen.

## B. Determinants of Blood Rheology

The flow properties of blood are determined by the interaction of its cellular and noncellular components and may be assessed by measuring the viscosity of whole blood, serum, or plasma. Whole blood viscosity is determined to a major extent by the number of cells present. Thus, when the number of RBCs is abnormally high (polycythemia) blood is hyperviscous. White blood cells are much fewer in number than red cells, but they are larger and stiffer than red cells. Therefore, they may influence whole blood viscosity in conditions where white cell number is much greater than normal (as in leukemia), although it is difficult to quantify their contribution. Whole blood viscosity is also influenced by any reduction in deformability of RBCs. The pathogenic potential of this type of change is epitomized by sickle cell disease and hereditary spherocytosis. Chien et al.[5] showed that blood viscosity increased when RBCs were stiffened with glutaraldehyde. The mechanism of red cell stiffening *in vivo* is not understood. Although RBCs become less deformable in response to a variety of changes in their environment such as acidosis, hyperglycemia, or hyperlipidemia, it is not known whether reduced deformability is due to membrane changes, altered internal viscosity, some changes in the cytoskeleton, or to a shape change which alters surface area-to-volume ratio. The picture is confused further by the fact that a variety of pharmaceutical agents of widely different chemical composition appear to have the ability to improve RBC deformability which reduces blood viscosity and enhances blood flow in the microcirculation.

Plasma viscosity is determined principally by the concentration of fibrinogen. In conditions where the rate of blood flow is reduced below an undefined limit, fibrinogen is believed to form temporary cell-protein-cell bonds which result in three-dimensional aggregates.[6,7] This type of interaction is possibly an electrostatic attraction between negatively charged RBCs and positively charged fibrinogen molecules.

Serum viscosity reflects the concentration of proteins, particularly albumin and globulins, and serum hyperviscosity syndromes are well described.[8] Albumin plays an important role in the maintenance of normal blood viscosity. This may be due to the electrostatic interaction of its multiple negative charges with negatively charged RBCs in preventing red cell aggregation, as hypoalbuminemia is associated with increased blood viscosity.

In addition to viscosity the effects of two other physical factors must be recognized if the determinants of capillary blood flow are to be understood. These are (1) capillary size and (2) the thixotropic nature of blood.

**Capillary size** — In 1840 Poiseuille showed that rate of flow through narrow tubes was directly proportional to the fourth power of the tube radius, but inversely proportional to fluid viscosity and tube length. According to Fishberg,[9] "While the laws of Poiseuille do not hold exactly for the branching circulation . . . they furnish valuable approximations which are at least of the correct order." Therefore, it seems likely that when blood viscosity is increased, flow in the smallest capillaries and postcapillary venules (where flow rate is lowest normally) may be jeopardized. Cessation of blood flow (stasis) may be overcome as long as there is sufficient reserve of vasodilation in arterioles to sustain a pressure gradient across the capillary bed. If this does not occur then the capillary and the tissue it supplies will die. It should be emphasized that this example may typify the pathogenic potential of blood hyperviscosity which is manifested first according to the Poiseuille formula, in the smallest blood vessels. In addition, the example provides a basis for understanding the relatively poor predicting power of those aspects of capillary blood flow which can be measured. Thus, while it is possible to assess blood viscosity and red cell deformability it is not possible to obtain information about capillary size which could be the limiting factor in microvascular blood flow.

**Blood thixotropy** — Thixotropic systems exhibit a reversible reduction in viscosity with increasing shear rate (flow rate). The thixotropic nature of blood is well established[10,11] and this means that whenever there is a reduction in the rate of blood flow, blood viscosity will increase for that reason alone.

It is proposed to apply the foregoing concepts to an evaluation of hemorheological changes associated with moderate ethanol ingestion and prolonged ethanol abuse and of the possible role of altered blood rheology in the pathogenesis of some of the complications of alcoholism, for example, in hypertension and central nervous system disorders.

## II. MODERATE ETHANOL INGESTION AND BLOOD FLOW

With moderate alcohol intake peripheral blood flow is increased, the face reddens, and there is a rise in hand and toe temperatures. Such changes are more than a simple ethanol-induced vasodilation. Fewings et al.[12] showed that ethanol ingestion caused vasodilation of the superficial vessels in the hand and forearm, but reduced blood flow in forearm muscles. In contrast, intra-arterially injected ethanol caused vasoconstriction of both skin and muscle blood vessels, reducing blood flow in hand and forearm. It was considered that the dilator effects on skin vessels of ethanol taken by mouth were sympathetically mediated, but as the fall in muscle blood flow occurred also in sympathectomized limbs, it was concluded that the constrictor effect of ethanol was due to direct local action on muscle blood vessels. However, it has been reported that not all subjects respond to ethanol by vasodilation[13] and in some subjects the effects were opposite, although the mechanism is unexplained.

### A. The Effects of Ethanol Per Se

Couchman[14] has shown that the absorption of ethanol is a very variable phenomenon, and pointed out that most laboratory studies involved the relatively rapid consumption of large quantities of alcohol, usually after fasting. If food is consumed with alcohol, blood alcohol levels are lower than after fasting. This is probably due to a delay in gastric emptying as has been shown experimentally.[15]

According to Horrobin,[16] in *in vitro* studies ethanol in the range of 30 to 400 mg% converted dihomogammalinolenic acid (DHGLA) stored in platelet membrane phospholipids to prostaglandin $E_1$ ($PGE_1$). $PGE_1$ has a direct effect on RBC deformability[17,18] and $PGE_1$ infusions have been found to be beneficial in conditions with impaired microcirculatory blood flow.[19,20] In the expectation that alcohol ingestion would be associated with a $PGE_1$-induced improvement in RBC deformability and reduced high shear rate blood viscosity, we studied the changes occurring in the blood rheology of 14 subjects (7 male, 7 female) after drinking sufficient vodka in fruit juice to provide 1.2 g ethanol per kilogram of body weight. Controls (4 males, 3 females) drank the same volume of fruit juice. Changes in whole blood viscosity and filterability showed individual variability similar to the variability in alcohol absorption reported by Couchman.[14] Although some individuals showed improved blood rheology in the first 100 min after begining their first drink, the majority showed either no change or a negative effect on the flow properties of the blood. The conclusions from that unpublished study were (1) that drinking fruit juice with or without alcohol caused slight reductions in high shear rate blood viscosity in 15/21 subjects; and (2) that there were no corresponding changes in RBC filterability which would indicate either an alcohol-induced increase or decrease in RBC deformability. Our protocol and results were similar to those of Hillbom et al.[21] who concluded that ethanol "...failed to induce any marked changes in blood rheology." That conclusion is in agreement with the results of a study of the *in vitro* effects of ethanol on RBC filterability. Shand[103] failed to find any change in the filterability of suspensions of washed RBCs after exposure for 6 min to ethanol concentrations up to 300 mg/dl. The observation by Galea and Davidson[22] that platelet hypoaggregability was evident at 2 h and maximal at 3 h after alcohol intake is the only evidence that alcohol might have stimulated DHGLA release and, thus, increased plasma PGE[1] levels. Increased blood viscosity attributed to hemoconcentration with increases in total plasma protein concentration, and hematocrit occurred after 2 h of social drinking and eating, with mean alcohol loads of 1.2 g ethanol per kilogram of body weight.[23]

## B. The Consequences of Ethanol Metabolism

According to Couchman,[14] in normal volunteers alcohol was eliminated from the blood in a near-linear fashion at average rates of 20.4 and 18.8 mg/dl/h for females and males, respectively. However, the rate of elimination diminishes as blood alcohol concentrations are reduced. Hillbom et al. showed graphically that when most alcohol had been eliminated from the blood at 12 to 16 h after the first drink, both erythrocyte rigidity and whole blood viscosity had increased.[21] Galea and Davidson[22] studied four subjects whose maxium blood alcohol levels were comparable to those in the study of Hillbom et al.[21] It was reported that whole blood viscosity increased progressively at all shear rates, but the increase reached statistical significance only in samples obtained 15 h after taking alcohol.

The increase in high shear rate blood viscosity at times when most alcohol would have been metabolized[21,22] draws attention to the possibility that a metabolite of ethanol could be responsible for the change in blood viscosity. As acetaldehyde is a metabolite of ethanol,[24] it is a likely suspect. Korsten et al.[25] showed that after intravenously administered ethanol, blood acetaldehyde levels plateaued for several hours, then fell sharply at blood ethanol concentrations between 15 and 25 m$M$. It was found that after the same alcohol loading significantly higher acetaldehyde levels were observed in alcoholic than in nonalcoholic subjects. Such a result is explicable on the basis of impaired intrahepatic oxidation of acetaldehyde in alcoholics.[26] Eriksson[27] has drawn attention to the possibility that the high levels of acetaldehyde reported by Korsten et al.[25] were artefactual. Lindros[28] points out that after alcohol ingestion by nonalcoholics, plasma concentrations of acetaldehyde do not reach high levels. However, it was noted that because of ethical considerations, information is lacking about the blood acetaldehyde levels after heavy alcohol intoxication. Baraona et al.[29] have reported that after modest alcohol ingestion most of the acetaldehyde is reversibly bound to red cells, and intracellular concentrations were about ten times that of the plasma. The authors considered that in this way acetaldehyde was transported from the liver to exert its toxic influence in other tissues. Palmer and Jenkins[26] reviewed the literature in this rather controversial area and confirmed the observation of Korsten et al.[25] by showing that alcoholics had higher concentrations of acetaldehyde in venous blood than normal subjects after ingesting similar quantities of ethanol. They concluded that the higher levels of acetaldehyde were not due to enhanced synthesis, but more probably were a consequence of reduced oxidation in the alcoholic liver, either as a primary abnormality or secondary to liver damage. Finally, they suggested that reduced hepatic aldehyde dehydrogenase activity might represent a primary and potentially inheritable abnormality, at least in some alcoholics. That suggestion may have been prescient as von Wartburg and Buhler[30] have introduced the concept of ''aldehydism'' as a consequence of the genetically determined enzyme pattern of ''chronic aldehydism''.

Further insight into the involvement of acetaldehyde in the adverse effects of drinking ethanol has been provided in studies on the phenomenon of facial flushing in Japanese. (See Volume II, Chapter 16.) Mizoi et al.[31] found a number of differences between a ''flushing'' and a ''nonflushing'' group after the consumption of a small amount of ethanol. Although there were no differences in the maximum levels of blood alcohol or in the rate of alcohol elimination, the ''flushing'' group showed a ''conspicuous rise in the blood acetaldehyde levels''. In addition, increased urinary concentrations of epinephrine and norepinephrine (NE) were noted. In a study of alcohol and aldehyde dehydrogenases in autopsy livers from Japanese people, Harada et al.[32] found that 52% of the livers contained an unusual type of aldehyde dehydrogenase. This observation led to the speculation that because of the delayed oxidation of acetaldehyde, individuals with the atypical enzyme would be exposed to elevated concentrations of acetaldehyde. Subsequently, Adachi and Mizoi[33] investigated the possibility that the cardiovascular response to ethanol may represent the catecholamine-mediated effects of acetaldehyde. They reasoned that as both the aldehydes from alcohol and from biogenic amines require aldehyde dehydrogenase, then the metabolic pathway of catecholamines would be changed in the presence of the unusual form of aldehyde dehydrogenase. They found that facial flushing occurred in the unusual aldehyde dehydrogenase group accompanied by marked and sustained increases in plasma and urinary NE. In the group with the usual aldehyde dehydrogenase, blood acetaldehyde did not increase. The observed changes in urinary metabolies of NE were in agreement with the findings of Gitlow et al.[34] who noted that ethanol ingestion increased both synthesis and turnover of NE. As Disulfiram inhibits aldehyde dehydrogenase with a consequent rise in plasma acetaldehyde, conditions comparable to those occurring in "flushing" Japanese will develop after ingestion of ethanol by subjects taking Disulfiram. (See Volume II, Chapter 15.)

In such circumstances it can be anticipated that blood rheology will be affected adversely, although the mechanisms involved could be complex. Chien et al.[35] have shown that prolonged exposure of RBCs to acetaldehyde reduced deformability, although the concentration and time involved were not physiological. Whether or not the reversible binding of acetaldehyde reported by Baraona et al.[29] influences RBC deformability is unknown. As both Hillbom et al.[21] and Galea and Davidson[22] reported changes in blood rheology more than 12 h after the first drink, intraerythrocytic acetaldehyde could be involved. Alternatively, the alcohol-stimulated synthesis of NE[34] could be the cause of the change in blood rheology. NE has been shown to reduce RBC filterability.[18] As there are reports of high levels of plasma NE when blood flow is impaired, it seems reasonable to suspect this catecholamine of contributing to any change in blood rheology which might occur when its rate of secretion is enhanced. However, in a study of the filterability of suspensions of washed red cells, Shand[103] did not detect a reduction in the filterability of RBCs exposed to NE. Hawley et al.[36] cited six references to support a proposal that in the release of catecholamines the primary releasing action was caused by acetaldehyde. Thus, the rise in high shear rate blood viscosity can be interpreted as the consequence of either a direct effect of acetaldehyde or an acetaldehyde-initiated NE-mediated reduction in RBC deformability. The exact changes which result in reduced RBC deformability remain undefined, although there is evidence that changes in the bilipid layer, for example, in cholesterol content, can increase membrane stiffness. Ethanol has been shown by spin labeling to increase the fluidity of RBC membranes, probably through its action on the lipid rather than the protein components of the membrane.[37] By means of a fluorescence polarization technique Hrelia et al.[38] claimed that ethanol improved RBC membrane fluidity in chronic alcoholism in "a dose-dependent manner". In contrast, Benedetti et al.[39] used diphenylhexatriene as a probe to show decreased fluidity of the membrane of RBCs from chronic alcoholics. However, this occurred in the presence of increased cholesterol/phospholipid ratios and probably reflects associated liver damage.

Thus, in normal individuals the intake of ethanol initially improves blood rheology by enhancing $PGE_1$ synthesis. This will continue until the stores of membrane-bound DHGLA are exhausted and/or high blood alcohol levels impair the efficiency of the enzyme delta-6-desaturase which converts linoleic acid to DHGLA.[16] Both $PGE_1$[17] and ethanol[37] improved RBC membrane fluidity, as asssessed by spin labeling. However, if the cell membrane stores of DHGLA are not replenished by adequate food intake or because of an ethanol block at delta-6-desaturase, $PGE_1$ synthesis will be suboptimal. In such conditions RBC deformability may be reduced. In contrast to the direct effects of ethanol, its metabolites may have adverse effects on blood rheology. Acetaldehyde appears to play a major role, although the mechanisms involved remain undefined. It is possible that RBC deformability is affected by reversibly bound acetaldehyde in erythrocytes as reported by Baraona et al.[29] Although Hawley et al.[36] believe that acetaldehyde stimulated the release of NE, Gitlow et al.[34] reported that increased NE synthesis and turnover were due to ethanol, but the observations are not necessarily mutually exclusive. As NE has been shown to reduce RBC filterability,[18] it could be the effective agent in causing the reduced RBC deformability observed more than 12 h after alcohol ingestion.[21,22] Poorly deformable RBCs increase blood viscosity through their disturbance of streamlines in flow. In the microcirculation flow in the smallest capillaries will be impaired first, but because of the thixotropic nature of blood, a viscosity-induced reduction in the rate of flow will cause a further increase in viscosity. In such conditions the rate of capillary blood flow will be reduced, impairing the delivery of oxygen and nutrients to the tissues and allowing the accumulation of tissue metabolites. Normal rates of blood flow may be restored if there is an autoregulated vasodilation upstream to increase intracapillary pressure sufficiently to overcome the resistance of the viscous blood. While a rise in capillary pressure might restore blood flow, it would also increase the rate of fluid passing out through the capillary walls. This rise in transudation rate which shifts fluid from intravascular to extravascular sites will reduce blood volume, and the resulting hemoconcentration will further increase blood viscosity.

As the hemorheological changes will have systemic effects and because poorly deformable RBCs and increased blood viscosity have been reported 12 to 18 h after the first drink, it is possible that the symptoms of hangover are due to hemorheological changes. In a study of the pathogenesis of hangover Ylikahri et al.[40] noted that as peak acetaldehyde did not correlate with the severity of hangover, it could not be a cause of the symptoms. However, the observation of Baraona et al.[29] of erythrocyte-bound acetaldehyde may lead to a revision of that conclusion. Furthermore, two factors which could influence blood rheology showed good time-related correlations with the severity of hangover. Plasma triglyceride levels which rose and fell after 12 h would increase plasma viscosity. Blood lactate rose to a peak at a time when physical signs were most abnormal. Shand[41] has shown that RBC incubated with physiological concentrations of lactate showed reduced filterability with increasing concentration.

## C. Conclusion

Future studies of the acute effects of drinking alcohol will need to take cognizance of the need to select panels of subjects with similar rates of ethanol absorption and metabolism. In addition, such panels will need to comprise only those who vasoconstrict or who vasodilate in response to ethanol. If panel homogeneity is to be achieved it may be necessary also to take cognizance of the genetic nature of the enzymes alcohol dehydrogenase and aldehyde dehydrogenase. All or some of these potential sources of variation will have contributed to the differences among the results of published studies.

There is evidence that small amounts of ethanol improve blood rheology in those with the "usual" aldehyde dehydrogenase. The rise in blood acetaldehyde following ethanol ingestion by those with an "unusual" aldehyde dehydrogenase, and the similarities of their symptoms to those manifested by subjects of Disulfiram after alcohol, point to acetaldehyde being involved causally. Whether or not this is due to a direct effect through erythrocyte binding or an indirect effect through NE remains to be determined. The symptoms of hangover which follow a drinking episode with high levels of blood alcohol can be explained on rheological grounds, and is an area which needs further investigation. Finally, if hangover does have a hemorheological basis, then it might be prevented by taking evening primrose oil to enhance $PGE_1$ formation or fish oil rich in eicosapentaenoic acid which has been shown to improve RBC membrane fluidity.

## III. PROLONGED ETHANOL INTAKE AND BLOOD RHEOLOGY

Although several studies of the acute effects of alcohol on blood rheology have been referred to, only one paper dealing with the hemorheological consequences of chronic alcoholism was identified in a computerized literature search. Therefore, the proposals in this and the following sections are extrapolations from the known effects on blood rheology of acutely raised levels of blood alcohol and the numerous reports of impaired microcirculatory blood flow in alcoholics. In addition to the acute effects which ethanol induces in the flow properties of blood, prolonged heavy drinking, frequently accompanied by inappropriate nutrition, induces further changes. Such changes may be due to intensification of those conditions already described, and/or to effects which may be secondary to ethanol-related organ dysfunction.

### A. Changes in Blood Cells

Larkin and Watson-Williams[42] provided a detailed account of the effects of chronic alcoholism on hemopoietic tissue and RBCs. They emphasized the difficulties of discriminating between primary and secondary and direct or indirect effects of alcohol which present as a variety of anemias. To a large extent such changes were due to inadequate nutrition and/or the consequences of increasing liver dysfunction. Possibly because of increasing abnormalities of lipid metabolism, alcoholic liver disease is associated with a variety of changes in red cell morphology. The most frequently reported change is an increase in red cell size (macrocytosis), although it has been shown that increased mean cell volume is not sufficiently specific to be used as a marker for alcoholism.[43] Macrocytic target cells with mean corpuscular volume ranging from 100 to 106 fl (normal range 76 to 96) develop by acquiring increased amounts of cholesterol and phosphatidylcholine (predominantly the former) from plasma lipoprotein cholesterol and RBC membrane. The enlarged surface area was considered to confer a survival advantage, and target cells had increased osmotic resistance which reverted to normal osmotic fragility after the cells had been transfused into normal recipients for a few days.[42] Cooper[44] reported that sera from patients with severe alcoholic liver disease and "spur cells" (erythrocytes with branched extensions of the cell membrane) readily transferred free cholesterol to normal red cells which developed "thorny projections". Similar observations have been reported by others.[45] As increased cholesterol content of cell membranes is associated with reduced flexibility, it is likely that erythrocytes having an enhanced cholesterol content would exhibit reduced deformability.

Despite the hemorheological implications of the changes in red cell morphology which have been reported, there is a surprising lack of information concerning blood rheology and alcohol abuse. In addition to macrocytic target cells, spur cells, stomatocytes[42,46] (cup-shaped red cells), and knizocytes[47] (red cells with triconcave indentations) have been described. As it has been found that such cells could not be produced by incubation with ethanol or acetaldehyde, it has been proposed that the presence of such abnormal RBCs may be a marker for alcohol abuse.[48] It is unlikely that such morphological changes can occur without some change in their ability to deform. Evidently, the ethanol-induced changes in morphology are associated with an alteration in membrane fluidity, as Chin and Goldstein[49] found that RBCs from mice chronically treated with ethanol did not fluidize their membranes when exposed to ethanol *in vitro*. The hyperlipoproteinemia resulting from alcoholic liver disease could be expected to influence plasma viscosity. Goebel et al.[50] studied the blood rheology of 12 alcoholic patients during the acute phase and while in remission from chronic alcoholism. While the patients exhibited the major signs of alcoholic liver disease (hemolytic anemia, jaundice, hyperlipoproteinemia) RBCs revealed markedly impaired deformability when compared with RBCs obtained after the symptoms of liver disease had subsided. They concluded that the marked changes in the rheological properties of RBCs might impair the flow of blood in the peripheral circulation.

## B. The Rheological Consequences

In spite of lack of specific studies there are good reasons to believe that because of the combined influence of a number of factors, blood rheology in alcoholics will be altered. Thus, the ethanol-induced depletion of DHGLA stores reduces $PGE_1$ synthesis; acetaldehyde-related release of NE may result in a reduction in RBC deformability; a state of "chronic aldehydism"[30] may develop in those with genetically determined enzyme patterns; and the changes in blood lipids due to liver dysfunction would have a cumulatively adverse effect on blood rheology in general and RBC deformability in particular. Changes in the flow properties of blood resulting from the consumption of alcohol have been assessed by studying blood flow in the capillaries of the conjunctiva.[51] The subjects were "periodic obsessive drinkers" hospitalized to recover from an episode of heavy drinking. It was not possible to ascertain how long the drinking bout had lasted, but at the time of the study blood alcohol levels ranged from 0 to 328 mg/dl. Reduced rates of blood flow were observed in association with blood alcohol concentrations of 200 mg/dl and higher. In addition, at higher concentrations, stasis and vascular occlusions were seen frequently and microscopic hemorrhages were observed at blood alcohol levels in excess of 225 mg/dl. The authors expressed the view that the sludging of blood could cause damage to brain, liver, and heart by initiating locally defined areas of hypoxia or anoxia. Since then it has been shown that during the first 2 weeks of withdrawal, alcoholics had significant global reduction in cerebral blood flow,[52] and that there was a linear relationship between ethanol consumption and the degree of impairment of gray matter blood flow.[53]

## C. Conclusion

Given the extent of the reported changes in RBC morphology and the probable effects on plasma viscosity of abnormal lipid levels due to alcohol-related liver dysfunction, it is unlikely that blood rheology will remain unaltered in chronic alcoholism. Furthermore, it would seem highly improbable that blood flow in the microcirculation could be impaired to the extent that localized regions of stasis were observed, without pathological consequences.

## IV. ALTERED BLOOD RHEOLOGY AS A PATHOGENIC AGENT IN THE COMPLICATIONS OF ALCOHOL ABUSE

Armed with the knowledge that prolonged heavy drinking results in liver dysfunction and changes in RBC morphology and deformability, it may be possible to unravel the fabric of alcohol-induced pathology, thus, enabling the thread of altered blood rheology which permeates the complications of alcohol abuse to be recognized. Goebel et al.[50] expressed the view that RBCs with changed rheological characteristics might impair flow in the peripheral circulation. Although Moskow et al.[51] did not consider their observations from a hemorheological point of view, their conclusions are of such significance as to justify their repetition. They concluded that the effects of ethanol on circulating blood cells reduced the forward rates of blood flow in capillaries. The reduction in flow rate led to edema, red cell diapedesis, plugging of vessels, congestion, possible stasis, and hemorrhages. As a result anatomically defined minute areas of severe hypoxia and anoxia developed. In nerve tissue such anoxia led to the destruction of neurons and it was pointed out that it was possible that anoxic areas could develop around "minute cardiac venules". This concept is compatible with a suggested pathogenic potential of rheologically impaired blood flow in the microcirculation. It should be emphasized that localized anoxia with congested capillaries could lead to focal ischemic necrosis, with the significance of such lesions being determined by their anatomical location.

The pathogenesis of some of the more important complications of alcoholism will be discussed from the point of view that the essential element is impaired microcirculatory blood flow due to altered blood rheology. Although it is possible that other metabolites of ethanol may also have deleterious effects on blood rheology, in alcoholics the major factor may be the effect on RBC deformability of an enhanced secretion of NE, the release of which is triggered either by ethanol or acetaldehyde.

## A. Alcohol and Hypertension

Many authors have made the observation that there is a relationship between the regular consumption of alcohol and raised blood pressure, while emphasizing the fact that the mechanisms involved are unknown. Saunders et al.[54] studied blood pressures of alcoholics while they were drinking, during withdrawal, and after a period of abstinence. They found that blood pressures were high initially, fell to normal in most cases after detoxification, and rose again in those who recommenced their old drinking habits. A conclusion from that study was that many of the symptoms of alcohol withdrawal were suggestive of a hyper-adrenergic state. Arkwright et al.[55] found that there was a close relationship between the quantity of alcohol consumed regularly and the level of blood pressure in a group of civil servants. In a companion paper[56] they reported the results of a study of matched pairs of men who were drinkers or nondrinkers in which pressor hormone responses to standardized physiologic stresses were assayed. Although it was not possible to distinguish drinkers from nondrinkers on the basis of circulating NE or other pressor hormones, the authors did not exclude the possibility that those agents had been involved earlier in causing the observed differences in blood pressure. There are many reports of the levels of circulating catechol-amines at various times after alcohol loading in normal or alcoholic subjects or in experimental animals. Carlsson and Haggendal[57] found that in alcoholic subjects, NE levels were highest 13 to 19 h after they stopped drinking alcohol, i.e., when reduced RBC deformability and increased blood viscosity has been noted.[21,22] However, plasma NE levels did not differ between individuals of the same age with high or low alcohol intake,[58] nor were plasma NE levels uniformly raised in association with high alcohol consumption. In an attempt to define the role of NE in essential hypertension, Goldstein[59] analyzed the results of 32 studies reporting plasma NE concentrations. He found that the differences arose from three factors: the normotensive control values, type of assay used, and age of the patient population. In a comprehensive examination of the possible mechanisms of alcohol-associated hypertension, Potter and Beevers[60] summarized their views as follows: "Alcohol dependent persons have a high incidence of hypertension which is a common clinical problem. It is probably due to alcohol withdrawal and is mediated via increased cortisol and catecholamine production." According to Davis et al.,[61] ethanol ingestion alters the metabolism of both NE and epinephrine from the normal oxidative route to a reductive pathway which results in a marked shift in the nature of the metabolites. Gitlow et al.,[34] by means of a radiolabeling technique, confirmed the observation of Davis et al.[61] that ethanol could modify the metabolism of NE. It was noted that ethanol enhanced the synthesis and turnover of NE, but withdrawal of alcohol was followed by a prompt decrease in turnover rate. This implies that high levels of NE would be sustained for some time after drinking ceased. Gitlow et al.[34] noted that the turnover of the labeled pool of NE "which was most rapid during ethanol ingestion, rebounded to its slowest rate during withdrawal before returning to normal." Borg et al.[62] showed that in patients with alcohol dependence the cerebrospinal fluid content of 3-methoxy-4-hydroxy-phenylethyleneglycol, a metabolite of NE by the reductive pathway, was maximal during alcohol intoxication, had fallen markedly 24 h later, and by 21 d showed a similar distribution to that in healthy volunteers. Such reports substantiate the idea that in alcoholics abnormally high levels of NE may prevail for some time. In addition to any vasoactive effects, the high NE concentrations could reduce RBC deformability. This would increase high shear rate blood viscosity, impairing microcirculatory blood flow and cause a commensurate rise in total peripheral resistance. There seems to be a close parallelism between the role of NE in the hypertension of alcoholics and in nonalcoholics. Messerli et al.[63] found that in borderline hypertensives, circulating NE increased with age with a corresponding increase in total peripheral resistance, which they suggested might give rise to established hypertension. In a later study involving 30 patients more than 65 years of age with established essential hypertension, it was found that high arterial blood pressure was associated with an hypertrophied heart working against a high total peripheral resistance.[64] Corea et al.[65] suggested that NE might be associated with pressure factors in regulating the development of left ventricular hypertrophy in patients with hypertension. As Devereux et al.[66] have proposed that whole blood viscosity might be a determinant of cardiac hypertrophy in hypertension, it is possible that the ''pressure factors'' of Corea et al.[65] could involve altered blood rheology. This suggestion is given credibility by other reports of increased blood viscosity in patients with hypertension.[67-69]

We conclude that altered blood rheology may have a causal role in the pathogenesis of alcoholic hypertension. The reduction in RBC deformability due to the ethanol-enhanced synthesis of NE would increase high shear rate blood viscosity and enhance total peripheral resistance. The extent to which blood pressure was raised would depend upon the pressure needed to overcome the peripheral resistance, and would fall when the RBC stiffening influence of NE waned. Such a proposal would account for the observations of Saunders et al.[54] that blood pressure fell during abstinence and rose again after further ethanol ingestion.

## B. Alcohol and the Central Nervous System

Because of its overall importance, this topic will be considered in some detail. The extensive literature dealing with alcohol-related changes will be subdivided into two broad categories for the purpose of this discussion: first, those studies which deal with the mechanisms of brain damage manifested as altered behavior and/or functional disorders; second, those studies dealing with the association of alcohol and acute cerebrovascular accidents (CVAs) and stroke.

### 1. Studies Dealing with the Mechanisms of Brain Damage

There is substantial evidence that the aging process is associated with a progressive reduction in the number of cerebral cortical neurons. It has been suggested that the observed age-related reduction in regional cerebral blood flow[70] could be a response to a reduction in the need for metabolites from the smaller cell population. On the other hand, aging is associated with a reduction in RBC deformability and it is possible that the loss of cerebral neurons is the result of anoxic-ischemic death because of inefficient perfusion. In relatively young subjects, alcohol-induced reduction in RBC deformability may be associated with brain changes similar to those induced by age alone. For this reason it seems more likely that loss of cerebral neurons is the result rather than a cause of a decline in regional cerebral blood flow. Several workers comment that changes seen in alcohol-damaged brains are suggestive of accelerated aging. Therefore, in elderly alcoholics it may be impossible to separate the contributions which age and alcohol make to their brain damage.

### a. Cerebral and Cerebellar Changes

Perhaps the most striking effects on the brain of chronic alcohol abuse is the enlargement of the ventricles as revealed by pneumoencephalography[71,72] or computerized tomography,[73,74] with an associated thinning of the cerebral cortical gray matter[75] and atrophy of the anterior superior part of the cerebellar vermis.[76,77] (See also this volume, Chapter 3.) Goldman et al.[78] showed that rats given an alcohol load of 2.4 g/kg had reduced blood flow to the cerebellum and hippocampus, and it was suggested that both regions ''...may be selectively vulnerable to ethanol.'' In a comprehensive study of the degeneration which occurs in the cerebellar cortex of alcoholics, Victor et al.[79] refuted the idea that the degenerative changes were restricted to Purkinje cells and noted that all neural elements appeared to be affected equally. Furthermore, by concluding that the localization of the lesions could be due to "...some factor of circulation" they implied an ischemic event as the cause of the degenerative lesions. However, in discussing the etiology of cerebellar lesions they reached no firm conclusions concerning causal mechanisms. It was, however, pointed out that the cerebellar lesions correspond to the "leg area" of experimental animals. This provided a basis for understanding the poor leg coordination which was the most obvious feature of their alcoholic patients. In contrast to the permanent effects of the cerebellar lesions there are several reports indicating that the cerebral deficits may improve in the absence of alcohol. The time relationships indicate that such improvements are more than a simple absence of alcohol. Carlen et al.[80] showed partially reversible "cerebral atrophy" by repeated computer tomography scans of abstinent alcoholics who also showed functional improvement.

The etiology of the degenerative changes in the brains of alcoholics has yet to be explained fully, but there seems little doubt that blood flow is impaired and capillary dysfunction plays an important role. Xenon washout studies have shown that alcohol consumption was inversely related to blood flow in cerebral gray matter.[53] As the decline in regional cerebral blood flow was detectable in social drinkers, it was proposed that reduced blood flow might be due to the neurotoxic effects of chronic alcohol consumption as part of a continuum, with the Wernicke-Korsakoff syndrome as the end point.[53] Earlier it had been shown that the major abnormality in patients with the Wernicke-Korsakoff syndrome was a reduction in total cerebral blood flow[81] associated with increased cerebrovascular resistance and marked reduction in blood flow to the gray matter. Lynch[75] quoted Courville (not seen) as stating that " . . . some interference with capillary blood supply lies at the root of the process" (of chronic nerve cell loss). In addition, Courville was quoted as pointing out the pericapillary distribution of brain lesions in alcoholics " . . . which could be due to ischaemia produced by blocking of the vessel." The findings of Harper[76] of perivascular tissue necrosis are in accord with Courville's observations. This evidence of alcohol-related reduction in cerebral blood flow with concurrent pericapillary brain lesions parallels the conditions which have been reported as relating to the brain changes in multiple sclerosis (MS). Swank et al.[82] showed that cerebral blood flow in MS patients was much reduced in comparison with normal subjects, and Putnam and Alexander[83] and Macchi[84] have described pericapillary changes and capillary congestion as a feature of plaque formation in MS brain. Furthermore, just as the alcoholic disorders are associated with altered blood rheology and poorly deformable RBCs, increased high shear rate blood viscosity and morphologically abnormal RBCs have been found in patients with clinically definite MS.[85] Until there are more studies of blood rheology in chronic alcoholics before, during the development of brain lesions, and after their effects have become obvious, it is possible only to speculate about the role of altered blood rheology in the pathogenesis of cerebral and cerebellar lesions. However, there are sufficient indications to suggest that hemorheological studies in chronic alcoholics could be rewarding. For example, Rogers et al.[53] concluded that in contrast to his age-matched counterparts, the heavy drinker with alcohol-impaired blood rheology was predisposed to other disorders, not linked to alcohol, including cerebrovascular accidents.

### b. Changes in the Cerebrospinal Fluid

Neuronal loss is associated with the appearance of the breakdown products of nucleic acids accumulating in the cerebrospinal fluid (CSF). Farstad et al.[86] showed that in psychiatric patients without cerebral atrophy CSF uric acid levels were within the normal range. In a companion paper[87] it was shown that similar levels of CSF uric acid occurred in patients without progressive cerebral atrophy, but significantly increased CSF uric acid levels were found in those with progressing atrophy. As there was no dependence of CSF uric acid on serum uric acid, it was considered that the CSF acid was derived from a "nonserum" source, most likely from the catabolism of cerebral neuronal nucleic acids. Carlsson and Dencker[88] found high levels of uric acid in CSF (but not in serum) of chronic alcoholics following a bout of drinking and after they had spent some time in the hospital. Those observations provide a basis for understanding the CSF acidosis which was found in chronic alcoholics many weeks after their last drink, in the absence of systemic acidosis.[89] The acidosis did not occur in controls with other neurological disorders. A general review of the relationship between CSF pH and neurologic symptoms concluded that CSF acidosis was likely to affect brain function adversely irrespective of the level of systemic pH.[90]

### c. Conclusion

On the basis of the studies referred to above it is suggested that alcohol-related brain changes have impaired blood flow as their primary pathogenic agent. Cerebellar lesions appear to be irreversible and could be the consequence of impaired blood supply resulting in ischemic necrosis of the anterior superior region of the cerebellum. Evidence of impaired middle cerebral artery blood flow and reduced regional cerebral blood flow in the elderly[70] provides a basis for proposing an ischemic cause of neuronal death in alcoholics. Catabolism of the nucleic acids of dead neurons might produce sufficient uric acid to increase CSF acidity with deleterious effects on central nervous system function.[87,88] Because the pH of the CSF changes relatively slowly, the effects of abstinence from alcohol will become manifest only with reduced CSF acidity leading to improved brain function with progressive improvement spanning several weeks. Abstinence may not only result in improved blood rheology, but also give rise to a partial decrease in cerebral atrophy.[80]

### 2. Alcohol, Cerebrovascular Accidents, and Stroke

A reduction in RBC deformability may alter the flow properties of blood sufficiently to impair microcirculatory blood flow. When other factors such as reduction in vessel lumen dimensions by vasospasm, or vessel wall pathology or a sudden lowering of systemic blood pressure occur concurrently with altered blood rheology, the consequences for blood flow may be beyond the compensatory limits of autoregulation and result in localized stasis. If the stasis is irreversible, some tissue will become hypoxic, anoxic, and die of ischemic necrosis. Whether or not such lesions have physiological significance will be determined by their anatomical location. An example of such changes in the brain are the "silent" plaques which occur in MS.

Applying these basic assumptions to CVAs we would anticipate that the conditions which might give rise to an alcohol-related CVA would be as follows: (1) an unusually high level of alcohol consumption; (2) a reduction in RBC deformability due to the combined effects of intraerythrocyte acetaldehyde and the ethanol-enhanced, acetaldehyde-mediated release of NE; (3) the persistence of high levels of plasma NE for 12 to 18 h due to the reduction in NE turnover rate following the cessation of drinking; (4) the presence of some other factor which reduces the flow properties of blood, i.e., and aberrant vasospasm, a fall in central blood pressure, cigarette smoking, or the use of oral contraceptives.

### a. CVAs, Alcoholics, and "Binge" Drinking

There are a number of reports of CVAs in alcoholics, e.g., References 91 and 92, and according to Walbran et al.,[92] alcoholics with cerebral infarcts had average survivals about 10 years less than nonalcoholics with infarcts. Hillbom and Kaste[93] reported that in 76 successive cases of ischemic brain infarction in subjects less than 40 years old, 20% had been on a drinking spree within 24 h of the onset of symptoms. Ethanol-related infarcts were more common in the youngest age group (40% of 16 to 19 year olds) than in the oldest group (13% of 30 to 39 year olds). In a subsequent paper they confirmed that ethanol intoxication might precipitate brain infarction in young people and showed that most cases occurred as a consequence of drinking on Friday and Saturday nights.[94] Although the authors referred to studies on the hemorheological effects of alcohol, they concluded that it was unclear why ethanol increased the risk of cerebral infarction. To some extent their results were corroborated by an American study[95] which had the disadvantage of small numbers and problems in obtaining appropriate controls. It was pointed out that stroke patients were not only more likely to be drinkers than controls, but also that they had been drinking within 24 h of the first symptoms of stroke. The possibility of a relationship between "binge" drinking and stroke was raised in a *Lancet* editorial[96] and discussed in a rather sceptical manner. Another report involving stroke after "binge" drinking[97] led to the conclusion that excessive consumption of alcohol may be a risk factor for stroke. In the ensuing correspondence Gill et al.[98] expressed doubt, as they had done earlier,[99] that "binge" drinking was as important a predisposing factor for stroke as regular, moderate, to heavy drinking. Other contributors[100] put forward the proposition that marked hypotension after the acute ingestion of a large amount of alcohol could be a major factor in stroke. Although none of those reports referred to the pathogenic potential of alcohol-induced impairment of blood rheology, in view of what has been proposed about the hemorheological effects of ethanol and its metabolite acetaldehyde, there is no doubt that because of the thixotropic nature of blood an episode of hypotension would have disastrous effects on cerebral blood flow. It has been reported that rat pial arterioles and venules responded by contraction to graded concentrations of alcohol.[101] "Threshold constriction of arterioles" was induced with local alcohol concentrations between 10 and 15 mg/dl, while the constrictions following exposure to alcohol concentrations up to 200 mg/dl showed a dose-related relaxation " . . . within 5 to 40 minutes." When perivascular or blood alcohol concentrations exceeded 300 mg/dl, irreversible spasm, often followed by rupture with local hemorrhage, occurred within 5 to 10 min. Similar changes and effects were observed in cortical venules where ethanol concentrations of 0.01 to 1% caused a reduction in vessel diameter of from 8 to 60%, while 3% or higher concentrations of ethanol "...often resulted in irreversible spasm and rupture." It was noted also that "blood flow in cortical microvessels was markedly curtailed." Altura et al.[101] proposed that their results supported an earlier suggestion that impaired cerebral blood flow due to vasoconstriction resulted in brain hypoxia.[78] Furthermore, it was proposed that the results might explain the hypertension of alcoholics and the CVAs of "binge" drinkers. However, it must be emphasized that hypertension and CVAs occur hours after drinking and are much more likely to be responses to the secondary effects of alcohol rather than to alcohol per se, as explained earlier in this paper. A more important aspect of that paper[101] is that it took no account of the effects of sustained vasoconstriction on the flow properties of the blood, especially in vessels distal to the constriction. During the 5 to 40-min period of vasoconstriction induced by ethanol concentrations up to 200 mg/dl, changes will take place which affect the flow properties of the blood. Simply because of the reduction in flow rate and, therefore, shear rate, thixotropic amplification of blood viscosity will occur and increase resistance to flow. Should stasis occur with the formation of three-dimensional aggregates, clotting phenomena would not be involved, but an adequate pressure gradient to disrupt the bonding of the temporary aggregates would be needed to initiate reflow. As very high intravascular pressure would be needed to disperse such aggregates, the stasis may become irreversible. Finally, in such conditions the lack of oxygen could lead to ischemic necrosis. If ethanol-induced vasoconstriction occurred in human beings, then CVAs should occur at the time of peak blood alcohol concentration. The fact that this does not occur tends to diminish the relevance of this frequently quoted paper.

In an editorial on cerebral ischemia Astrup et al.[102] pointed out that "complete arrest of the cerebral circulation leads within seconds to cessation of neuronal electrical activity and within a few minutes to deterioration of the energy state and ion homeostasis." Should this persist for longer than 5 to 10 min it is likely that irreversible cell damage will occur. Although such changes were the inevitable outcome of cessation of blood flow in the brain, the outcome of reduced blood flow is more difficult to predict and is largely dependent on the extent to which residual perfusion can continue to supply oxygen to the tissue involved. It seems that irrespective of the initial cause of the stoppage or reduction in blood flow, a variety of factors, either intrinsic to the blood itself or arising as a consequence of impaired blood flow, will act together to make reinstatement of flow difficult or impossible.

### b. Conclusion

As alcoholic-related CVAs occur 12 to 18 h after the end of a drinking "binge", it is more likely that the cerebral infarct relates to the ethanol-induced increase in plasma NE and its persistence after drinking than to ethanol per se, as assayed by blood alcohol levels. About the times that CVAs occur, RBCs containing reversibly bound acetaldehyde would have been exposed to the adverse influence of NE for some hours and for those reasons their deformability would be reduced. This change has been reported[21] together with an associated increase in blood viscosity.[21,22] Capillary blood flow would be impaired commensurately with the degree of hemorheological change. Two types of pathology could arise from such circumstances. First, in the absence of a sufficient autoregulated rise in blood pressure irreversible stasis could develop, resulting in a localized region of ischemic necrosis. Alternatively, high blood pressure developed by an autoregulated vasodilation to overcome the resistance to flow of viscous blood could rupture a blood vessel at a junction or at a previously weakened point and result in a cerebral hemorrhage. This proposal of a hemorheological basis for alcohol-related stroke is not entirely without supporting evidence. Since it provides a basis for the logical explanation of the mechanisms of cerebral infarction after excessive alcohol consumption, it is a proposal which deserves investigation. A corollary to the proposal is that it implies that preventive measures such as the consumption of oils which improve RBC deformability, prior to going on a drinking spree, might reduce the hemorheological consequences of ethanol ingestion.

## V. EPILOGUE

In the introduction we stated the objective of drawing to the attention of readers the effects which alcohol consumption might have on the flow properties of blood. By using hypertension and brain damage as examples of the complications of alcohol abuse we have endeavored to demonstrate the pathogenic potential of altered blood rheology. Limitation on space prevents detailed examination of other important complications in the chronic alcoholic. Unless there is recognition of impaired blood flow as a consequence of alcohol ingestion, major problems will confront investigators trying to unravel the tangled threads of primary and secondary events which can occur in the liver, the placenta, and in skeletal muscles, in which reduced blood flow disrupts normal metabolic processes. Therefore, it is essential that alcohol-related change in blood rheology be recognized as a systemic factor which has the potential to initiate ischemic necrosis in any organ in the body.

## REFERENCES

1. **Landis, E. M.,** Capillary pressure and capillary permeability, *Physiol. Rev.,* 14, 404, 1934.
2. **Simpson, L. O.,** Basement membranes and biological thixotropy: a new hypothesis, *Pathology,* 12, 377, 1980.

3. **Simpson, L. O.**, Biological thixotropy of basement membranes: the key to the understanding of capillary permeability, in *Progress in Microcirculation Research,* Garlick, D., Ed., Committee in Postgraduate Medical Education, The University of New South Wales, Sydney, Australia, 1981, 55.

4. **Ohsaka, A., Suzuki, K., and Ohashi, M.,** The spurting of erythrocytes through junctions of the vascular endothelium treated with snake venom, *Microvasc. Res.,* 10, 208, 1975.

5. **Chien, S.,** Principles and techniques for assessing erythrocyte deformability, *Blood Cells,* 3, 71, 1977.

6. **Wells, R. E., Gawronski, T. H., Cox, P. J., and Perea, R. D.,** Influence of fibrinogen on the flow properties of erythrocytes suspensions, *Am. J. Physiol.,* 207, 1035, 1964.

7. **Chien, S., Usami, S., Dellenback, R. J., and Gregersen, M. I.,** Shear dependent interaction of plasma proteins with erythrocytes in blood rheology, *Am. J. Physiol.,* 219, 143, 1970.

8. **Wells, R. E.,** Syndromes of hyperviscosity, *N. Engl. J. Med.,* 283, 183, 1970.

9. **Fishberg, A. M.,** *Hypertension and Nephritis,* 5th ed., Lea & Febiger, Philadelphia, 1954, 296.

10. **Huang, C. R., Siskovic, N., Robertson, R. W., Fabisiak, W., Smitherberg, E. H., and Copley, A. L.,** Quantitative characterisation of thixotropy of whole human blood, *Biorheology,* 12, 279, 1975.

11. **McMillan, D. E., Utterback, N. G., and Baldridge, J. B.,** Thixotropy of blood and red cell suspensions, *Biorheology,* 17, 445, 1980.

12. **Fewings, J. D., Hanna, M. J. D., Walsh, J. A., and Whelan, R. F.,** The effects of ethyl alcohol on the blood vessels of the hand and forearm in man, *Br. J. Pharmacol. Chemother.,* 27, 93, 1966.

13. **Dengerink, H. A., Mead, J. D., and Bertilson, H. S.,** Individual differences in response to alcohol. Vasoconstriction and vasodilation, *J. Stud. Alcohol,* 39, 12, 1978.

14. **Couchman, K. G., Ethanol Metabolism in Humans, Ph.D.** Thesis, Massey University, New Zealand, 1979.

15. **McFarlane, A., Pooley, L., Welch, I. McL., Rumsey, R. D. E., and Read, N. W.,** How does dietary lipid lower blood alcohol concentrations?, *Gut,* 21, 15, 1986.

16. **Horrobin, D. F.,** A biochemical basis for alcoholism and alcohol-induced damage including the fetal alcohol syndrome and cirrhosis: interference with essential fatty acid and prostaglandin metabolism, *Med. Hypotheses,* 6, 929, 1980.

17. **Kury, P. G., Ramwell, P. W., and McConnell, H. M.,** The effects of prostaglandins $E_1$ and $E_2$ on the human erythrocyte as monitored by spin labels, *Biochem. Biophys. Res. Commun.* 56, 478, 1974.

18. **Rasmussen, H., Lake, W., and Allen, J. E.,** The effect of catecholamines and prostaglandins upon human and rat erythrocytes, *Biochim. Biophys. Acta,* 411, 63, 1975.

19. **Dowd, P. M., Kovacs, I. B., Bland, C. J. H., and Kirby, J. D. T.,** Effect of prostaglandins $I_2$ and $E_1$ on red cell deformability in patients with Raynaud's phenomenon and systemic sclerosis, *Br. Med. J.,* 283, 350, 1981.

20. **Kyle, V. and Hazleman, B.,** Prostaglandin $E_1$ and peripheral vascular disease, *Lancet,* 2, 282, 1983.

21. **Hillbom, M. E. Kaste, M., Tarssanen, L., and Johnsson, R.,** Effect of ethanol on blood viscosity and erythrocyte flexibility in healthy men, *Eur. J. Clin. Invest.,* 13, 45, 1983.

22. **Galea, G. and Davidson, R. J. L.,** Some haemorheological and haematological effects of alcohol, *Scand. J. Haematol.,* 30, 308, 1983.

23. **Hamazaki, T. and Shishido, H.,** Increase in blood viscosity due to alcohol drinking, *Thromb. Res.,* 30, 587, 1983.

24. **Lieber, C. S.,** Metabolism and metabolic effects of alcohol, *Semin. Haematol.,* 17, 85, 1980.

25. **Korsten, M. A., Matsuzaki, S., Feinman, L., and Lieber, C. S.,** High blood acetaldehyde levels after ethanol administration. Difference between alcoholic and nonalcoholic subjects, *N. Engl. J. Med.,* 292, 386, 1975.

26. **Palmer, K. R. and Jenkins, W. J.,** Impaired acetaldehyde oxidation in alcoholics, *Gut,* 23, 729, 1982.

27. **Eriksson, C. J. P.,** Elevated blood acetaldehyde levels in alcoholics and their relatives: a reevaluation, *Science,* 207, 1383, 1980.

28. **Lindros, K. O.,** Human blood acetaldehyde, (Vol. 2, chap. 13 of these volumes).

29. **Baraona, E., Padova, C. D., Tabasco, J., and Lieber, C. S.,** Red blood cells: a new modality for acetaldehyde transport from liver to other tissues, *Life Sci.,* 40, 253, 1987.

30. **von Wartburg, J.-P. and Buhler, R.,** Biology of disease. Alcoholism and aldehydism: new biomedical concepts, *Lab. Invest.,* 50, 5, 1984.

31. **Mizoi, Y., Ijiri, I., Tatsuno, Y., Kijima, T., Fujiwara, S., and Adachi, J.,** Relationship between facial flushing and blood acetaldehyde levels after alcohol intake, *Pharmacol. Biochem. Behav.,* 10, 303, 1979.

32. **Harada, S., Misawa, S., Agarwal, D. P., and Goedde, H. W.,** Liver alcohol dehydrogenase and aldehyde dehydrogenase in the Japanese: isozyme variation and its possible role in intoxication, *Am. J. Hum. Genet.,* 32, 8, 1980.

33. **Adachi, J. and Mizoi, Y.,** Acetaldehyde-mediated alcohol sensitivity and elevation of plasma catecholamine in man, *Jpn. J. Pharmacol.,* 33, 531, 1983.

34. **Gitlow, S. E., Dziedzic, L. M., Dziedzic, S. W., and Wong, B. L.,** Influence of ethanol on human catecholamine metabolism, *Ann. N.Y. Acad. Sci.,* 273, 263, 1976.

35. **Chien, S., Usami, S., Dellenback, R. J., Bryant, G. A., and Gregersen, M. I.,** Change of erythrocyte deformability during fixation in acetaldehyde, in *Theoretical and Clinical Haemorheology,* Hartert, H. H. and Copley, A. L., Eds., Springer-Verlag, Berlin, 1971, 136.

36. **Hawley, R. J., Major, L. F., Schulman, E. A., and Lake, C. R.,** C.S.F. levels of norepinephrine during alcohol withdrawal, *Arch. Neurol.,* 38, 289, 1981.

37. **Logan, B. J., Laverty, R., and Peake, B. M.,** ESR measurements on the effects of ethanol on the lipid and protein conformation in biological membrane, *Pharmacol. Biochem. Behav.,* 18, 31, 1983.

38. **Hrelia, S., Lercker, G., Biagi, P. L., Bordoni, A., Stefanini, F., Zunarelli, P., and Rossi, C. A.,** Effect of ethanol intake on human erythrocyte membrane fluidity and lipid composition, *Biochem. Int.,* 12, 741, 1986.

39. **Benedetti, A., Birarelli, A. M., Brunelli, E., Curatola, G., Feretti, G., Jezequel, A. M., and Orlandi, F.,** Effect of chronic ethanol abuse on the physicochemical properties of erythrocyte membranes in man, *Pharmacol. Res. Commun.,* 18 (Abstr.) 1003, 1986.

40. **Ylikahri, R. H., Huttunen, M. D., Eriksson, C. J. P., and Nikkila, E. A.,** Metabolic studies on the pathogenesis of hangover, *Eur. J. Clin. Invest.,* 4, 93, 1974.

41. **Shand, B. I.,** Changes in blood rheology induced by lactic acid, *Proc. Univ. Otago Med. Sch.,* 64, 71, 1986.

42. **Larkin, E. C. and Watson-Williams, E. J.,** Alcohol and the blood, *Med. Clin. North Am.,* 68, 105, 1984.

43. **Stimmel, B., Korts, D., Jackson, G., and Gilbert, H. S.,** Failure of mean red cell volume to serve as a biological marker for alcoholism in narcotic dependence, *Am. J. Med.,* 74, 369, 1983.

44. **Cooper, R. A.,** Anaemia with spur cells: a red cell defect acquired in serum and modified in the circulation, *J. Clin. Invest.,* 48, 1820, 1969.

45. **Powell, L. W., Halliday, J. W., and Knowles, B. R.,** The relationship of red cell membrane lipid content to red cell morphology and survival in patients with liver disease, *Aust. N. Z. J. Med.,* 5, 101, 1975.

46. **Wisloff, F. and Boman, D.,** Acquired stomatocytosis in alcoholic liver disease, *Scand. J. Haematol.,* 23, 43, 1979.

47. **Homaidan, F., Kricka, L. J., and Whitehead, T. P.,** Morphology of red blood cells in alcoholics, *Lancet,* 1, 913, 1984.

48. **Homaiden, F. R., Kricka, L. J., Bailey, A. R., and Whitehead, T. P.,** Red cell morphology in alcoholics: a new test for alcohol abuse, *Blood Cells,* 11, 375, 1986.

49. **Chin, J. B. and Goldstein, D. B.,** Drug tolerance in biomembranes: a spin label study of the effects of ethanol, *Science,* 196, 684, 1977.

50. **Goebel, K. M., Goebel, F. D., and Lanser, K. G.,** Einfluss alkoholtoxischer Hyperlipoproteinamien auf die Verformbarkeit und Viskositatder Erythrozytosen, *Schweiz. Med. Wochenschr.,* 114, 1386, 1984.

51. **Moskow, H. A., Pennington, R. C., and Knisely, M. H.,** Alcohol, sludge, and hypoxic areas of nervous sytem, liver and heart, *Microvasc. Res.,* 1, 174, 1968.

52. **Berglund, M. and Risberg, J.,** Regional cerebral blood flow during alcohol withdrawal, *Arch. Gen. Psychiatry,* 38, 351, 1981.

53. **Rogers, R. L., Meyer, J. S., Shaw, T. G., and Mortel, K. F.,** Reductions in regional cerebral blood flow associated with chronic consumption of alcohol, *J. Am. Geriatr. Soc.,* 31, 540, 1983.

54. **Saunders, J. B., Beevers, D. G., and Paton, A.,** Alcohol-induced hypertension, *Lancet,* 2, 65, 1981.

55. **Arkwright, P. D., Beilin, L. J., Rouse, I., Armstrong, B. K., and Vandongen, R.,** Effects of alcohol use and other aspects of lifestyle on blood pressure levels and the prevalance of hypertension in a working population, *Circulation,* 66, 60, 1982.

56. **Arkwright, P. D., Beilin, L. J., Vandongen, R., Rouse, I. A., and Lalor, C.,** The pressor effect of moderate alcohol consumption in man: a search for mechanisms, *Circulation,* 66, 515, 1982.

57. **Carlsson, C. and Haggendal, J.,** Arterial noradrenaline levels after ethanol withdrawal, *Lancet,* 2, 889, 1967.

58. **Ibsen, H., Christensen, N. J., Rasmussen, S., Hollnagel, H., Nielsen, M. D., and Giese, J.,** The influence of chronic high alcohol intake on blood pressure, plasma noradrenaline concentration and plasma renin concentration, *Clin. Sci.,* 61 (Suppl. 7), 377s, 1981.

59. **Goldstein, D. S.,** Plasma norepinephrine in essential hypertension. A study of the studies, *Hypertension,* 3, 48, 1981.

60. **Potter, J. F. and Beevers, D. G.,** The possible mechanisms of alcohol associated hypertension, *Ann. Clin. Res.,* 16 (Suppl. 43), 97, 1984.

61. **Davis, V. E., Cashaw, J. L., Huff, J. A., Brown, H., and Nicholas, N. L.,** Alteration of endogeneous catecholamine metabolism by ethanol ingestion, *Proc. Soc. Exp. Biol. Med.,* 125, 1140, 1967.

62. **Borg, S., Liljeberg, P., and Mossberg, D.,** Clinical studies on central noradrenergic activity in alcohol abusing patients, *Acta Psychiatr. Scand.,* 73 (Suppl. 327), 43, 1986.

63. **Messerli, F. H., Frohlich, E. D., Suarez, D. H., Reisin, E., Dreslinski, G. R., Dunn, F. G., and Cole, F. E.**, Borderline hypertension: relationship between age, haemodynamics and circulating catecholamines, *Circulation*, 64, 760, 1981.

64. **Messerli, F. H., Sundergaard-Riise, K., Ventura, H. O., Dunn, F. G., Glade, L. B., and Frohlich, E. D.**, Essential hypertension in the elderly: haemodynamics, intravascular volume, plasma renin activity and circulating catecholamine levels, *Lancet*, 2, 983, 1983.

65. **Corea, L., Bentivoglio, M., Verdecchia, P., and Motolese, M.**, Plasma norepinephrine and left ventricular hypertrophy in systemic hypertension, *Am. J. Cardiol.*, 53, 1299, 1984.

66. **Devereux, R. B., Drayer, J. I. M., Chien, S., Pickering, T. G., Letcher, R. L., De Young, J. L., Sealey, J. E., and Laragh, J. H.**, Whole blood viscosity as a determinant of cardiac hypertrophy in systemic hypertension, *Am. J. Cardiol.*, 54, 592, 1984.

67. **Tibblin, G., Bergentz, S.-E., Bjure, J., and Wilhelmsen, L.**, Haematocrit, plasma protein, plasma volume, and viscosity in elderly hypertensive disease, *Am. Heart J.*, 72, 165, 1966.

68. **Letcher, R. L., Chien, S., Pickering, T. G., Sealey, J. E., and Laragh, J. H.**, Direct relationship between blood pressure and blood viscosity in normal and hypertensive subjects, *Am. J. Med.*, 70, 1195, 1981.

69. **Letcher, R. L., Chien, S., Pickering, T. G., and Laragh, J. H.**, Elevated blood viscosity in patients with borderline hypertension, *Hypertension*, 5, 757, 1983.

70. **Naritomi, H., Meyer, J. S., Sakai, F., Yamaguchi, F., and Shaw, T.**, Effect of advancing age on regional cerebral blood flow, *Arch. Neurol.*, 36, 410, 1979.

71. **Haug, J. O.**, Pneumoencephalographic evidence of brain damage in chronic alcoholics. A preliminary report, *Acta Psychiatr. Scand.*, 44 (Suppl. 203), 135, 1968.

72. **Carlsson, C., Claeson, L.-E., Karlsson, K.-I., and Petterson, L.-E.**, Clinical, psychometric and radiological signs of brain damage in chronic alcoholism, *Acta Neurol. Scand.*, 60, 85, 1979.

73. **Carlen, P. L., Wilkinson, D. A., Wortzman, G., Holgate, R., Cordingley, J., Lee, M. A., Huszar, L., Moddel, G., Singh, R., Kiraly, L., and Rankin, J. G.**, Cerebral atrophy and functional deficits in alcoholics without clinically apparent liver disease, *Neurology*, 31, 377, 1981.

74. **Ron, M. A., Acker, W., Shaw, G. K., and Lishman, W. A.**, Computerised tomography of the brain in chronic alcoholism. A survey and follow-up study, *Brain*, 105, 497, 1982.

75. **Lynch, M. J. G.**, Brain lesions in chronic alcoholism, *Arch. Pathol.*, 69, 342, 1960.

76. **Harper, C.**, The incidence of Wernicke's encephalopathy in Australia — a neuropathological study of 131 cases, *J. Neurol. Neurosurg. Psychiatry*, 46, 593, 1983.

77. **Pollock, M.**, Alcohol and the nervous system, *N. Z. Med. J.*, 79, 602, 1974.

78. **Goldman, H., Sapirstein, L. A., Murphy, S., and Moore, J.**, Alcohol and regional blood flow in the brains of rats, *Proc. Soc. Exp. Biol. Med.*, 144, 983, 1973.

79. **Victor, M., Adams, R. D., and Moncall, E. L.**, A restricted form of cerebellar cortical degeneration occurring in alcoholic patients, *Arch. Neurol.*, 1, 579, 1959.

80. **Carlen, P. L., Wortzman, G., Holgate, R. C., Wilkinson, D. A., and Rankin, J. G.**, Reversible cerebral atrophy in recently abstinent chronic alcoholics measured by computed tomography scans, *Science*, 200, 1076, 1970.

81. **Shimojyo, S., Steinberg, P., and Reinmuth, O.**, Cerebral blood flow and metabolism in the Wernicke-Korsakoff syndrome, *J. Clin. Invest.*, 46, 849, 1967.

82. **Swank, R. L., Roth, J. G., and Woody, D. C.**, Cerebral blood flow and red cell delivery in normal subjects and in multiple sclerosis, *Neurol. Res.*, 5, 37, 1983.

83. **Putman, T. J. and Alexander, L.**, Disseminated encephalomyelitis. A histological syndrome associated with thrombosis of small cerebral vessels, *Arch. Neurol. Psychiatry*, 41, 1087, 1939.

84. **Macchi, G.**, The pathology of the blood vessels in multiple sclerosis, *J. Neuropathol. Exp. Neurol.*, 13, 378, 1954.

85. **Simpson, L. O., Shand, B. I., and Olds, R. J.**, Red cell and haemorheological changes in multiple sclerosis, *Pathology*, 19, 51, 1987.

86. **Farstad, M., Skaug, O. E., and Solheim, D. M.**, Uric acid in the cerebrospinal fluid in psychiatric disorders, *Acta Neurol. Scand.*, 41, 59, 1965.

87. **Farstad, M., Haug, J. O., Lindbak, H., and Skaug, O. E.**, Uric acid in the cerebrospinal fluid in cerebral atrophy, *Acta Neurol. Scand.*, 41, 52, 1965.

88. **Carlsson, C. and Dencker, S. J.**, Cerebrospinal uric acid in alcoholics, *Acta Neurol. Scand.*, 49, 39, 1973.

89. **Carlen, P.L., Kapur, B., Huszar, L. A., Lee, M. A., Moddel, G., Singh, R., and Wilkinson, D. A.**, Prolonged cerebrospinal fluid acidosis in recently abstinent chronic alcoholics, *Neurology*, 30, 956, 1980.

90. **Posner, J. B. and Plum, F.**, Spinal-fluid pH and neurologic symptoms in systemic acidosis, *N. Engl. J. Med.*, 277, 605, 1967.

91. **Lee, K.**, Alcoholism and cerebrovascular thrombosis in the young, *Acta Neurol. Scand.*, 59, 270, 1979.

92. **Walbran, B. B., Nelson, J. S., and Taylor, J. R.,** Association of cerebral infarction and chronic alcoholism: an autopsy study, *Alcoholism Clin. Exp. Res.,* 5, 531, 1981.
93. **Hillbom, M. and Kaste, M.,** Does ethanol intoxication promote brain infarction in young adults?, *Lancet,* 2, 1181, 1978.
94. **Hillbom, M. and Kaste, M.,** Ethanol intoxication: a risk factor for ischaemic brain infarction in adolescents and young adults, *Stroke,* 12, 422, 1981.
95. **Taylor, J. R., Combs-Orme, T., Anderson, D., Taylor, D. A., and Koppenol, C.,** Alcohol, hypertension and stroke, *Alcoholism Clin. Exp. Res.,* 8, 283, 1984.
96. **Anon.,** "Binge" drinking and stroke, *Lancet,* 2, 660, 1983.
97. **Wilkins, M. R. and Kendall, M. J.,** Stroke affecting young men after alcoholic binges, *Br. Med. J.,* 291, 1342, 1985.
98. **Gill, J. S., Zezulka, A. V., and Beevers, D. G.,** Stroke affecting young men after alcoholic binges (letter), *Br. Med. J.,* 291, 1645, 1985.
99. **Gill, J. S., Beevers, D. G., and Tsementzis, S. A.,** Strokes and alcohol (letter), *Lancet,* 2, 1142, 1983.
100. **Stott, D. J., Ball, S. G., Connell, J. M. C., and McInnes, G. T.,** Stroke affecting young men after alcoholic binges (letter), *Br. Med. J.,* 291, 1645, 1985.
101. **Altura, B. M., Altura, B. T., and Gebrewold, A.,** Alcohol-induced spasms of cerebral blood vessels: relation to cerebrovascular accidents and sudden death, *Science,* 220, 331, 1983.
102. **Astrup, J., Siesjo, B. K., and Symon, L.,** Thresholds in cerebral ischaemia — the ischaemic penumbra, *Stroke,* 12, 723, 1981.
103. **Shand, B. I.,** unpublished.

Chapter 6

# ALCOHOL INTERACTIONS WITH THE REPRODUCTIVE SYSTEM

**Judith S. Gavaler and David H. Van Thiel**

## TABLE OF CONTENTS

## I. BACKGROUND AND CONCEPTUALIZATION

### A. Recognition of Alcohol Effects on Endocrine Function

It has been recognized for a long time that alcoholic beverage use and abuse can produce reproductive dysfunction. Well before even the vaguest recognition of either the role of gametes or the concept of hormones, abuse of alcoholic beverages was noted to have an adverse effect on reproductive capability. The consumption of alcoholic beverages by young married couples was prohibited in ancient Greek and Old Testament writings. In more recent times, Shakespeare noted that alcohol "provokes the desire but it takes away the performance" (*Macbeth*: Act II, Scene 3).

Not surprisingly, the endocrine effects of chronic alcoholic beverage abuse were first systematically reported in alcoholic males with Laennec's (alcohol-induced) cirrhosis. Essentially, all of these early reports were based on observational studies in males, and later experimental studies were also performed exclusively in males. This early reliance on study populations composed only of males simply reflects two realities: males, but not females, were readily available for both observational and experimental studies; the external aspects of male, but not female, reproductive function are such that the endocrine sequelae of alcoholic beverage abuse were easily detectable in males even before the advent of such technological advances as hormone radioimmunoassay.

There was little disagreement among early investigators that alcoholic cirrhotic males manifested signs of endocrine dysfunction. However, during the last half of the 19th century and first half of this century, the malnutrition associated with alcoholism was presumed to be the cause of the observed cirrhosis. Further, the presence of cirrhosis, rather than alcoholic beverage abuse per se, was considered to be the major factor responsible for both the hypogonadism and feminization commonly observed in alcoholic cirrhotic males.

### B. Duality of Alcohol Effects: Hypogonadism and Feminization

Even with the demonstration that cirrhosis was the end stage in a spectrum of liver damage produced by alcohol even in the absence of malnutrition, little progress was made in elucidating the effects of alcoholic beverage abuse on endocrine function until the syndromes of hypogonadism and feminization were conceptually separated. In the male, hypogonadism includes both hypoandrogenization and reproductive failure, while the physical findings of feminization include gynecomastia, a female escutcheon, palmar erythema, and spider angiomata. Because the early observational studies evaluated alcoholic cirrhotic males in whom the signs of both feminization and hypogonadism were present, it is not surprising that these two syndromes were considered to be a single entity.

With the conceptual dissociation of hypogonadism and feminization, hypotheses were more clearly formulated and study protocols were designed to evaluate specific components of testable hypotheses. The development of animal models and the use of ethanol as the experimental approximation of alcoholic beverages provided powerful experimental tools. With the use of these tools, it was possible to dissect out and separate the direct effects of alcohol on endocrine function from the sequelae of liver injury produced by alcohol abuse. The implementation of animal models provided a framework within which both the reproductive failure and the endocrine disruption and dysfunction associated with alcoholic beverage abuse could be studied.

In males, the effects of alcohol on the hypothalamic pituitary-gonadal axis have been extensively examined and a variety of intermeshing components of the mechanisms involved have been described. In females, there has been far less progress. This differential in the extent to which alcohol endocrine interactions have yielded to understanding may reflect the difference between the sexes in the complexity of gonadal function and regulation of the hypothalamic pituitary gonadal axis.

# II. HYPOGONADISM: DISRUPTION OF GONADAL FUNCTION BY ALCOHOL

## A. The Normal Hypothalamic Pituitary Gonadal Axis

Although the overall function of the gonads is to produce gametes for reproductive purposes and to synthesize sex steroids for the development and maintenance of secondary sex characteristics, the mechanisms by which these functional goals are achieved in each sex are quite different. The gonads of both males and females contain two distinct functional compartments. The reproductive compartment of the male is composed of the seminiferous tubules with their contained developing and mature sperm; this process of active germ cell production occupies up to 95% of the volume of the testes. Millions of mature germ cells (spermatozoa) are produced daily and this daily production usually continues until late adult life and may be unabated until death. The endocrine compartment of the testes is composed of Leydig cells which synthesize and secrete androgens. The Leydig cells occupy less than 5% of total testicular volume. At the hypothalamic pituitary level, the male is characterized neuroendocrinologically by regular bursts of gonadotropin-releasing hormone (GnRH) release and rather stable plasma levels of both luteinizing hormone (LH) and follicle-stimulating hormone (FSH).

In contrast to the large volume of the male gonad committed to germ cell production, less that 10% of the ovarian volume of adult females during the reproductive years consists of germ cells available for gamete production. Further, unlike the male, the absolute number of germ cells per ovary is fixed even before birth, and that number diminishes steadily thereafter as a result of either ovulation or atresia. The endocrine compartment of the ovary occupies a considerable volume of the adult ovary; it consists of three component cell types, each of which is functional as part of the various follicular structures produced during and after the development of a mature ovum. Each of the components of the ovarian endocrine compartment functions at a different rate during the menstrual cycle, and as much as 90% of the biologically active estradiol secreted by the ovary during any given cycle originates from the unique primordial follicle containing the primary oocyte destined for ovulation during that cycle. As the primordial follicle becomes the primary follicle, the granulosa cells of this developing structure synthesize the estrogen which is required for both follicular maturation and the maintenance of female sexual characteristics. As the primary follicle matures to become a graafian follicle, the contained thecal cells, which produce the androgen precursors of estrogen, augment estrogen production to provide for the burst of gonadotropin secretion essential for ovulation. Following ovulation, the ruptured follicle develops into the component of the endocrine compartment responsible for the production of progesterone, the corpus luteum.

In contrast to the male, the female is characterized by a complex neuroendocrine clock such that gonadotropins are not secreted synchronously and plasma levels vary dramatically in a regular manner which is characteristic of the menstrual cycle. During the normal cycle, plasma LH levels show a slight rise in the late follicular phase, followed by a marked preovulatory surge, and then a decline during the luteal phase. The target sites for LH are theca, stromal, and luteal cells, as well as granulosa cells; the target for FSH is exclusively the granulosa cell. Plasma FSH rises during the late luteal and early follicular phases followed by a decline which is interrupted at midcycle by a small surge coinciding with the preovulatory LH peak.

## B. Perturbation of Gonadal and Hypothalamic Pituitary Regulation by Alcohol

### 1. The Male

If one were to guess that the adverse effects of alcohol in the male might ultimately result in failure of both the endocrine compartment and the reproductive compartment, then one might expect to see the classic signs of testicular failure in alcoholic males. Failure of the male reproductive compartment is commonly manifested as testicular atrophy and hypo- or azoospermia. Testicular endocrine failure is manifested by the loss of secondary sex characteristics; in response to reduced testosterone levels, such men become less muscular, lose sexual hair on the face, axilla, and pubic areas, have reduced size of prostate, seminal vesicles, and phallus, manifest decrements in scrotal wrinkling, and complain of impotence and loss of libido.

Indeed, such were the findings reported in the earliest studies describing the endocrine abnormalities observed in cirrhotic alcoholic males.[1] With the advent of steroid hormone assays, it became clear that the testicular atrophy, impotence, and loss of libido in alcoholic males with alcohol-induced liver disease occurred in the presence of decreased circulating testosterone levels;[2-11] it is also to be noted that many of these studies also concomitantly reported findings of feminization in these cirrhotic alcoholics. With the application of animal models and the use of ethanol as the experimental approximation of alcoholic beverages, the fact that testicular failure could be produced in the absence of advanced biochemical and/or histologic liver disease emerged. Specifically, administration of ethanol to mice and rats resulted in reduced numbers of progeny, testicular atrophy, and decrements in plasma testosterone levels occurring with hepatic histology no more severe than steatosis.[12-18]

With the separation of the concepts of hypogonadism and feminization, followed by the subsequent conceptual dissociation of hypogonadism and putatively obligate liver disease, hypogonadism emerged as an individual entity to be elucidated with respect to alcohol rather than liver disease as the operative variable. Important information was provided via the application of three differing clinical approaches. First, studies comparing alcoholic cirrhotic males to males with cirrhosis of a nonalcohol etiology demonstrated that the testicular atrophy and reductions in testosterone were, indeed, a function of alcohol rather than of cirrhosis per se.[19] Second, clinical studies in which ethanol was administered to normal male volunteer subjects reported that ethanol administered acutely resulted in a fall in testosterone which occurred synchronously with an appropriate rise in LH, while chronic ethanol administration for 4 weeks to normal males resulted in a continued progressive decline of testosterone accompanied by a decrease (rather than an appropriate increase) in LH levels.[20-22] The third clinical approach evaluated abstaining alcoholic males and demonstrated reduced levels of testosterone, but ''normal'' to moderately increased LH concentrations.[23] Taken together, these clinical data suggested that alcohol ingestion is followed by an initial Leydig cell injury resulting in a reduction in plasma testosterone and an appropriate compensatory increase in gonadotropin levels; with continued alcohol abuse, testosterone levels continue to fall while gonadotropin concentrations return towards paradoxically ''normal'' levels.

With the link between alcohol and decrements in gonadal endocrine function demonstrated in clinical studies, the animal models could then be exploited to evaluate hypotheses as to the mechanisms involved. Studies using an isolated perfused rat testis preparation demonstrated that ethanol is a direct testicular toxin producing dose-dependent decreases in testicular testosterone production and synthesis.[24] Studies using Leydig cells obtained from both rats and humans, maintained in tissue culture and exposed to stimulatory levels of gonadotropin, demonstrated that testosterone secretion is markedly reduced by the addition of either ethanol or its first metabolic product, acetaldehyde.[25,26] Ethanol at levels as low as 25 mg/dl (5.4 m$M$) inhibits rat Leydig cell synthesis and secretion of testosterone by as much as 44%. In passing, it is worth noting not only that the legal limit defining intoxication is 100 mg/dl, but also that such a finding provides biochemical confirmation for Shakespeare's pungent observation. Acetaldehyde is an even more toxic substance and significantly inhibits rat Leydig cell synthesis and secretion of testosterone when added to the culture medium at concentrations as low as 5 $\mu M$. These studies have provided clear evidence of the toxic effect of both ethanol and of its first metabolic product, acetaldehyde, on the male gonadal endocrine compartment.

However, even with the above studies, the question of the hypothalamic pituitary failure of gonadotropin release observed in chronic alcoholic men with advanced testicular failure remained. Studies using an inhibitor of ethanol metabolism to maintain sustained high blood alcohol levels have provided additional information.[27] In male rats given the standard alcohol diet to which was added 4-methylpyrazole, an inhibitor of alcohol dehydrogenase, plasma gonadotropin levels do not rise to counteract the decrease in testosterone, but rather fall to levels below the normal range; these data suggest that, in this particular model at least, both a Leydig cell injury and a central hypothalamic pituitary failure of gonadotropin secretion can occur with alcohol abuse if sustained high levels of blood alcohol are maintained.

Morphologic studies have contributed to our knowledge concerning the intracellular mechanisms of alcohol-induced testicular injury. Morphologically, the testes of chronic alcoholic men and of chronic alcohol-fed male rats appear quite similar. Both show advanced germ cell injury with few, if any, normal residual germ cells within the seminiferous tubules, which are markedly decreased in diameter and are surrounded by a peritubular fibrotic process.[16,18,28,29] At the light microscopic level, the Leydig cells appear to be normal and possibly increased in number. This apparent Leydig cell hyperplasia probably represents the collapse of normal numbers of Leydig cells into groups around residual seminiferous tubules occurring as a consequence of the loss of seminiferous tubular volume. At the electron microscopic level the Leydig cells of rats chronically fed alcohol appear to be reduced in size, to have a reduced cytoplasmic volume, and to demonstrate a relative hypertrophy of the mitochondria, such that individual mitochondria are larger and the fractional volume of the cytopolasmic mass committed to mitochondrial function is increased significantly.[30]

Probably even more important for testosterone biosynthesis, however, is the reduction in the volume of the Leydig cell committed to microsomal function. Moreover, even when one corrects for the reduction in microsomal protein seen in the Leydig cells obtained from rats fed alcohol, ethanol feeding can be shown to reduce the activity of 3-β-hydroxysteroid dehydrogenase/Δ-5-4 isomerase, desmolase, and 17-α-oxidoreductase, the three microsomal enzymes which are important in steroidogenesis.[31-39] Not only is the activity of 3-β-hydroxysteroid dehydrogenase/Δ-5-4 isomerase reduced as a consequence of alcohol abuse when tested under ideal conditions of excess cofactors and substrates, but ethanol feeding also alters the testicular redox state in a direction disadvantageous for the activity of this enzyme. More importantly, increases in NADH per se, even in the presence of equimolar increases in $NAD^+$ such that the net ratio of $NAD^+$ to NADH is not disturbed, also inhibit this enzyme complex.[40]

Relevant to this latter issue of NADH increases and altered redox states occurring as a consequence of alcohol abuse, as well as the effects of acetaldehyde upon Leydig cells, is the relatively recent observation that the testes contain an alcohol dehydrogenase distinct from hepatic alcohol dehydrogenase.[41-43] Much, although not all, of the activity of the testicular alcohol dehydrogenase is contained within the Leydig cells.[42,43] The demonstration of this enzyme within the testes provides a mechanism whereby the metabolism of alcohol locally within the testes could thereby alter the testicular redox state by generating locally increased concentrations of the even more toxic product acetaldehyde. Indeed, recent studies have demonstrated that chronic ethanol administration to male rats results in measurable changes in testicular mitochondrial diene conjugates, polyenoic fatty acid composition, malonaldehyde formation, and reduced glutathione levels; these findings suggest that lipid peroxidation may also contribute to the testicular injury observed.[44]

Finally, the presence of a testicular alcohol dehydrogenase provides an additional mechanism by which germ cell injury can occur as a consequence of alcohol abuse. Specifically, retinol is essential for normal spermatogenesis. It reaches the testes via the circulation as the retinyl ester and is converted to the free alcohol and then to the aldehyde, retinal, which is the form of the vitamin essential for normal spermatogenesis. Testicular alcohol dehydrogenase not only oxidizes ethanol to acetaldehyde, but also converts retinol to retinal.[41,43,45]

Moreover, ethanol inhibits the generation of retinal from retinol in a noncompetitive manner. Thus, alcohol metabolism by testicular alcohol dehydrogenase contributes importantly to both the Leydig cell and the germ cell injury associated with alcohol abuse in the male.

### 2. The Female: The Reproductive Years

If one were to guess that the adverse effects of alcoholic beverage abuse in females during the reproductive years might ultimately result in ovarian failure, then one might expect to see disruption of cyclic ovarian function as mainifested by oligomenorrhea, hypomenorrhea, and amenorrhea, reduced luteal phase levels of progesterone, and the consequences of hypoestrogenization which include vaginal dryness, labial atrophy, and loss of fat mass of breast and buttocks in alcoholic women. Indeed, observational studies of endocrine effects of alcoholic beverage abuse by women during the reproductive years have reported an increased prevalence of menstrual dysfunction, and amenorrhea;[46-50] with alcohol abstinence, the amenorrhea may be reversible, while with continued abuse the amenorrhea progresses to early menopause.[51] It has further been reported that the ovaries of women dying during their reproductive years of Laennec's cirrhosis have been found to contain no corpora lutea or corpora hemorrhagica, suggesting ovulatory failure for at least several months prior to death.[52]

In contrast to males, acute administration of ethanol to normal women under controlled conditions at specific phases of the menstrual cycle has been reported to produce no consistent effect on levels of either sex steroids or gonadotropins.[53-56] It has been suggested that such findings are consistent with a hypothesis that a single episode of intoxication is not sufficient to suppress basal hormone levels, during either the follicular or luteal phase of the menstrual cycle.[57] It is further possible, given the dynamics of ovarian hormone production vis-à-vis the phases of follicular maturation, that to disrupt a particular cycle it is necessary that the ethanol insult be delivered at the earliest developmental phase of the particular follicle containing the oocyte destined for ovulation.

Studies in animals have provided data which have replicated the findings of studies in humans describing the effects of chronic abuse of alcoholic beverages on female reproductive function; such animal studies have evaluated ethanol effects on the estrous cycle,[58-64] on ovulatory function,[18,60,61,63-67] and on fertility.[58,65,68]

In mice, rats, and monkeys, it has been demonstrated that as alcohol exposure is prolonged over time or increased in dose, an increased number of female animals show a total loss of estrous cyclicity or a disruption of existing cycles.[58-65] Similarly, in rodents and monkeys, an increased incidence of ovulatory failure with ethanol exposure has been documented by findings of absence of ova in the fallopian tubes, absence of ovarian corpora hemorrhagica and corpora lutea, loss of the midcycle ovulatory gonadotropin surge, and failure of plasma estradiol and progesterone levels to increase in the latter half of the cycle.

That abuse of alcoholic beverages by women during the reproductive years can produce disruption of the menstrual cycle, which may result in amenorrhea and ultimately in early menopause, has been repeatedly demonstrated in observational clinical studies. Further, such findings have been replicated and extended in various animal models. In spite of such progress, however, the specific mechanisms for the effect of ethanol on the ovarian hypothalamic pituitary axis, in general, or on follicular maturation and development, in particular, remain unelucidated.

### 3. The Female: The Menopause

As contrasted to women during the reproductive years, postmenopausal women are characterized by a relatively simple pattern of hypothalamic pituitary gonadal axis regulation. By definition, cyclic follicular function has ceased to occur in the menopause; the postmenopausal ovarian stromal cells, however, continue to produce androgens which are then available for conversion to estrogens at peripheral sites. Although endocrine effects of alcoholic beverage use and abuse in postmenopausal women have been studied to even a lesser degree than such effects in women during the reproductive years, there are a variety of potential pathways by which ethanol might be postulated to influence postmenopausal endocrine function.[69,70] Given that the degree of complexity of hypothalamic pituitary regulation in postmenopausal women is somewhat comparable to that in males, rather than to that in women with reproductive capability, it might be expected that ethanol effects on postmenopausal endocrine function might be readily detectable.

Such is not the case with respect to effects of acute administration of ethanol to normal postmenopausal women. Three recent studies have reported that comparison of the responses obtained following the administration of placebo to those obtained after ethanol administration have demonstrated no change in circulating levels of either LH or FSH.[55,71,72] These studies in normal postmenopausal women have been confirmed in monkeys.[73] Unfortunately, none of these studies report concomitantly obtained and assayed levels of any steroid hormones, and, thus, these reports provide no clue as to acute ethanol effects on postmenopausal levels of androgens or estrogens.

Interestingly, the findings of the few studies of endocrine effects of chronic alcoholic beverage abuse in postmenopausal women with Laennec's cirrhosis echo the observations reported in chronic alcoholic cirrhotic males. None of these four reports is entirely adequate with respect either to sample size and/or to the completeness of the endocrine characterization provided for the postmenopausal women studied; nevertheless, these four studies provide generally consistent data: gonadotropins are often decreased while estrogen concentrations are increased.[48,49,74,75]

The question of whether or not moderate consumption of alcoholic beverages by healthy postmenopausal women might increase circulating estrogen has recently been raised. It is known that acute ethanol administration to normal male volunteers results in an increase in the conversion of testosterone to estradiol.[22] It has also been demonstrated in animal studies that even moderate doses of ethanol can stimulate adrenal production of corticosterone, progesterone, and androstenedione.[76,77] Given that the major source of estrogens in normal postmenopausal women is the conversion of available androgen precursors to estrogens, and in view of the above findings,[48,49,74-77] it is somewhat surprising that alcoholic beverage consumption has not previously been evaluated as a determinant of circulating levels of estrogens in normal postmenopausal women. The majority of postmenopausal women consume alcohol beverages to at least a moderate degree.[78-82] Indeed, a recent study in normal postmenopausal women provides evidence that even low to moderate consumption levels of alcohol result in significantly increased circulating concentrations of estradiol.[83]

Studies in bilaterally oophorectomized rats have confirmed several of the findings observed in the human studies. Administration of graded doses of alcohol in drinking water (up to 5.5% ethanol) for prolonged periods of time (up to 10 weeks) results in increases in circulating estradiol levels;[84] further, although concentrations of estradiol and LH were significantly correlated in this group of 60 ovariectomized rats, no significant effect of chronic administration of low to moderate doses of ethanol on LH levels was detected.[85]

## III. FEMINIZATION

Despite numerous similarities between the sexes in terms of the end stage of effects of alcohol abuse on the hypothalamic pituitary gonadal axis, a major difference between alcoholic men and women exists. When one examines the issue of estrogenization, chronic alcoholic women during the reproductive years, like alcoholic men, become progressively hypogonadal with continued alcohol abuse; as a result, they lose their secondary sex characteristics and the evidence of adequate estrogenization. In contrast, chronic alcoholic men, particularly those with advanced liver disease characterized by portal hypertension and portal systemic shunting, become progressively feminized (hyperestrogenized), as manifested by the physical findings of gynecomastia, palmar erythema, spider angiomata, a female escutcheon, and changes in the distribution of body fat and hair, as well as by the biochemical findings of increases in estrogen-responsive proteins such as sex steroid binding globulin and neurophysin.

Part of the mechanism responsible for these apparent different responses between the sexes probably reflects the marked difference in normal androgen production rather than the differences in estrogen production rates between the two sexes. For example, primary alcohol-induced gonadal injury reduces estrogen biosynthesis in the female and androgen biosynthesis in the male. The net result is a clear reduction in plasma estrogen levels in the alcoholic female, but markedly reduced testosterone levels in the alcoholic male. As noted earlier, alcohol is a potent inducer of microsomal aromatase activity.[22,86,87] Aromatase is the enzyme complex responsible for the conversion of androgens and proandrogens to estrogens, principally estrone, and to a lesser extent estradiol. Thus, a minor (<1%) alcohol-associated change in the fraction of proandrogens (androstenedione, dehydroepiandrostrone, and dehydroepiandrostrone sulfate) and androgen (testosterone), which are produced in millimolar amounts per day by the male, has a profound effect upon estrogen levels which are normally produced in picomolar amounts. Thus, despite reduced testicular androgen biosynthesis, the level of plasma estrogens in alcoholic men is either normal (estradiol) or increased (estrone) as a result of a slight increase in the peripheral conversion of androgens to estrogens as a result of alcohol-associated aromatase induction.

In the normal female the production rate of androgens is much lower than in males, and normally a much larger fraction of secreted proandrogens and testosterone is converted to estrogens; thus, a slight increase in the conversion rate of androgens to estrogens as a result of aromatase induction is not sufficient for the maintenance of the female secondary sex characteristics. As a result, alcoholic men become feminized while alcoholic women during the reproductive years become progressively defeminized.

In males, the stimulatory effect of alcohol on adrenal steroid production also contributes directly to the increased circulating levels of estrogens seen in alcoholic cirrhotic males.[1-11,19,88,89] Not only has adrenal overproduction of weak androgens and estrogen precursors been reported to regularly occur in chronic alcoholic men,[90,91] but also the adrenals of chronic alcoholic men have been reported to directly produce estrone.[11] Specifically, under experimental conditions of adrenal suppression subsequent to gonadal suppression, levels of estrone, estradiol, testosterone, and androstenedione are reduced; subsequent gonadal stimulation produces no change in either estrone or androstenedione, while further subsequent adrenal stimulation results in increases in levels of both androstenedione and estrone.[11] Thus, at least in chronic alcoholic males, chronic alcohol stimulation of the adrenals results not only in increased production of the androgen substrate available for conversion, but also in increased direct production of estrone.

The problem of portal systemic shunting which occurs as a consequence of advanced alcoholic liver disease with portal hypertension only compounds the problem of feminization of the male, and reduces but does not eliminate the defeminization of the alcoholic woman.[91,92] Specifically, as a result of portal systemic shunting of proandrogens (progestens and weak adrenal androgens) large masses of sterols in the millimolar range escape hepatic clearance and are presented to peripheral tissues (fat, muscle, skin, etc.) where alcohol-induced aromatase activity converts a slight but significant fraction to estrogens.

Yet another factor contributing importantly to the differences between the feminization observed in the male, and the lack of feminization observed in the female, is the effect of alcohol upon cytosolic estrogen binding proteins and receptors.[91,93,94] Alcohol increases the level of cytosolic estrogen receptors in both males and females. However, the net change is minimal in the female and marked in the male. Moreover, the male is normally protected from estrogen because male hepatocytes contain a cytosolic estrogen binding protein which is not a receptor, but which competes for cytosolic estrogen with the receptor. Under normal conditions only 5 or 10% of the cytosolic estrogen binding protein in the male cell is estrogen receptor and the remaining bulk of the estrogen binding protein is nonreceptor protein. Under conditions of alcohol abuse (or feeding of ethanol in animal experiments), the level of estrogen receptor increases slightly, although the level of nonreceptor protein is reduced dramatically. As a consequence, the ratio of receptor to nonreceptor is shifted markedly in favor of the receptor and, as a result, feminization of the male is observed. Such is not the case in the female; estrogen binding in female hepatic cytosol occurs solely via receptors, as the female lacks entirely the male-specific nonreceptor estrogen binding protein.

Additional factors that may contribute to alcohol-associated gonadal failure are the hyperprolactinemia and the changes in the plasma levels of albumin and sex steroid binding globulin seen in advanced alcoholic liver disease.[3,7-9] Chronic alcoholics, particularly those with cirrhosis, have been found to have pituitary prolactin cell hyperplasia at autopsy and some data exist to suggest that prolactin cell secretion may become autonomous in a few alcoholics.[52,95-97] These observations are of interest in that hyperprolactinemia is associated with hypogonadism, and deficient gonadotropin secretion and reduced numbers of gonadotropin receptors at the level of the gonad. Equally important is the fact that prolactin may act as a stimulant for adrenal secretion of proandrogens. Such stimulation only enhances the "feminization" of the male and masculinizes the female alcoholic.

Serum albumin is principally responsible for the transport of plasma estrogens, while sex steroid binding globulin is principally responsible for testosterone transport. Chronic alcoholism, as a consequence of malnutrition, liver disease, and its various gonadal effects, results in a reduction in plasma albumin levels and a marked increase in the plasma levels of sex steroid binding globulin. The net result for the male is more feminization and less masculinization for any given level of plasma estrogens and androgens. In the case of the female, the net result is a reduction in circulating estrogen levels, probably because less androgen is available for peripheral conversion to estrogen; as a result, progressive defeminization occurs.

Combining the above components of the mechanism responsible for feminization into a unified whole has great appeal. The role of systemic shunting, which would permit steroids to escape the enterohepatic circulation and thus result in increased peripheral tissue exposure, fits well with evidence. Specifically, signs of feminization are seen only in alcoholic males who have developed cirrhosis and the accompanying portal hypertension. Similarly, decreases in hepatic albumin secretion, which would result in increased circulating levels of unbound estrogens, are a common finding in cirrhotics. The reports of alcohol-associated decreases in the hepatic male-specific nonreceptor estrogen binding protein and concomitant increases in the hepatic content of estrogen receptors provide a mechanism whereby even small increases in circulating levels of estrogen might result in an amplification of estrogen effects. Thus, for example, in response to only moderately elevated circulating levels of estrogen which are reported to occur in alcoholic cirrhotic males, the reported increases in sex hormone binding globulin are predictable.

The findings in alcoholics of increases in the adrenal production of weak androgens and proandrogens, the substrate needed for subsequent conversion to estrogens, along with the findings of increases in the conversion of such precursors to estrogens, are consistent with the feminization theory as a whole, but fit the evidence less well. Specifically, while levels of estrone in cirrhotic alcoholic males have consistently been reported to be significantly elevated,[6,7,11,89] levels of the five times more potent estradiol have been reported to be somewhat increased,[8,18,89] or more usually to be not significantly different from levels in normal controls.[3,4,6,9,11,19] Given the degree of hyperestrogenization observed in feminized alcoholic cirrhotic males, perhaps it is possible that such men are exposed not only to the steroidal estrogens such as estrone and estradiol, but also to nonsteroidal estrogens as well. In passing, it is worth noting that if nonsteroidal estrogens were present, they would not be detected by a radioimmunoassay in which the antibody recognizes estrone or estradiol, but does not cross-react with nonsteroidal estrogenic substances.

Very recent reports provide data which fit a hypothesis which includes not only all of the components of the feminization mechanism discussed above, but which also incorporates exposure to biologically active nonsteroidal estrogens as well. Specifically, recent studies provide evidence that bourbon, a whiskey made from corn, contains the biologically active nonsteroidal phytoestrogen biochanin A.[98-100] In addition to isolating and identifying this estrogen of plant origin in an extract of bourbon, these reports also demonstrate that the bourbon concentrate not only produces two measurable estrogenic dose responses in an appropriate animal model, but also interacts with estrogen receptors in a dose-dependent manner. Although it yet remains to be demonstrated that other alcoholic beverages contain biologically active phytoestrogens, at the very least, these reports suggest that the mechanisms responsible for the feminization observed in some alcoholic cirrhotic men have not yet been fully elucidated.

## ACKNOWLEDGMENT

This work was supported in part by grants AA06772 and AA04425 from the National Institute on Alcohol Abuse and Alcoholism.

## REFERENCES

1. **Lloyd, C. W. and Williams, R. H.,** Endocrine changes associated with Laennec's cirrhosis of the liver, *Am. J. Med.,* 4, 315, 1948.
2. **Chopra, I. J., Tulchinsky, D., and Greenway, F. L.,** Estrogen-androgen imbalance in hepatic cirrhosis, *Ann. Intern. Med.,* 79, 198, 1973.
3. **Galvao-Teles, A., Anderson, D. C., Burke, C. W., Marshall, J. C., Corker, C. S., Brown, R. L., and Clark, M. L.,** Biologically active androgens and oestradiol in men with chronic liver disease, *Lancet,* 1, 173, 1973.
4. **Kent, J. R., Scaramuzzi, R. J., Lauwers, W., Parlow, A. F., Hill, M., Penardi, R., and Hilliard, J.,** Plasma testosterone, estradiol and gonadotropin in hepatic insufficiency, *Gastroenterology,* 64, 111, 1973.
5. **Gordon, G. G., Olivo, J., Raffi, F., and Southren, A. L.,** Conversion of androgens to estrogens in cirrhosis of the liver, *J. Clin. Endocrinol. Metab.,* 40, 1018, 1975.
6. **Kley, H. K., Nieschlag, E., Wiegelmann, W., Solbach, M. G., and Kruskemper, H. L.,** *Acta Endocrinol.,* 79, 275, 1975.
7. **Van Thiel, D. H., Gavaler, J. S., Lester, R., Loriaux, D. L., and Braunstein, G. D.,** Plasma estrone, prolactin, neurophysin, and sex steroid binding protein in chronic alcoholic men, *Metabolism,* 24, 1015, 1975.
8. **Baker, H. W. G., Burger, H. G., Dekretser, D. M., Dulmanis, A., Hudson, B., O'Connor, S., Paulsen, C. A., Purcell, N., Rennie, G. C., Seah, C. S., Taft, H. P., and Wang, C.,** A study of the endocrine manifestations of hepatic cirrhosis, *Q. J. Med.,* 45, 145, 1976.
9. **Lindholm, J., Fabricius-Bjerre, N., Bahnsen, M., Boiesen, P., Hagen, C., and Christensen, T.,** Sex steroids and sex-steroid binding globulin in males with chronic alcoholism, *Eur. J. Clin. Invest.,* 8, 273, 1978.
10. **Van Thiel, D. H., Lester, R., and Vaitukaitis, J.,** Evidence for a defect in pituitary secretion of luteinizing hormone in chronic alcoholic men, *J. Clin. Endocrinol. Metab.,* 47, 499, 1978.
11. **Van Thiel, D. H. and Loriaux, D. L.,** Evidence for an adrenal origin of plasma estrogens in alcoholic men, *Metabolism,* 28, 536, 1979.
12. **Badr, F. M., and Bartke, A.,** Effect of ethyl alcohol on plasma testosterone levels in mice, *Steroids,* 23, 921, 1974.

13. **Cicero, T. J. and Badger, T. M.,** Effects of alcohol in the hypothalamic pituitary gonadal axis in the male rat, *J. Pharmacol. Exp. Ther.,* 201, 427, 1977.
14. **Cicero, T. J., Bernstein, D., and Badger, T. M.,** Effects of acute alcohol administration on reproductive endocrinology in the male rat, *Alcoholism Clin. Exp. Res.,* 2, 249, 1978.
15. **Van Thiel, D. H., Gavaler, J. S., Lester, R., and Goodman, M. D.,** Alcohol-induced testicular atrophy: an experimental model for hypogonadism occurring in chronic alcoholic men, *Gastroenterology,* 69, 326, 1975.
16. **Van Thiel, D. H., Gavaler, J. S., Cobb, C. F., Sherins, R., and Lester, R.,** Alcohol-induced testicular atrophy in the adult male rat, *Endocrinology,* 105, 888, 1979.
17. **Badr, R. M., Bartke, A. Dalterio, S., and Bugler, W.,** Suppression of testosterone production by ethyl alcohol: possible mode of action, *Steroids,* 30, 647, 1977.
18. **Gavaler, J. S., Van Thiel, D. H., and Lester, R. R.,** Ethanol, a gonadal toxin in the mature rat of both sexes: similarities and differences, *Alcoholism Clin. Exp. Res.,* 4, 271, 1980.
19. **Van Thiel, D. H., Gavaler, J. S., Spero, J. A., Egler, K. M., Wight, C., Sanghvi, A., Hasiba, U., and Lewis, J. H.,** Patterns of hypothalamic pituitary gonadal dysfunction in men with liver disease due to differing etiologies, *Hepatology,* 1, 39, 1981.
20. **Mendelson, J. H., Mello, N. K., and Ellingboe, J.,** Effects of acute alcohol intake on pituitary gonadal hormones in normal human males, *J. Pharmacol. Exp. Ther.,* 202, 676, 1977.
21. **Ylikahri, R., Huttunen, I. M., Harkonen, M., Seuderling, U., Onikki, S., Karonen, S. I., and Aldercruetz, H.,** Low plasma testosterone values in men during hangover, *J. Steroid Biochem.,* 5, 655, 1974.
22. **Gordon, G. G., Altman, K., Southren, A. L., Rubin, E., and Lieber, C. S.,** Effect of alcohol (ethanol) administration on sex hormone metabolism in men, *N. Engl. J. Med.,* 295, 793, 1976.
23. **Van Thiel, D. H.,** Ethanol: its adverse effects upon the hypothalamic pituitary gonadal axis, *J. Lab. Clin. Med.,* 101, 21, 1983.
24. **Cobb, C. F., Ennis, M. F., Van Thiel, D. H., Gaveler, J. S., and Lester, R.,** Isolated testes perfusion: a method using a cell- and protein-free perfusate useful for the evaluation of potential drug and/or metabolic injury, *Metabolism,* 30, 71, 1980.
25. **Van Thiel, D. H., Gavaler, J. S., Cobb, C. F., Santucci, L., and Graham, T. O.,** Ethanol, a Leydig cell toxin: evidence obtained in vivo and in vitro, *Pharmacol. Biochem. Behav.,* 18, 317, 1983.
26. **Santucci, L., Graham, T. O., and Van Thiel, D. H.,** Inhibition of testosterone production by rat Leydig cells with ethanol and acetaldehyde: prevention of ethanol toxicity with 4-methylpyrazole, *Alcoholism Clin. Exp. Res.,* 7, 135, 1983.
27. **Gavaler, J. S., Gay, V., Egler, K. M., and Van Thiel, D. H.,** Evaluation of the differential *in vivo* toxic effects of ethanol and acetaldehyde on the hypothalamic pituitary gonadal axis using 4-methylpyrazole, *Alcoholism Clin. Exp. Res.,* 7, 332, 1983.
28. **Van Thiel, D. H., Gavaler, J. S., Herman, G. B., Lester, R., Smith, W. I., Jr., and Gay, V.,** An evaluation of the respective roles of liver disease and malnutrition in the pathogenesis of the hypogonadism seen in alcoholic rats, *Gastroenterology,* 79, 533, 1980.
29. **Van Thiel, D. H., Cobb, C. F., Herman, G. B., Perez, H. A., Estes, L., and Gavaler, J. S.,** An examination of various mechanisms for ethanol-induced testicular injury: studies utilizing the isolated perfused rat testes, *Endocrinology,* 109, 2009, 1981.
30. **Gavaler, J. S., Perez, H. A., Estes, L., and Van Thiel, D. H.,** Morphologic alterations of rat Leydig cells induced by ethanol, *Pharmacol. Biochem. Behav.,* 18 (Suppl. 1), 341, 1983.
31. **Chiao, Y.-B. and Van Thiel, D. H.** Biochemical mechanisms that contribute to alcohol-induced hypogonadism in the male, *Alcoholism Clin. Exp. Res.,* 7, 131, 1983.
32. **Munono, E. P., Lin, T., Osterman, J., and Nankin, H. R.,** Direct inhibition of testosterone synthesis in rat interstitial cells by ethanol: possible sites of action, *Steroids,* 36, 619, 1980.
33. **Cicero, T. J., Newman, K. S., and Meyer, E. R.,** Ethanol induced inhibitions of testicular steroidogenesis in the male rat: mechanisms of action, *Life Sci.,* 28, 871, 1981.
34. **Ellingboe, J. and Varanelli, C. C.,** Ethanol inhibits testosterone biosynthesis by direct action on Leydig cells, *Res. Commun. Chem. Pathol. Pharmacol.,* 28, 87, 1979.
35. **Cicero, T. J., Bell, R. D., Meyer, E. R., and Badger, T. M.,** Ethanol and acetaldehyde directly inhibit testicular steroidogenesis, *J. Pharmacol. Exp., Ther.,* 213, 228, 1980.
36. **Johnston, D. E., Chiao, Y.-B., Gavaler, J. S., and Van Thiel, D. H.,** Inhibition of testosterone synthesis by ethanol and acetaldehyde, *Biochem. Pharmacol.,* 30, 1827, 1981.
37. **Chiao, Y.-B., Johnston, D. E., Gavaler, J. S., and Van Thiel, D. H.,** Effect of chronic ethanol feeding on testicular content of enzymes required for testosteronogenesis, *Alcoholism Clin. Exp. Res.,* 5, 230, 1981.
38. **Cicero, T. J. and Bell, R. D.,** Effects of ethanol and acetaldehyde on the biosynthesis of testosterone in the rodent testes, *Biochem. Biophys. Res. Commun.,* 94, 814, 1980.
39. **Cicero, T. J. and Bell, R. D.,** Acetaldehyde directly inhibits the conversion of androstenedione to testosterone, in *Alcohol and Acetaldehyde Metabolizing Systems,* Vol. 4, Thurman, R. G., Ed., Plenum Press, New York, 1980.

40. **Gordon, G. G., Vittek, J., Southren, A. L., Munnang, P., and Lieber, C. S.,** Effect of chronic alcohol ingestion on the biosynthesis of steroids in rat testicular homogenates in vitro, *Endocrinology,* 106, 1880, 1980.
41. **Van Thiel, D. H., Gavaler, J. S., and Lester, R.,** Ethanol inhibition of vitamin A in the testes: possible mechanism for sterility in alcoholics, *Science,* 186, 941, 1974.
42. **Messiha, F. S.,** Subcellular fractionation of alcohol and aldehyde dehydrogenase in the rat testicle, *Prog. Biochem. Pharmacol.,* 18, 155, 1981.
43. **Chaio, Y.-B. and Van Thiel, D. H.,** Characterization of rat testicular alcohol dehydrogenase, *Alcohol Alcoholism,* 21, 9, 1986.
44. **Rosenblum, E. R., Gavaler, J. S., and Van Thiel, D. H.,** Lipid peroxidation: a mechanism for ethanol associated testicular injury in rats, *Endocrinology,* 116, 311, 1985.
45. **Rosenblum, E. R., Gavaler, J. S., and Van Thiel, D. H.,** Vitamin A at the pharmacologic doses ameliorates the membrane lipid peroxidation injury and testicular atrophy which occur with chronic ethanol feeding in rats, *Alcohol Alcoholism,* 22, 241, 1987.
46. **Moskovic, S.,** Effect of chronic alcohol intoxication on ovarian dysfunction, *Sel. Transl. Int. Alcoholism Res.,* 20, 2, 1975.
47. **Jones-Saumty, D. J., Fabian, M. S. and Parsons, D. A.,** Medical status and congitive function in alcoholic women, *Alcoholism Clin. Exp. Res.,* 5, 372, 1981.
48. **Hugues, J. N., Perret, G., Adessi, G., Coste, T., and Modigliani, E.,** Effect of chronic alcoholism on the pituitary gonadal function of women during menopausal transition and in the postmenopausal period, *Biomedicine,* 29, 279, 1978.
49. **Hugues, J. N., Coste, T., Perret, G., Jayle, M. F., Sebaqun, J., and Modigliani, E.,** Hypothalamic pituitary ovarian function in thirty-one women with chronic alcoholism, *Clin. Endocrinol.,* 12, 543, 1980.
50. **Wilsnack, S. C., Klassen, A. D., and Wilsnack, R. W.,** Drinking and reproductive dysfunction among women in a 1982 national survey, *Alcoholism Clin. Exp. Res.,* 8, 451, 1984.
51. **Ryback, R. S.,** Chronic alcohol consumption and menstruation, *J.A.M.A.,* 238, 2143, 1977.
52. **Jung, Y. and Russfield, A. B.,** Prolactin cells in the hypophysis of cirrhotic patients, *Arch. Pathol.,* 94, 265, 1972.
53. **McNamee, B., Grant, J., Ratcliffe, J., Ratcliffe, W., and Oliver, J.,** Lack of effect of alcohol on pituitary gonadal hormones in women, *Br. J. Addict.,* 74, 316, 1979.
54. **Mendelson, J. H., Mello, N. K., Barli, S., Ellingboe, J., Bree, M., Harvey, K., King, N., and Seghal, R.,** Alcohol effects on female reproductive hormones, in Ethanol Tolerance and Dependence: Endocrinologic Aspects, Cicero, T. J., Ed., NIAAA Res. Monogr., DHHS Publ. No. (ADM) 83-1285, Department of Health and Human Services, Washington, D. C., 1983.
55. **Mendelson, J. H., Mello, N. K., and Ellingboe, J. H.,** Acute alcohol intake and pituitary gonadal hormones in normal human females, *J. Pharmacol. Exp. Ther.,* 218, 23, 1981.
56. **Valimaki, M., Harkonen, M., and Ylikahri, R.,** Acute effects of alcohol on female sex hormones, *Alcoholism Clin. Exp. Res.,* 7, 289, 1983.
57. **Mello, N. K.,** Effects of alcohol abuse on reproductive function in women, in *Recent Developments in Alcoholism,* Vol. 6, Galanter, M., Ed., Plenum Press, New York, 1988, 254.
58. **Cranston, E. M.,** Effect of tranquilizers and other agents on the sexual cycle of mice, *Proc. Soc. Exp. Biol. Med.,* 98, 320, 1958.
59. **Aron, E., Flanzy, M., Combescot, C., Puisais, J., Demaret, J., Reynouard-Brault, F., and Igbert, C.,** L'alcool est-il dans le vin l'element qui perturbe chez la ratte le cycle vaginal?, *Bull. Acad. Natl. Med.,* 149, 112, 1965.
60. **Kieffer, J. D. and Ketchel, M.,** Blockade of ovulation in the rat by ethanol, *Acta Endocrinol.,* 65, 117, 1970.
61. **Van Thiel, D. H., Gavaler, J. S., and Lester, R.,** Alcohol-induced ovarian failure in the rat, *J. Clin. Invest.,* 61, 624, 1978.
62. **Eskay, R. L., Ryback, R. S., Goodman, M., and Majchrowicz, E.,** Effect of chronic ethanol administration on plasma levels of LH and the estrus cycle in the female rat, *Alcoholism Clin. Exp. Res.,* 5, 204, 1981.
63. **Bo, W. J., Krueger, W. A., Rudeen, P. K., and Symmes, S. K.,** Ethanol-induced alterations in the morphology and function of the rat ovary, *Anat. Rec.,* 202, 255, 1982.
64. **Mello, N. K., Bree, M. P., Mendelson, J. H., and Ellingboe, J.,** Alcohol self-administration disrupts reproductive function in female Macaque monkeys, *Science,* 211, 677, 1983.
65. **Chaudhury, R. R. and Matthews, M.,** Effects of alcohol on the fertility of female rabbits, *J. Endocrinol.,* 34, 275, 1966.
66. **Blake, C. A.,** Paradoxical effects of drugs acting on the central nervous system on the preovulatory release of pituitary luteinizing hormone in pro-oestrous rats, *J. Endocrinol.,* 79, 319, 1978.

67. **Gavaler, J. S.,** Sex related differences in ethanol induced hypogonadism and sex steroid responsive tissue atrophy: analysis of the weanling ethanol-fed rat model using epidemiologic methods, in Ethanol Tolerance and Dependence: Endocrinologic Aspects, Cicero, T. J., Ed., NIAAA Res. Monogr., DHHS Publ. No. (ADM) 83-1285, Department of Health and Human Services, Washington, D. C., 1983.
68. **Merari, A., Ginton, A., Heifez, T., and Lev-Ran, T.,** Effects of alcohol on the mating behavior of the female rat, *Q. J. Stud. Alcohol,* 34, 1095, 1973.
69. **Gavaler, J. S.,** Effects of alcohol on endocrine function in postmenopausal women: a review, *J. Stud. Alcohol,* 46, 495, 1985.
70. **Gavaler, J. S.,** Effects of moderate consumption of alcoholic beverages on endocrine function in post-menopausal women: bases for hypotheses, in *Recent Developments in Alcoholism,* Vol. 6, Galanter, M., Ed., Plenum Press, New York, 1988, 229.
71. **Mendelson, J. H., Mello, N. K., Ellingboe, J., and Bavli, S.,** Alcohol effects on plasma luteinizing hormone levels in postmenopausal women, *Pharmacol. Biochem. Behav.,* 22, 233, 1985.
72. **Valimaki, M., Penkonen, K., and Ylikahri, R.,** Acute ethanol intoxication does not influence gonado-tropin secretion in postmenopausal women, *Alcohol Alcoholism,* in press.
73. **Mello, N. K., Mendelson, J. H., Bree, M. P., and Skupny, A. S.,** Alcohol effects on LHRH stimulated LH and FSH in ovariectomized female rhesus monkeys, *J. Pharmacol. Exp. Ther.,* 236, 590, 1986.
74. **Valimaki, M., Penkonen, K., Salaspuro, M., Harkonen, M., Hirvonen, E., and Ylikahri, R.,** Sex hormones in amenorrheic women with alcoholic liver disease, *J. Clin. Endocrinol. Metab.,* 59, 133, 1984.
75. **James, V. H. T., Green, J. R. B., Walker, J. G., Goodall, A., Short, F., Jones, D. L., Noel, C. T., and Reed, M. J.,** The endocrine status of postmenopausal women, in *The Endocrines and the Liver,* Langer, M., Chiandussi, L., Chopra, I. J., and Martini, L., Eds., Academic Press, New York, 1982, 417.
76. **Cobb, C. F., Van Thiel, D. H., Gavaler, J. S., and Lester, R.,** Effects of ethanol and acetaldehyde on the rat adrenal, *Metabolism,* 30, 537, 1981.,
77. **Cobb, C. F., Van Thiel, D. H., and Gavaler, J. S.,** Isolated rat adrenal perfusion: a new method to study adrenal function, *J. Surg. Res.,* 31, 247, 1984.
78. **Goodwin, J. S., Sanchez, C. J., Thomas, P., Hunt, C., Garry, P. J., and Goodwin, J. M.,** Alcohol intake in a healthy elderly population, *Am. J. Public Health,* 77, 173, 1987.
79. **Wechsler, H.,** Summary of the literature: epidemiology of male and female drinking over the last century, in Alcoholism and Alcohol Abuse Among Women, NIAAA Res. Monogr. No. 1, DHEW Publ. No. (ADM) 80-835, Department of Health, Education and Welfare, Washington, D. C., 1980, 3.
80. **Graham, K.,** Identifying and measuring alcohol abuse among the elderly: serious problems with instru-mentation, *J. Stud. Alcohol* 47, 322, 1986.
81. **Clark, W. B. and Widanik, L.,** Alcohol use and alcohol problems among U.S. adults: results of the 1979 national survey, in Alcohol Consumption and Related Problems, NIAAA Alcohol and Health Monogr. No. 1, DHHS Publ. No. (ADM) 82-1190, Department of Health and Human Services, Washington, D.C., 1982, 3.
82. **Gomberg, E. S. L.,** Alcohol use and alcohol problems among the elderly, in Special Population Issues, NIAAA Alcohol and Health Monogr. No. 4, DHHS Publ. No. (ADM) 82-1193, Department of Health and Human Services, Washington, D.C., 1982, 263.
83. **Gavaler, J. S., Belle, S., and Cauley, J.,** Effects of moderate alcoholic beverage consumption on estradiol levels in normal postmenopausal women, *Alcoholism Clin. Exp. Res.,* 11, 199, 1987.
84. **Gavaler, J. S. and Rosenblum, E. R.,** Exposure dependent effects of ethanol administered in drinking water on serum estradiol and uterus mass in sexually mature oophorectomized rats: a rat model for bilaterally ovariectomized/postmenopausal women, *J. Stud. Alcohol,* 48, 295, 1987.
85. **Van Thiel, D. H., Rosenblum, E. R., Pohl, C., and Gavaler, J. S.,** Lack of an effect of ethanol administered in drinking water to sexually mature oophorectomized rats on LH levels, *Alcoholism Clin. Exp. Res.,* 9, 194, 1985.
86. **Gordon, G. G., Southren, A. L., Vittek, J., and Lieber, C. S.,** The effect of alcohol ingestion on hepatic aromatase activity and plasma steroid hormones in the rat, *Metabolism,* 28, 20, 1979.
87. **Gordon, G. G., Southren, A. L., and Lieber, C. S.,** Hypogonadism and feminization in the male: a triple effect of alcohol, *Alcoholism Clin. Exp. Res.,* 3, 210, 1979.
88. **Pincus, I. J., Rakoff, A. E., Cohn, E. M., and Tumen, H. J.,** Hormonal studies in patients with chronic liver disease, *Gastroenterology,* 19, 735, 1951.
89. **Pentikainen, P. J., Pentikainen, L. A., Azarnoff, D. L., and Dujovne, C. A.,** Plasma levels and excretion of estrogens in urine in chronic liver disease, *Gastroenterology,* 69, 20, 1975.
90. **Frajria, R. and Anojeli, A.,** Alcohol induced pseudo-Cushings Syndrome, *Lancet,* 1, 1050, 1977.
91. **Van Thiel, D. H.,** Feminization of chronic alcoholic men: a formulation, *Yale J. Biol. Med.,* 52, 219, 1979.
92. **Van Thiel, D. H., Gavaler, J. S., Slone, F. L., Cobb, C. F., Smith, W. I., Jr., Bron, K. M., and Lester, R.,** Is feminization in alcoholic men due in part to portal hypertension: a rat model, *Gastroenterology,* 78, 81, 1980.

93. **Eagon, P. K., Porter, L. E., Gavaler, J. S., Egler, K. M., and Van Thiel, D. H.,** Effect of ethanol feeding upon levels of a male-specific hepatic estrogen binding protein: a possible mechanism for feminization, *Alcoholism Clin. Exp. Res.,* 5, 183, 1981.
94. **Eagon, P. K., Zdunek, R. J., Van Thiel, D. H., Singletary, B., Egler, K., Gavaler, J. S., and Porter, L. E.,** Alcohol-induced changes in hepatic estrogen binding proteins, *Arch. Biochem. Biophys.,* 211, 48, 1981.
95. **Van Thiel, D. H., McClain, C. J., Elson, M. K., McMillin, M. J., and Lester, R.,** Evidence for autonomous secretion of prolactin in some alcoholic men with cirrhosis and gynecomastia, *Metabolism,* 27, 1178, 1978.
96. **Van Thiel, D. H., McClain, C. J., Elson, M. K., and McMillin, M. J.,** Hyperprolactinemia and thyrotropin releasing factor (TRH) responses in men with alcoholic liver disease, *Alcoholism Clin. Exp. Res.,* 2, 344, 1978.
97. **Tarquini, B., Gheri, R., and Anichini, P.,** Circadian study of immunoreactive prolactin in patients with cirrhosis of the liver, *Gastroenterology,* 73, 116, 1977.
98. **Gavaler, J. S., Imhoff, A. F., Pohl, C., Rosenblum, E. R., and Van Thiel, D. H.,** Alcoholic beverages: a source of estrogenic substances?, in *Advances in Biomedical Alcohol Research,* Lindros, K. O., Ylikahri, R., and Kiianmaa, K., Eds., Pergamon Press, 1987.
99. **Rosenblum, E. R., Van Thiel, D. H., Campbell, I. M., Eagon, P. K., and Gavaler, J. S.,** Separation and identification of phytoestrogenic compounds isolated from bourbon, in *Advances in Biomedical Alcohol Research,* Lindros, K. O., Ylikahri, R., and Kiianmaa, K., Eds., Pergamon Press, Oxford, 1987, 551.
100. **Gavaler, J. S., Rosenblum, E. R., Van Thiel, D. H., Eagon, P. K., Pohl, C. R., Campbell, I. M., and Gavaler, J.,** Biologically active phyto-estrogens are present in bourbon, *Alcoholism Clin. Exp. Res.,* 11, 399, 1987.

# SECTION II
## Interactions of Alcohol and Its Metabolism with Biochemical and Physiological Functions

Chapter 7

INTERACTIONS OF ETHANOL AND LIPID METABOLISM

**Jens Kondrup, Niels Grunnet, and John Dich**

TABLE OF CONTENTS

# I. INTRODUCTION

Ethanol intake has a marked effect on nearly all aspects of lipid metabolism in humans; this review will focus on some common clinical problems: alcoholic fatty liver, alcoholic ketoacidosis, and alcoholic hyperlipidemia. In a biochemical sense these clinical entities are only poorly understood. This is, in part, due to the natural restriction of experiments with human volunteers, but also, in part, due to the rapid development of basic lipid research which makes many hitherto accepted views outdated. For extensive reviews the reader is referred to previous publications.[1-3] The recent literature is reviewed in context with a brief summary of earlier work, and special emphasis is put on issues that, according to our opinion, are unsolved or controversial.

# II. FATTY LIVER

Fat infiltration is common in liver biopsies from patients with alcoholic liver disease. The main lipid accumulating is triacylglycerol. The mean content of triacylglycerol is reportedly increased four- to tenfold compared to control liver biopsies.[4-6] In selected biopsies, in which all hepatocytes contained lipid vacuoles, the mean content of triacylglycerol was increased 30-fold compared to control values, corresponding to 350 μmol triacylglycerol per gram wet weight; in extreme cases about 50% of the wet weight of the liver biopsy was triacylglycerol.[7]

In human volunteers, the hepatic content of triacylglycerol increased sixfold, as a mean, after an intake of 100 g ethanol per day for 8 to 14 d; ethanol was given in a diet that otherwise was nutritionally sufficient.[8] Triacylglycerol also increased in the liver of volunteers given 200 to 300 g ethanol per day for only 1 to 2 d.[8,9] These results indicate that the fatty liver of alcoholics is caused by the intake of ethanol per se and that other factors such as the poor nutritional habits encountered among alcoholics are probably not a prerequisite for the development of fatty liver.

The etiology of the fatty liver is still being investigated; it is not even known with certainty to what extent hepatic vs. extrahepatic effects of ethanol are responsible for the fatty liver. This section will concentrate on the effects of ethanol on hepatic fatty acid uptake, fatty acid synthesis, fatty acid oxidation, and synthesis of triacylglycerol, as well as possible extrahepatic effects of ethanol contributing to the fatty liver.

## A. Fatty Acid Uptake

In humans, an intake of 300 to 400 g ethanol per day (giving a blood ethanol concentration of about 30 mM) increases the fatty acid concentration of blood, probably reflecting increased peripheral lipolysis.[10] This could contribute to fatty liver by increasing hepatic fatty acid uptake.[11] Ethanol in large doses increases the concentrations of glucocorticoids and epinephrine in blood,[12-15] and adrenalectomy and hypophysectomy prevent the development of fatty liver after a single large dose of ethanol.[12,16] However, fatty liver is experimentally induced in man by an ethanol intake of only about 100 g/d, a dose that does not increase the blood fatty acid concentration.[10] Further, the fatty acids accumulating in hepatic triacylglycerol of human volunteers have a composition that resembles the food much more than the fatty acid composition of adipose tissue.[17] In rat experiments corrected for the different turnover rates of triacylglycerol in liver and adipose tissue, it was calculated that about 90% of the fatty acids accumulated in hepatic triacylglycerol originated from the diet — the remainder originated from adipose tissue.[18] These observations suggest that peripheral lipolysis does not contribute significantly to the development of fatty liver except during periods of excessive drinking.

Hepatic fatty acid uptake may increase by mechanisms other than increased peripheral lipolysis. In perfused rat liver, fatty acid uptake increases with increased flow of the perfusion medium.[19] After injection of ethanol, Abrams and Cooper[20,21] found an increased hepatic uptake of fatty acids *in vivo*. They ascribed this to the increased hepatic blood flow caused by the injection of ethanol. In man, infusion of ethanol to give blood ethanol concentrations of 2 to 6 m$M$ did not increase hepatic blood flow,[22,23] but infusion of ethanol to give concentrations of 10 to 25 m$M$ did increase hepatic blood flow by about 30%.[24] Therefore, hepatic fatty acid uptake may be increased in humans when the hepatic blood flow is increased by large doses of ethanol, but this has not been shown experimentally.

Ethanol may also increase hepatic fatty acid uptake in a more direct way. In perfused rat liver ethanol increased fatty acid uptake by 15%;[25] this effect was reproduced with hepatocytes in culture[26] and the increase was not alleviated by 4-methylpyrazole,[27] indicating that it is independent of the metabolism of ethanol. In the investigations cited above,[22,23] infusion of ethanol to volunteers did not increase fatty acid uptake, indicating that this effect of ethanol is not important in man, at least not at low concentrations of ethanol.

### B. Fatty Acid Synthesis

In rats a number of studies employing different preparations — perfused liver, freshly isolated hepatocytes, hepatocytes in culture, and experiments *in vivo* with prolonged ethanol feeding — have failed to show increased fatty acid synthesis during ethanol metabolism.[28-31] Ethanol also failed to increase fatty acid synthesis in tissue from human liver biopsies.[32] When feeding volunteers ethanol in a diet with an extremely low fat content, palmitate was the main fatty acid accumulating in liver triacylglycerol. Linoleate was the main fatty acid of the diet and it was concluded that increased fatty acid synthesis could, in part, be responsible for the accumulation of triacylglycerol.[17] However, the observation could as well be due to increased hepatic storage of newly synthesized fatty acids, rather than to increased synthesis of fatty acids.

### C. Fatty Acid Oxidation

When the oxidation of fatty acids is inhibited in perfused rat liver, the fatty acids will accumulate as triacylglycerol.[33-35] Inhibition of fatty acid oxidation by ethanol could, therefore, contribute to the accumulation of triacylglycerol, as suggested by Lieber et al.[36] In human volunteers, infusion of ethanol decreased the conversion of a radioactively labeled fatty acid to oxidative products (ketone bodies and $CO_2$) by about 50%.[22] Similar results were observed in human liver slices.[37] A number of rat experiments have confirmed these observations (see Reference 38). These investigators agree that the inhibition of fatty acid oxidation could contribute to the development of fatty liver.

However, results obtained in experiments with a radioactive fatty acid cannot by themselves be taken to reflect a decreased *total* fatty acid oxidation (ketogenesis + $CO_2$ formation from endogenous and exogenous fatty acids). Endogenous fatty acids contribute to total oxidation of fatty acids, and in the presence of ethanol the contribution of endogenous fatty acids may increase and thereby lower the specific radioactivity of acylcarnitine and acyl CoA, the immediate precursors for β-oxidation; this would lead to a reduced conversion of the exogenous-labeled fatty acids to oxidative products without any change in the total rate of oxidation of fatty acids. Further, the radioactivity of acetyl CoA formed during β-oxidation could be diluted by acetyl CoA formed from acetate during ethanol oxidation, leading to a reduced appearance of radioactivity in $CO_2$ and ketone bodies without any change in flux through the tricarboxylic acid cycle or rates of ketogenesis. In addition, changes in the asymmetric labeling of carbon atoms 1 and 3 in acetoacetate may lead to erroneous estimates of ketogenesis.[38]

Alternatively, fatty acid oxidation has been estimated by stoichiometric calculations, based on chemical measurement of ketone body production and assuming that the oxygen uptake not accounted for by ketogenesis reflects oxidation to $CO_2$ of acetyl CoA originating from fatty acid β-oxidation.[39,40] This approach is inaccurate since it ignores the formation of ketone bodies from pyruvate and, in the presence of ethanol, also from acetate arising from ethanol metabolism. In the fed state and in the absence of exogenously added fatty acids, this may lead to a gross overestimation of the effect of ethanol.

When correcting experimentally for these flaws, it was found that ethanol inhibits total fatty acid oxidation (at 1 m$M$ palmitate) by 25 and 20% in isolated hepatocytes from fed and starved rats, respectively,[38,41] and by 30% in rat hepatocyte cultures.[26] These results are not very different from those obtained by others using less elaborate methods, but a quantitatively accurate calculation of the effect of ethanol on overall hepatic fatty acid metabolism is now possible (see below). At 0.3 m$M$ palmitate, which is a physiological concentration in the fed state, there was, surprisingly, no inhibition of total fatty acid oxidation in isolated hepatocytes.[41]

All investigators agree that ethanol inhibits $CO_2$ formation from fatty acids by inhibition of the tricarboxylic acid cycle and that the inhibition is caused by the increased NADH/NAD$^+$ ratio in the mitochondria.[36,42,43] With the inhibition of the tricarboxylic acid cycle it might be expected that acetyl CoA would simply be diverted to ketone bodies in the mitochondria, as happens during starvation.[44] However, ketogenesis from fatty acids is either slightly reduced or unchanged during ethanol metabolism.[26,38,41] This means that β-oxidation must be inhibited to a degree that corresponds approximately to the reduced $CO_2$ formation from fatty acids. The mechanism of this inhibition is not entirely understood. Possible regulatory factors in this pathway include the levels of the substrate acyl CoA; the cofactor carnitine; the allosteric inhibitor of carnitine-acyltransferase, malonyl CoA; the acetyl CoA thiolase inhibitor, acetyl CoA; and the NADH/NAD$^+$ ratio.

The level of acyl CoA in the liver is reduced by 30% after acute ethanol treatment *in vivo*,[45] but it is difficult to evaluate this finding because of the intracellular compartmentation of the compound. The concentration of carnitine is increased[45] and malonyl CoA is decreased[46] after acute ethanol exposure *in vivo*, in fed rats, and both effects would be expected to increase β-oxidation of fatty acids.[47]

The mitochondrial NADH/NAD$^+$ ratio is increased two- to fourfold by ethanol and this has been suggested to be responsible for the decreased β-oxidation;[40,43,48,49] an increased NADH/NAD$^+$ ratio is supposed to cause a shift towards 3-hydroxyacyl CoA in the reaction catalyzed by 3-hydroxyacyl CoA dehydrogenase, thus, enhancing the concentrations of 3-hydroxyacyl CoA and 3-enoyl CoA, both inhibitors of acyl-CoA dehydrogenase. This role of the increased NADH/NAD$^+$ ratio is not supported by the lack of effect of ethanol at low concentrations of long-chain fatty acids.[41] The experiments of Ryle et al.[50] also seem to indicate a minor role for the increased NADH/NAD$^+$ ratio; they added methylene blue to the diet of rats fed ethanol for 16 d. Methylene blue prevented the increase in the mitochondrial NADH/NAD$^+$ ratio in the liver, but it did not prevent the accumulation of triacylglycerol, suggesting that the accumulation of triacylglycerol is not caused by an NADH/NAD$^+$-induced decrease in fatty acid oxidation; however, in the absence of data on fatty acid metabolism the conclusion is somewhat speculative.

Acetyl CoA accumulates after acute ethanol administration *in vivo*,[45,51] and this may inhibit acyl CoA thiolase.[52] This inhibition may cause an increase in the concentration of 3-oxoacyl CoA intermediates, which will inhibit long-chain acyl CoA dehydrogenase.[53] Short-chain acyl CoA dehydrogenase is not inhibited by these intermediates, which could explain the lack of inhibition of the oxidation of octanoate.[33] However, these intermediates have apparently not been measured after ethanol administration.

In conclusion, it is well established that ethanol inhibits the oxidation of fatty acids *in vitro*, but the mechanism is still uncertain and it is also unknown to what extent ethanol decreases hepatic fatty acid oxidation *in vivo*.

## D. Triacylglycerol Synthesis

During prolonged ethanol treatment, there is an increase in hepatic cytoplasmic lipid droplets.[8] The increase in hepatic triacylglycerol after prolonged ethanol treatment *in vivo* is confined to the cytoplasmic lipid droplets, with no increase in triacylglycerol associated with the endoplasmic reticulum.[54] In the perfused liver, however, ethanol did not increase the synthesis or deposition of triacylglycerol in cytoplasmic lipid droplets. Ethanol increased the incorporation of radioactively labeled palmitate into triacylglycerol in the endoplasmic reticulum. This was not due to an increased accumulation of triacylglycerol, but rather to an increased turnover of this pool.[25] Apparently, ethanol does not significantly increase the accumulation of triacylglycerol *in vitro* in short-term studies; previous claims to the opposite[36,43] are probably due to inadequate analysis of tracer studies.

Subsequent experiments with rat hepatocytes in culture showed that ethanol does increase the content of triacylglycerol *in vitro*, but only after about 6 h exposure to ethanol.[30] After 6 h exposure to ethanol there was a 50% increase in the rate of accumulation of triacylglycerol, measured chemically; this was quantitatively accounted for by increased fatty acid uptake, decreased fatty acid oxidation, and decreased very low-density lipoprotein (VLDL) secretion (each accounting for 44, 49, and 7% of the increase in triacylglycerol, respectively). After 30 h exposure to ethanol, there was a 75% increase in the rate of accumulation of triacyl-glycerol; of this increase 34% was accounted for by increased fatty acid uptake, 34% was accounted for by decreased fatty acid oxidation, and 32% was accounted for by decreased VLDL secretion.[26] When adding 4-methylpyrazole together with ethanol the effect of ethanol on fatty acid oxidation and VLDL secretion was reversed, while the increased fatty acid uptake was unaffected, suggesting that the latter effect was not caused by the metabolism of ethanol.[27] It now remains to be investigated whether the same changes in fatty acid metabolism are responsible for the development of fatty liver *in vivo*, in the rat as well as in humans. In a following section the effect of ethanol on VLDL secretion will be dealt with in more detail.

## E. Role of Glucocorticoids

It was previously mentioned that ethanol in large doses increases cortisol production. In addition to increased peripheral lipolysis, this may affect the liver directly. As reviewed by Brindley,[55] cortisol is known to increase the activity of hepatic phosphatidate phosphohydrolase. The activity of this enzyme may contribute to the rate control of triacylglycerol synthesis and, therefore, an increased activity may contribute to the accumulation of triacylglycerol *in vivo*. The possible quantitative role of this effect has not been determined in relation to the other effects of ethanol.

## III. ACCUMULATION OF TRIACYLGLYCEROL IN OTHER TISSUES

Triacylglycerol also accumulates *in vivo* in nonhepatic tissues that do not oxidize ethanol to any significant extent. Some of the effects of ethanol that are not related directly to its metabolism (see above) are therefore likely to be operative *in vivo*.

Ethanol is apparently not metabolized by heart tissue; nevertheless, ethanol seems to have direct effects on the tissue. The triacylglycerol content in cardiac tissue of rats doubled after 3 weeks of ethanol treatment, and in heart tissue slices ethanol increased the incorporation of $^{14}C$ acetate into triacylglycerol.[56] This could be due to increased fatty acid synthesis or increased esterification of newly synthesized fatty acids. An increased incorporation of labeled palmitate was found in heart tissue of dogs and rats after prolonged ethanol treatment, suggesting an increased capacity for triacylglycerol synthesis.[57,58]

In alcoholic patients, the pancreas exhibits accumulation of lipid in acinar cells.[59] Ethanol given *in vivo* or *in vitro* increased the incorporation of $^{14}C$ acetate into triacylglycerol in rat pancreas tissue slices[60] which could reflect increased fatty acid synthesis or increased esterification of newly synthesized fatty acids.

Lung tissue of rats fed ethanol for 7 weeks exhibited a threefold increase in the content of triacylglycerol,[61] but no experiments have been carried out to elucidate the etiology.

## IV. FATTY ACID METABOLISM IN THE FATTY LIVER

In perfused liver from rats treated with ethanol for 4 weeks, there was an increase in the proportion of fatty acids oxidized and the rate of esterification was reduced.[54] This could, in part, be due to the increased hepatic content of carnitine after prolonged ethanol treatment,[45] but, in addition, β-oxidation has also been found to be increased in mitochondria isolated from ethanol-induced fatty liver.[62] The adaptation is probably not related to the accumulation of fat itself, since fatty acid oxidation and esterification were identical in two populations of hepatocytes isolated from fatty liver, separated according to their content of triacylglycerol.[63] When ethanol was added to the perfusion medium in the experiments cited above, the rates of oxidation and esterification in the fatty liver became similar to the rates observed in control livers in the absence of ethanol. This adaptation may explain why accumulation of triacylglycerol ceases after 4 weeks of ethanol treatment in rats.[64] In fatty liver of baboons, there was an increased activity of triacylglycerol synthesizing enzymes,[65] but the significance of this is uncertain since fatty acid metabolism was not investigated. Biopsy tissue from human fatty liver exhibited increased rates of incorporation of labeled fatty acids into triacylglycerol,[66] but this *in vitro* system is not yet characterized in any detail with respect to metabolic integrity.

Analysis of the composition of fatty acids in phospholipids of liver biopsy tissue from humans with alcoholic cirrhosis revealed an increased ratio of saturated to unsaturated fatty acids which, together with an increased cholesterol/phospholipid ratio, suggests a decreased fluidity of hepatocyte membranes; tissue from patients with fatty liver showed changes in the same direction, but the changes were not statistically significant.[6] A decreased membrane fluidity may have important effects on cell function (see this volume, Chapter 11). In mice, similar changes have been found in membrane lipids as early as 2 h after ethanol administration.[67] It is known that ethanol treatment causes inhibition of $\Delta^5$ and $\Delta^6$ desaturases which may be responsible for the increase in saturated fatty acids.[68]

## V. PATHOGENIC ROLE OF HEPATIC FAT ACCUMULATION

The etiology of severe alcoholic liver lesions (cirrhosis and alcoholic hepatitis) is still not understood. In this section, the possible role of fat accumulation will be reviewed. Liver biopsies from patients with severe obesity, taken before intestinal bypass operations, often exhibit fat accumulation. A number of these biopsies also exhibit necrosis, fibrosis, and inflammatory cells.[69-72] It has been suggested that these changes are caused by the fat accumulation and that the changes may predispose to cirrhosis.[69] Cirrhosis is, however, rarely encountered in these samples.[73] This could be due to exclusion of obese patients with known cirrhosis of the liver before preoperative liver biopsy and due to weight loss among obese patients after development of cirrhosis. Hidden alcoholism could contribute significantly to the obesity and to the changes in liver morphology, although obese children in Japan exhibit similar hepatic morphological abnormalities.[74]

In an 11- to 13-d study comparing i.v. feeding of glucose with isocaloric feeding of emulsified fat, the glucose-fed patients developed fatty liver and small increases in serum aminotransferase activities; these changes were, however, not accompanied by hepatocellular destruction or inflammatory infiltration.[75] This study suggests that fat accumulation leads to slight liver damage only, at least in a short-term study.

In a study of fatty liver of obese or alcoholic patients it was found that the hepatic content of free fatty acids was increased 10-fold in fatty liver of obese patients and 10- to 15-fold in fatty liver of alcoholic patients, depending on the degree of liver damage.[5] The authors suggested that the cytotoxic effect of free fatty acids may be responsible for the liver damage in obesity and alcoholic liver disease. These results are, however, contradicted by another study reporting no increase in free fatty acids in fatty liver from alcoholic patients.[6] One reason for the discrepancy may be that the mean hepatic triacylglycerol content in the latter study was only about 50% of that in the former.

There are no studies dealing experimentally with the possible harmful effects of an increased content of free fatty acids in liver cells. However, from studies with isolated adipocytes, it is known that when the intracellular content of free fatty acids is increased about sixfold, the content of ATP declines, the uptake of potassium and $\alpha$-aminoisobutyrate declines, and the cells release about 90% of the activity of cytosolic enzymes.[76,77] The studies suggest that accumulation of free fatty acids is deleterious to the cell, probably because of a detergent effect on membranes.

In summary, some studies support the concept that the accumulation of fat has a pathogenic role, but direct experimental evidence is still lacking.

## VI. ALCOHOLIC KETOACIDOSIS

Acidosis due to accumulation of ketone bodies is sometimes encountered in chronic alcoholics, most frequently after a period of excessive alcohol intake associated with reduced intake of regular food. The ketosis develops in most cases after ethanol intake has ceased and the subjects are starving.[78-86] The concentration of ketone bodies may increase to 20 mM.[82] In the most extensive biochemical study of the condition the ketosis was found to be associated with very high concentrations of free fatty acids in plasma (1 to 4 mM) and high concentrations of cortisol and growth hormone concomitant with low or normal concentrations of insulin.[82] It seems that glucagon and catecholamines have never been measured.

Volunteers given ethanol (150 g/d for 1 week) showed a 30-fold increase in fasting ketone body concentration compared to controls fed an ordinary diet.[87] This study suggests that ethanol intake augments the physiological stimulus of starvation to increase ketonemia.

The accumulation of ketone bodies may be due to increased production of ketone bodies by the liver and/or to decreased utilization of ketone bodies by extrahepatic tissues, but these rates have never been measured in this condition.

An increase in ketone body production by the liver may result from a combination of increased fatty acid supply to the liver, prolonged ethanol intake, and starvation. The increased blood fatty acid concentration associated with alcoholic ketoacidosis is probably due to excessive adipose tissue lipolysis caused by the high level of cortisol and perhaps a low ratio between insulin and glucagon. The rate of ketone body production by the isolated liver is linearly correlated with the fatty acid concentration.[11]

Addition of ethanol to hepatocytes does not increase ketogenesis,[33,40,41,43] but prolonged ethanol intake increases fatty acid oxidation at the expense of esterification when ethanol is no longer present (see above). In addition, the high concentration of acetate during ethanol intake may contribute, since acetate is known to reduce the hepatic phosphorylation potential.[88] Alcoholic ketoacidosis is consistently associated with hypophosphatemia and blood ketone bodies exhibit extremely high 3-hydroxybutyrate/acetoacetate ratios.[82] These findings are in accordance with a low hepatic phosphorylation potential which leads to an increased oxidation of fatty acids at the expense of esterification.[88]

Starvation leads to a reduction in the concentration of tricarboxylic acid cycle intermediates and acetyl CoA will, therefore, be diverted to ketone bodies. In addition, the inhibition by ethanol of fatty acid oxidation ceases when ethanol metabolism stops, allowing the expression of the full capacity of the increased fatty acid oxidation (see above); this may add to the predominance of alcoholic ketosis after ethanol consumption has ceased.

Ketosis could also be caused by decreased utilization of ketone bodies by extrahepatic tissues. However, addition of ethanol had no effect on the oxidation of 3-hydroxybutyrate in rat diaphragm tissue.[87] Ethanol is not oxidized in any significant amount by extrahepatic tissues and it is, therefore, improbable that a physiological concentration of ethanol by itself has any acute effects on ketone body metabolism in these tissues. Acetate arising from ethanol oxidation may, by competition, inhibit ketone body oxidation in peripheral tissues, but this has apparently not been examined experimentally.

## VII. HYPERLIPIDEMIA

Alcoholism is one of the most common causes of hyperlipidemia.[89] The abnormalities include increases in triacylglycerol, mainly in the VLDL fraction, and of cholesterol in the high-density lipoprotein (HDL) fraction of plasma lipoproteins. This review will focus on plasma triacylglycerol with only a brief discusson of hypercholesterolemia.

### A. Triacylglycerol
*1. Single Dose of Ethanol*
A single dose of 60 to 80 g ethanol given to fasting healthy subjects increases plasma triacylglycerol by 20 to 35%.[90-92] Smaller doses (30 g or less) do not increase plasma triacylglycerol.[93] Administration of a single dose of ethanol to fasting abstinent alcoholics also increases plasma triacylglycerol; the effect is moderate in most patients, but in some a three- to fourfold increase was seen.[94-96] Food intake markedly exaggerates the ethanol-induced increase in plasma triacylglycerol, increasing the response twofold in healthy subjects.[97-99] In alcoholics, the effect of food can be even larger.[100] A high-fat meal by itself increases plasma triacylglycerol much more in patients with alcoholic fatty liver than in healthy controls.[101,102] When adding ethanol to a meal, the additional increase in plasma triacylglycerol in patients with alcoholic fatty liver is only minor; in patients with alcoholic cirrhosis, the addition of ethanol decreases plasma triacylglycerol.[101]

In conclusion, a single dose of ethanol generally increases plasma triacylglycerol, but the response depends on, among other factors, the size of the dose, the duration of ethanol exposure, and the time points of sampling.[91,103,104] The maximal effect of ethanol occurs 6 to 8 h after ethanol administration.[90,98,100] The response is further dependent on concomitant food intake and on the presence or absence of liver disease.

*2. Prolonged Ethanol Intake*
Epidemiological studies have shown a correlation between regular ethanol consumption and an increase in plasma triacylglycerol.[105-107] Daily consumption of 50 to 90 g ethanol by healthy subjects for 1 to 4 weeks resulted in a 25 to 50% increase in plasma triacylglycerol.[108-110] Intake of 200 to 400 g ethanol daily by alcoholics for 2 to 3 weeks increased triacylglycerol in plasma two- to fivefold.[10,111] In these studies, the effect vanished after 1 to 2 weeks despite continued ethanol administration, but in another study the increase persisted.[112] Among alcoholics admitted to alcoholic wards only about one third exhibited increased plasma triacylglycerol.[113-117] In one study half of the patients exhibited a persistent increase in plasma triacylglycerol after 12 d of abstinence.[89] Factors other than ethanol metabolism probably contribute to the development of hyperlipidemia. Healthy controls with high fasting plasma triacylglycerol are known to be more susceptible to the effect of ethanol.[99,100,118,119] Alcoholics with preexisting hyperlipidemia (type IV) are prone to develop high levels of triacylglycerol;[120] also, coexisting diabetes may modify the effect of ethanol.[113,121,122] Baboons fed ethanol for years consistently exhibit elevated plasma con-

centrations of triacylglycerol.[65] It seems that prolonged ethanol intake causes hyperlipidemia, but individual predisposition to hyperlipidemia plays an important role in determining the degree of hyperlipidemia developed.

### 3. Very Low-Density Lipoprotein Secretion

The ethanol-induced increase in plasma triacylglycerol is due either to an increased secretion of VLDL and/or chylomicrons, a decreased degradation of these, or both. In man a single dose of ethanol increases the incorporation of radioactively labeled fatty acids into plasma triacylglycerol *in vivo*.[95,98,123] Similar results have been obtained in rats,[124] but others found no increase;[125] the difference may be due to different time points of sampling. After prolonged ethanol treatment there was also an increased labeling in rats;[125,126] similar studies seem not to have been carried out in humans. The increased incorporation does not by itself prove an increased synthesis, since it may merely reflect an increase in the specific radioactivity of the intrahepatic precursor pool for plasma triacylglycerol.[25] Further, the increased radioactivity of plasma triacylglycerol could be due to decreased degradation which was not determined in these studies.

Another experimental approach is to give ethanol-treated rats Triton WR 1339 which inhibits the degradation of plasma triacylglycerol. Changes in plasma triacylglycerol then reflect changes in secretion. In the starved state, the liver is responsible for 60 to 80% of total triacylglycerol secretion in the rat, the remaining part being secreted by the intestine.[127,128] With this method a time-dependent response to ethanol was discovered, the rate of secretion being decreased during the first 2 h after ethanol administration, followed by a period of increasing secretion, reaching control values after 6 h.[129] Inhibition of synthesis at early time points was reproduced by others[130] and 4 to 6 h after ethanol administration even an increased rate of triacylglycerol release has been found.[131,132] The initial decrease may be due to the high initial blood ethanol concentration[129] (see below), while the increase at later time points may be due to the developing fatty liver.[34,35] After prolonged ethanol treatment, there was an increased secretion of triacylglycerol,[126,130] and when a single dose of ethanol was given to these ethanol-treated rats there was a decreased secretion.[130] In conclusion, the immediate effect of ethanol *in vivo* seems to be an inhibition of the secretion, but at later time points and after prolonged ethanol treatment, there is an increased secretion. It is not known to what extent the liver and the intestine contribute to these changes.

The *apoproteins* of the lipoproteins may also be affected by ethanol ingestion. No human studies have been carried out in this field, but in rats, a single dose of ethanol *in vivo* decreased the secretion of VLDL-apoprotein.[133] After prolonged ethanol intake, there was an increase in the protein content of VLDL and an increase in the incorporation of labeled lysine into VLDL.[126] In isolated hepatocytes, the synthesis of VLDL apoprotein was inhibited in the presence of ethanol, the inhibition being dependent on the concentration of ethanol.[134] This is compatible with the general inhibitory effect of ethanol on the secretion of export proteins from hepatocytes.[135] In hepatocytes isolated 4 h after ethanol treatment there was an increased secretion of VLDL-apoprotein.[134] These studies are generally in agreement with the *in vivo* studies regarding VLDL secretion discussed above.

*Hepatic triacylglycerol secretion,* studied *in vitro,* is dependent on the concentration of ethanol. In perfused rat liver, concentrations of ethanol below 50 m$M$ had no effect on the secretion of triacylglycerol,[25,136,137] while at concentrations of 80 to 90 m$M$ there was an inhibition of 60 to 70%.[25,138] These studies are consistent with the immediate effect of ethanol in the *in vivo* studies mentioned above. In cultured hepatocytes 50 m$M$ ethanol had only a small effect on triacylglycerol secretion after 6 h of incubation, but after 30 h of incubation with ethanol there was a decrease of 50%.[26] After prolonged ethanol treatment *in vivo* the perfused liver exhibited an unchanged rate of triacylglycerol secretion, and addition of 80 m$M$ ethanol to the perfusion medium did not affect the secretion.[54] This study is not consistent with the *in vivo* study mentioned above,[130] and there is no obvious explanation for the discrepancy.

*Intestinal triacylglycerol secretion* may contribute to the alcoholic hyperlipidemia, since chylomicron-like particles appear in the blood of healthy controls and in alcoholics after ethanol ingestion,[96,110] but it appears that no human studies have been carried out to verify this suggestion. In rats, both a single dose of ethanol and prolonged ethanol administration increase the incorporation of radioactively labeled fatty acids into triacylglycerol in slices of intestinal tissue.[139] Ethanol added to intestinal slices *in vitro* also increased the incorporation of radioactivity.[139]

Lymph lipid output was not increased by a single intragastric gavage of ethanol.[140] Immediately after beginning intraduodenal infusion of ethanol to starved rats the intestinal lymph output of VLDL-triacylglycerol was unaffected, but after infusion for 8 or 16 h there was an increase of 30 to 50%;[141] the source of this lipid is unknown. When lipid was administered together with a single dose of ethanol, the lymph lipid output was increased compared to administration of lipid alone, suggesting increased absorption and transport of dietary lipid.[140] This effect was not seen in rats after prolonged ethanol treatment.[140] In these rats the intestinal lymph lipid output after administration of lipid alone was less than in control animals, suggesting that prolonged ethanol treatment decreases lipid absorption. The effect of a single dose of ethanol has been examined by a different method. Orotic acid was given to rats to inhibit hepatic triacylglycerol secretion and Triton WR 1339 was given to inhibit the degradation of plasma triacylglycerol. Administration of ethanol (without dietary fat) did not change the plasma concentration of triacylglycerol, indicating that the intestinal triacylglycerol secretion was not increased.[132] In conclusion, increased intestinal fat absorption and secretion may contribute to the postprandial hyperlipidemia after a single dose of ethanol, but the intestine does not seem to contribute to the hyperlipidemia seen after prolonged ethanol treatment.

### 4. Degradation of Triacylglycerol

The degradation rate of endogenously labeled plasma triacylglycerol has been found to be unchanged[142] or decreased[143] in healthy subjects after a single dose of ethanol. The degradation rate of Intralipid, an artificial fat emulsion resembling chylomicrons, was unchanged in healthy subjects when ethanol was given over a period of 3 h,[144] but decreased when ethanol was given in a slightly larger dose over a period of 5 h.[145] The clearance of Intralipid was normal in alcoholics soon after admission when plasma triacylglycerol was high and also after a period of abstinence when plasma triacylglycerol had declined.[89] In rats, a decreased degradation of labeled chylomicrons was found 3 h after a single dose of ethanol,[146] while 16 h after ethanol administration, when ethanol had disappeared from the blood, the rate of degradation was unchanged.[124] After prolonged ethanol intake in rats the degradation of labeled chylomicrons was unchanged when measured over a period of 10 min after injection,[126] but when measured over a period of 30 min the degradation rate was decreased.[147,148] The latter study also showed that VLDL-apoproteins are retained in the blood for a longer period in ethanol-treated rats. These results are supported by the finding of a decreased degradation rate of labeled VLDL in rats after prolonged ethanol treatment.[149]

The decreased degradation could be due to decreased activity of lipoprotein lipase. The available evidence is, however, confusing. In human studies, single doses of ethanol and prolonged ethanol intake have been reported to increase,[117,150] decrease,[110,113,143,144,151] or not change[90,109] the activity.

Available evidence points towards an inhibitory effect of ethanol on the degradation of plasma triacylglycerol being at least partially responsible for hyperlipidemia.

## B. Cholesterol

Alcohol consumption also increases the level of plasma cholesterol. This has been reported in alcoholics after long drinking periods[10,96,114,121] and in healthy subjects during regular ethanol intake.[105-108] The concentration of cholesterol normalizes a few weeks after ethanol withdrawal.[121] A single dose of ethanol, given either to abstinent alcoholics[94] or to healthy subjects,[90-92,118] does not, however, increase plasma cholesterol.

The increase in plasma cholesterol is mainly confined to the HDL fraction;[107,117,152-154] experimentally, the level of HDL cholesterol is increased after administration of ethanol for several weeks to healthy subjects[109,110,155-157] and to alcoholics.[154,158] HDL cholesterol normalizes during abstinence.[117,159] The level of the apoproteins of HDL (A-I, A-II, and C) is also increased by long-term alcohol ingestion[160,161] and the level normalizes after 2 weeks of abstinence.[161] In alcoholics, the content of cholesterol is increased both in subclass $HDL_2$ and in subclass $HDL_3$.[117,162] In moderately drinking healthy subjects only subclass $HDL_3$ declined during abstinence, and resumption of drinking was followed by a rise in this subclass only.[163] It is, therefore, difficult to explain a possible correlation between moderate drinking and a low incidence of coronary heart disease, since it is an increase in the $HDL_2$ subclass that is associated with a low incidence of coronary heart disease.[164] It is suggested that the increased content of cholesterol and apolipoproteins in HDL is due to increased hepatic synthesis and/or altered degradation, but these rates have not yet been measured.[117]

## VIII. CONCLUDING REMARKS

Ethanol intake increases the content of triacylglycerol in liver, blood, and all other tissues examined. The effect is most marked in liver where metabolism of ethanol takes place, and it has been shown that the metabolism of ethanol is largely responsible for the hepatic accumulaton of triacylglycerol *in vitro*. However, it has also been shown *in vitro* that ethanol, in addition, increases hepatic triacylglycerol accumulation significantly by increasing hepatic fatty acid uptake, an effect that is independent of the metabolism of ethanol. *In vivo*, extrahepatic effects of ethanol, such as increased glucocorticoid production, may contribute to the development of fatty liver. It remains to be quantitated *in vivo* to what extent the fatty liver is caused by hepatic effects of ethanol not related to its metabolism vs. effects caused by its metabolism, and to what extent hepatic vs. extrahepatic effects of ethanol contribute to the fatty liver. The same questions remain to be solved for accumulation of triacylglycerol in other tissues. In clinical research it remains to be investigated whether the same effects as described for experimental animals are involved in the development of fatty liver. The possible pathogenic role of the fatty liver needs to be clarified, especially the possible harmful effects of accumulation of free fatty acids intracellularly and of the altered composition of membrane lipids.

Experimental research in humans is insufficient to explain alcoholic hyperlipidemia in any detail, and little is known about the clinical significance of this condition. Experimental research in rats is also limited in that there are very few comprehensive studies available. Parallel studies with simultaneous measurements of secretion and degradation *in vivo* are required and a more accurate quantitation of hepatic vs. intestinal triacylglycerol secretion is also needed. In addition, the bimodal, time-dependent effect of ethanol *in vivo* needs further clarification to establish whether the initial inhibition of VLDL secretion is due to the initially high blood ethanol concentration and to what extent the developing fatty liver is associated with an increased VLDL secretion. Finally, studies of apoprotein catabolism and lipoprotein lipase are required to elucidate the effect of ethanol on the degradation of plasma triacylglycerol.

# REFERENCES

1. **Reitz, R. C.,** The effects of ethanol ingestion on lipid metabolism, *Prog. Lipid Res.,* 18, 87, 1979.
2. **Kondrup, J.,** In vitro effect of ethanol on the synthesis of triacylglycerol in hepatic cytoplasmic lipid droplets, *Dan. Med. Bull.,* 31, 109, 1984.
3. **Lieber, C. S. and Savolainen, M.,** Ethanol and lipids, *Alcoholism Clin. Exp. Res.,* 8, 409, 1984.
4. **Laurell, S. and Lundquist, A.,** Lipid composition of human liver biopsy specimens, *Acta Med. Scand.,* 189, 65, 1971.
5. **Mavrelis, P. G., Ammon, H. V., Gleysteen, J. J., Komorowski, R. A., and Charaf, U. K.,** Hepatic free fatty acids in alcoholic liver disease and morbid obesity, *Hepatology,* 3, 226, 1983.
6. **Cairns, S. R. and Peters, T. J.,** Biochemical analysis of hepatic lipid in alcoholic and diabetic and control subjects, *Clin. Sci.,* 65, 645, 1983.
7. **Lundquist, A., Wiebe, T., and Belfrage, P.,** Liver lipid content in alcoholics, *Acta Med. Scand.,* 194, 501, 1973.
8. **Rubin, E. and Lieber, C. S.,** Alcohol-induced hepatic injury in nonalcoholic volunteers, *N. Engl. J. Med.,* 290, 128, 1968.
9. **Wiebe, T., Lundquist, A., and Belfrage, P.,** Time-course of liver fat accumulation in man after a single load of ethanol, *Scand., J. Clin. Lab. Invest.,* 27, 33, 1971.
10. **Lieber, C. S., Jones, D. P., Mendelson, J., and DeCarli, L. M.,** Fatty liver, hyperlipemia and hyperuricemia produced by prolonged alcohol consumption, despite adequate dietary intake, *Trans. Assoc. Am. Physicians,* 76, 289, 1963.
11. **Heimberg, M., Weinstein, I., and Kohout, M.,** The effects of glucagon, dibutyryl cyclic adenosine 3',5'-monophosphate, and concentration of free fatty acid on hepatic lipid metabolism, *J. Biol. Chem.,* 244, 5131, 1969.
12. **Brodie, B. B. and Maickel, R. P.,** Role of the sympathetic nervous system in drug-induced fatty liver, *Ann. N.Y. Acad. Sci.,* 104, 1049, 1963.
13. **Brodie, B. B., Butler, W. M., Horning, M. G., Maickel, R. P., and Maling, H. M.,** Alcohol-induced triglyceride deposition in liver through derangement of fat transport, *Am. J. Clin. Nutr.,* 9, 432, 1961.
14. **Maickel, R. P. and Brodie, B. B.,** Interaction of drugs with the pituitary-adrenocortical system in the production of the fatty liver, *Ann. N.Y. Acad. Sci.,* 104, 1059, 1963.
15. **Pritchard, P. H., Cooling, J., Burditt, S. L., and Brindley, D. N.,** Can the alterations in serum glucocorticoid concentrations explain the effects of ethanol and benfluorex on the synthesis of hepatic triacylglycerols?, *J. Pharm. Pharmacol.,* 31, 406, 1979.
16. **Soliman, K. F. A., Pinkston, J. N., and Owasoyo, J. O.,** The role of the adrenal gland in ethanol-induced triglyceride mobilization, *Pharmacology,* 29, 94, 1984.
17. **Lieber, C. S. and Spritz, N.,** Effects of prolonged ethanol intake in man: role of dietary, adipose, and endogenously synthesized fatty acids in the pathogenesis of the alcoholic fatty liver, *J. Clin. Invest.,* 45, 1400, 1966.
18. **Mendenhall, C. L.,** Origin of hepatic triglyceride fatty acids: quantitative estimation of the relative contributions of linoleic acid by diet and adipose tissue in normal and ethanol-fed rats, *J. Lipid Res.,* 13, 177, 1972.
19. **Morris, B.,** Some factors affecting the metabolism of free fatty acids and chylomicron triglycerides by the perfused rat's liver, *J. Physiol.,* 168, 584, 1963.
20. **Abrams, M. A. and Cooper, C.,** Quantitative analysis of metabolism of hepatic triglyceride in ethanol-treated rats, *Biochem. J.,* 156, 33, 1976.
21. **Abrams, M. A. and Cooper, C.,** Mechanism of increased hepatic uptake of unesterified fatty acids from serum of ethanol-treated rats, *Biochem. J.,* 156, 47, 1976.
22. **Wolfe, B. M., Havel, J. R., Marliss, E. B., Kane, J. P., Seymour, J., and Ahuja, S. P.,** Effects of a 3-day fast and of ethanol on splanchnic metabolism of FFA, amino acids, and carbohydrates in healthy young men, *J. Clin. Invest.,* 57, 329, 1976.
23. **Jorfeldt, L. and Juhlin-Dannfelt, A.,** The influence of ethanol on splanchnic and skeletal muscle metabolism in man, *Metabolism,* 27, 97, 1978.
24. **Stein, S. W., Lieber, C. S., Leevy, C. M., Cherrick, G. R., and Abelmann, W. H.,** The effect of ethanol upon systemic and hepatic blood flow in man, *Am. J. Clin. Nutr.,* 13, 68, 1963.
25. **Kondrup, J., Lundquist, F., and Damgaard, S. E.,** Metabolism of palmitate in perfused rat liver. Effect of low and high ethanol concentrations at various concentrations of palmitate in the perfusion medium, *Biochem. J.,* 184, 83, 1979.
26. **Grunnet, N., Kondrup, J., and Dich, J.,** Effect of ethanol on lipid metabolism in cultured hepatocytes, *Biochem. J.,* 228, 673, 1985.
27. **Grunnet, N., Kondrup, J., and Dich, J.,** Ethanol-induced accumulation of triacylglycerol in cultured hepatocytes: dependency on ethanol metabolism, *Alcohol Alcoholism,* Suppl. 1, 257, 1987.

28. **Brunengraber, H., Boutry, M., Lowenstein, L., and Lowenstein, J. M.,** The effect of ethanol on lipogenesis by the perfused liver, in *Alcohol and Aldehyde Metabolizing Systems*, Thurman, R. G., Yonetani, T., Williamson, J. R., and Chance, B., Eds., Academic Press, New York, 1974, 329.

29. **Selmer, J. and Grunnet, N.,** Ethanol metabolism and lipid synthesis by isolated liver cells from fed rats, *Biochim. Biophys. Acta*, 428, 123, 1976.

30. **Grunnet, N., Jensen, F., Kondrup, J., and Dich, J.,** Effect of ethanol on fatty acid metabolism in cultured hepatocytes: dependency on incubation time and fatty acid concentration, *Alcohol*, 2, 157, 1985.

31. **Savolainen, M. J., Hiltunen, J. K., and Hassinen, I. E.,** Effect of prolonged ethanol ingestion on hepatic lipogenesis and related enzyme activities, *Biochem. J.*, 164, 169, 1977.

32. **Venkatesan, S., Leung, N. W. Y., and Peters, T. J.,** Fatty acid synthesis in vitro by liver tissue from control subjects and patients with alcoholic liver disease, *Clin. Sci.*, 71, 723, 1986.

33. **McGarry, J. D. and Foster, D. W.,** The regulation of ketogenesis from oleic acid and the influence of antiketogenic agents, *J. Biol. Chem.*, 246, 6247, 1971.

34. **Ide, T. and Ontko, J. A.,** Increased secretion of very low density lipoprotein triglyceride following inhibition of long chain fatty acid oxidation in isolated rat liver, *J. Biol. Chem.*, 256, 10247, 1981.

35. **Fukada, N. and Ontko, J. A.,** Interactions between fatty acid synthesis, oxidation, and esterification in the production of triglyceride-rich lipoproteins by the liver, *J. Lipid Res.*, 25, 831, 1984.

36. **Lieber, C. S., Lefevre, A., Spritz, N., Feinman, L., and DeCarli, L. M.,** Difference in hepatic metabolism of long- and medium-chain fatty acids: the role of fatty acid chain length in production of the alcoholic fatty liver, *J. Clin. Invest.*, 46, 1451, 1967.

37. **Blomstrand, R., Kager, L., and Lantto, O.,** Studies on the ethanol-induced decrease of fatty acid oxidation in rat and human liver slices, *Life Sci.*, 13, 1131, 1973.

38. **Grunnet, N. and Kondrup, J.,** Effect of ethanol, noradrenalin and 3',5'-cyclic AMP on oxidation of fatty acids and lipolysis in isolated rat hepatocytes, *Pharmacol. Biochem. Behav.*, 18 (Suppl. 1), 245, 1983.

39. **Williamson, J. R., Scholz, R., Browning, E. T., Thurman, R. G., and Fukami, M. H.,** Metabolic effects of ethanol in the perfused rat liver, *J. Biol. Chem.*, 244, 5044, 1969.

40. **Fellenius, E. and Kiessling, K.-H.,** Effect of ethanol on fatty acid oxidation in the perfused livers of starved, fed and fat-fed rats, *Acta Chem. Scand.*, 27, 2781, 1973.

41. **Grunnet, N. and Kondrup, J.,** The effect of ethanol on the β-oxidation of fatty acids, *Alcoholism Clin. Exp. Res.*, 10, 64S, 1986.

42. **Forsander, O. A., Mäenpää, P. H., and Salaspuro, M. P.,** Influence of ethanol on the lactate/pyruvate and β-hydroxybutyrate/acetoacetate ratios in rat liver experiments, *Acta Chem. Scand.*, 19, 1770, 1965.

43. **Ontko, J. A.,** Effects of ethanol on the metabolism of free fatty acids in isolated liver cells, *J. Lipid Res.*, 14, 78, 1973.

44. **Krebs, H. A.,** The regulation of the release of ketone bodies by the liver, *Adv. Enzyme Regul.*, 4, 339, 1966.

45. **Kondrup, J. and Grunnet, N.,** The effect of acute and prolonged ethanol treatment on the contents of coenzyme A, carnitine and their derivatives in rat liver, *Biochem. J.*, 132, 373, 1973.

46. **Guynn, R. W., Veloso, D., Harris, R. L., Lawson, J. W., and Veech, R. L.,** Ethanol administration and the relationship of malonyl-coenzyme A concentrations to the rate of fatty acid synthesis in rat liver, *Biochem. J.*, 136, 639, 1973.

47. **McGarry, J. D. and Foster, D. W.,** Regulation of hepatic fatty acid oxidation and ketone body production, *Annu. Rev. Biochem.*, 49, 395, 1980.

48. **Davis, E. J. and Lumeng, L.,** Suppression of the mitochondrial oxidation of ($-$)palmitylcarnitine by the malate-aspartate and α-glycerophosphate shuttles: mechanisms for the inhibitory effect of ethanol on β-oxidation, in *Use of Isolated Liver Cells and Kidney Tubules in Metabolic Studies*, Tager, J. M., Söling, H. D., and Williamson, J. R., Eds., North-Holland, Amsterdam, 1976, 112.

49. **Latipää, P. M., Kärki, T. T., Hiltunen, J. K., and Hassinen, I. E.,** Regulation of palmitoylcarnitine oxidation in isolated rat liver mitochondria. Role of the redox state of NAD(H), *Biochim. Biophys. Acta*, 875, 293, 1986.

50. **Ryle, P. R., Chakraborty, J., and Thomson, A. D.,** The effect of Methylene Blue on the hepatocellular redox state and liver lipid content during chronic ethanol feeding in the rat, *Biochem. J.*, 232, 877, 1985.

51. **Forsander, O. A. and Lindros, K. O.,** Influence of ethanol on the acetyl-coenzyme A level of intact rat liver, *Acta Chem. Scand.*, 21, 2568, 1967.

52. **Olowe, Y. and Schultz, H.,** Regulation of thiolases from pig heart. Control of fatty acid oxidation in heart, *Eur. J. Biochem.*, 109, 425, 1980.

53. **Davidson, B. and Schulz, H.,** Separation, properties, and regulation of acyl coenzyme A dehydrogenases from bovine heart and liver, *Arch. Biochem. Biophys.*, 213, 155, 1982.

54. **Kondrup, J., Lundquist, F., and Damgaard, S. E.,** Metabolism of palmitate in perfused rat liver. Effect of ethanol in livers from rats fed on a high-fat diet with or without ethanol, *Biochem. J.*, 184, 89, 1979.

55. **Brindley, D. N. and Sturton, E. G.,** Phosphatidate metabolism and its relation to triacylglycerol biosynthesis, in *Phospholipids*, Hawthorne, J. N. and Ansell, G. B., Eds., Elsevier, Amsterdam, 1982, chap. 5.

56. **Somer, J. B., Colley, P. W., Pirola, R. C., and Wilson, J. S.,** Ethanol-induced changes in cardiac lipid metabolism, *Alcoholism Clin. Exp. Res.,* 5, 536, 1981.
57. **Lochner, A., Crowley, R., and Brink, A. J.,** Effect of ethanol on metabolism and function of perfused rat heart, *Am. Heart J.,* 78, 770, 1969.
58. **Regan, T. J., Khan, M. E., Ettinger, P. O., Harder, B., Lyons, M. M., and Oldewurtel, H. A.,** Myocardial function and lipid metabolism in the chronic alcoholic animal, *J. Clin. Invest.,* 54, 740, 1974.
59. **Dreiling, D. A. and Bordalo, O.,** A toxic-metabolic hypothesis of pathogenesis of alcoholic pancreatitis, *Alcoholism,* 1, 293, 1977.
60. **Somer, J. B., Thompson, G., and Pirola, R. C.,** Influence of ethanol on pancreatic lipid metabolism, *Alcoholism Clin. Exp. Res.,* 4, 341, 1980.
61. **Liau, D. F., Hashim, S. A., Pierson, R. N., III, and Ryan, S. F.,** Alcohol-induced lipid changes in the lung, *J. Lipid Res.,* 22, 680, 1981.
62. **Cederbaum, A. I., Dicker, E., Lieber, C. S., and Rubin, E.,** Ethanol oxidation by isolated hepatocytes from ethanol-treated and control rats: factors contributing to the metabolic adaption after chronic ethanol consumption, *Biochem. Pharmacol.,* 27, 7, 1978.
63. **Kondrup, J., Bro, B., Dich, J., Grunnet, N., and Thieden, H. I. D.,** Fractionaton and characterization of rat hepatocytes isolated from ethanol-induced fatty liver, *Lab. Invest.,* 43, 182, 1980.
64. **Lieber, C. S. and DeCarli, L. M.,** Quantitative relationship between amount of dietary fat and severity of alcoholic fatty liver, *Am. J. Clin. Nutr.,* 23, 474, 1970.
65. **Savolainen, M. J., Baraona, E., Pikkarainen, P., and Lieber, C. S.,** Hepatic triacylglycerol synthesizing activity during progression of alcoholic liver injury in the baboon, *J. Lipid Res.,* 25, 813, 1984.
66. **Leung, N. W. Y. and Peters, T. J.,** Palmitic acid oxidation and incorporation into triglyceride by needle biopsy specimens from control subjects and patients with alcoholic fatty liver disease, *Clin. Sci.,* 71, 253, 1986.
67. **Littleton, J. M., John, G. R., and Grieve, S. J.,** Alterations in phospholipid composition in ethano tolerance and dependence, *Alcoholism Clin. Exp. Res.,* 3, 50, 1979.
68. **Nervi, A. M., Peluffo, R. O., Brenner, R. R., and Leikin, A. I.,** Effect of ethanol administration on fatty acid desaturation, *Lipids,* 15, 263, 1980.
69. **Adler, M. and Schaffner, F.,** Fatty liver hepatitis and cirrhosis in obese patients, *Am. J. Med.,* 67, 811, 1979.
70. **Ludwig, J., Viggiano, T. R., McGill, D. B., and Ott, B. J.,** Nonalcoholic steatohepatitis. Mayo Clinic experiences with a hitherto unnamed disease, *Mayo Clin. Proc.,* 55, 434, 1980.
71. **Nasrallah, S. M., Wills, C. E., and Galambos, J. T.,** Hepatic morphology in obesity, *Dig. Dis. Sci.,* 26, 325, 1981.
72. **Andersen, T., Christoffersen, P., and Gluud, C.,** The liver in consecutive patients with morbid obesity: a clinical, morphological, and biochemical study, *Int. J. Obesity,* 8, 107, 1984.
73. **Andersen, T., and Gluud, C.,** Liver morphology in morbid obesity: A literature study, *Int. J. Obesity,* 8, 97, 1984.
74. **Kinugasa, A., Tsunamoto, K., Furukawa, N., Sawada, T., Kusonoki, T., and Shimada, N.,** Fatty liver and its fibrous changes found in simple obesity of children, *J. Pediatr. Gastroenterol. Nutr.,* 3, 408, 1984.
75. **Tulikoura, I. and Huikuri, K.,** Morphological fatty changes and function of the liver, serum free fatty acids, and triglycerides during parenteral nutrition, *Scand. J. Gastroenterol.,* 17, 177, 1982.
76. **Rodbell, M.,** The metabolism of isolated fat cells. IV. Regulation of release of protein by lipolytic hormones and insulin, *J. Biol. Chem.,* 241, 3909, 1966.
77. **Heindel, J. J., Cushman, S. W., and Jeanrenaud, B.,** Cell-associated fatty acid levels and energy-requiring processes in mouse adipocytes, *Am. J. Physiol.,* 226, 16, 1974.
78. **Jenkins, D. W., Eckel, R. E., and Craig, J. W.,** Alcoholic ketoacidosis, *J.A.M.A.,* 217, 177, 1971.
79. **Levy, L. T., Duga, J., Girgis, M., and Gordon, E. E.,** Ketoacidosis associated with alcoholism in nondiabetic subjects, *Ann. Intern. Med.,* 78, 213, 1973.
80. **Cooperman, M. T., Davidoff, F., Spark, R., and Palotta, J.,** Clinical studies of alcoholic ketoacidosis, *Diabetes,* 23, 433, 1974.
81. **Fulop, M. and Hoberman, H. D.,** Alcoholic ketoacidosis, *Diabetes,* 24, 785, 1975.
82. **Miller, P. D., Heinig, R. E., and Waterhouse, C.,** Treatment of alcoholic acidosis, *Arch. Intern. Med.,* 138, 67, 1978.
83. **Johansen, K. and Theilade, P.,** Alcohol ketoacidosis, *Ugeskr. Laeg.,* 140, 226, 1978.
84. **Offenstadt, G., Amstutz, P., N'Guyen Dinh, F., Kindermans, C., and Hericord, P.,** Acidocétose alcoolique, *Nouv. Presse Med.,* 8, 585, 1979.
85. **Hasselbach, H., Selmer, J., and Kassis, E.,** Alcoholic ketoacidosis presenting as an acute abdomen, *Dan. Med. Bull.,* 28, 218, 1981.
86. **Williams, H. E.,** Alcoholic hypoglycemia and ketoacidosis, *Med. Clin. North Am.,* 68, 33, 1984.

87. **Lefèvre, A., Adler, H., and Lieber, C. S.,** Effect of ethanol on ketone metabolism, *J. Clin. Invest.,* 49, 1775, 1970.

88. **Veech, R. L.,** The toxic impact of parenteral solutions on the metabolism of cells: a hypothesis for physiological parenteral therapy, *Am. J. Clin. Nutr.,* 44, 519, 1986.

89. **Chait, A., February, A. W., Mancini, M., and Lewis, B.,** Clinical and metabolic study of alcoholic hyperlipidemia, *Lancet,* 11, 62, 1972.

90. **Verdy, M. and Gattereau, A.,** Ethanol, lipase activity and serum-lipid level, *Am. J. Clin. Nutr.,* 20, 997, 1967.

91. **Kaffernik, H. and Schneider, J.,** Zur kurzfristigen wirkung des äthylalkohols auf serumlipid gesundener fastender versuchspersonen, *Z. Gesamte Exp. Med.,* 152, 187, 1970.

92. **Markiewicz, K., Cholewa, M., Lutz, W., Walasek, L., and Zach, E.,** Influence of ethyl alcohol on serum lipid concentration in healthy persons, *Acta Med. Pol.,* 16, 281, 1975.

93. **Friedman, M., Rosenman, R. H., and Byers, S. O.,** Effect of moderate ingestion of alcohol upon serum triglyceride responses of normal and hyperlipemic subject, *Proc. Soc. Exp. Biol. Med.,* 120, 696, 1965.

94. **Jones, D. P., Losowsky, M. S., Davidson, C. S., and Lieber, C. S.,** Effects of ethanol on plasma lipids in man, *J. Lab. Clin. Med.,* 62, 675, 1963.

95. **Nestel, P. S. and Hirsch, E. Z.,** Mechanism of alcohol-induced hypertriglyceridemia, *J. Lab. Clin. Med.,* 66, 357, 1965.

96. **Avogaro, P. and Cazzolato, G.,** Changes in the composition and physico-chemical characteristics of serum lipoproteins during ethanol-induced lipemia in alcoholic subjects, *Metabolism,* 24, 1231, 1975.

97. **Brewster, A. C., Lankford, H. G., Schwartz, M. G., and Sullivan, S. F.,** Ethanol and alimentary lipemia, *Am. J. Clin. Nutr.,* 19, 255, 1966.

98. **Barboriak, J. J. and Meade, R. C.,** Enhancement of alimentary lipemia by preprandial alcohol, *Am. J. Med. Sci.,* 255, 245, 1968.

99. **Taskinen, M.-R. and Nikkilä, E. A.,** Nocturnal hypertriglyceridemia and hyperinsulinemia following evening intake of alcohol, *Acta Med. Scand.,* 202, 173, 1977.

100. **Wilson, D. E., Schreibman, P. H., Brewster, A. C., and Arky, R. A.,** The enhancement of alimentary lipemia by ethanol in man, *J. Lab. Clin. Med.,* 75, 264, 1970.

101. **Borowsky, S. A., Perlow, W., Baraona, E., and Lieber, C. S.,** Relationship of alcoholic hypertriglyceridemia to stage of liver disease and dietary lipid, *Dig. Dis. Sci.,* 25, 22, 1980.

102. **Avereginos, A., Chu, P., Greenfield, C., Harry, D. S., and McIntyre, N.,** Plasma lipid and lipoprotein response to fat feeding in alcoholic liver disease, *Hepatology,* 3, 349, 1983.

103. **Schubotz, R., Mühlfellner, G., Schneider, J., Mühlfellner, O., and Kaffernik, H.,** Verhalten der triglycerid-fettsäuren bei fastenden und isokalorisch ernährten versuchspersonen unter akuter äthanolbelastung, *Res. Exp. Med.,* 167, 139, 1976.

104. **Schneider, J., Tesdorfpf, M., Kaffernik, H., Hausmann, L., Zöfel, P., and Zilliken, F.,** Alteration of plasma lipids and intermediates of lipid metabolism in healthy fasting volunteers by ethanol and fructose, *Res. Exp. Med.,* 167, 159, 1976.

105. **Ostrander, L. P., Lamphiear, D. E., Block, W. D., Johnson, B. C., Ravenscroft, C., and Epstein, F. H.,** Relationship of serum lipid concentrations to alcohol consumption, *Arch. Int. Med.,* 134, 451, 1974.

106. **Stähelin, H. B., Nissen, C., Sommer, P., and Widmer, L. K.,** Alcoholic hyperlipidemia, *Int. J. Vitam. Nutr. Res.,* 44, 507, 1974.

107. **Castelli, W. P., Gordon, T., Hjortland, M. C., Kagan, A., Doyle, J. T., Hames, C. G., Hulley, S. B., and Zukel, W. J.,** Alcohols and blood lipids, *Lancet,* 11, 153, 1977.

108. **Barboriak, J. J. and Hogan, W. J.,** Preprandial drinking and plasma lipids in man, *Atherosclerosis,* 24, 323, 1976.

109. **Crouse, J. R. and Grundy, S. M.,** Effects of alcohol on plasma lipoproteins and cholesterol and triglyceride metabolism in man, *J. Lipid Res.,* 25, 486, 1984.

110. **Schneider, J., Liesenfeld, A., Mordasini, R., Schubotz, R., Zöfel, P., Kubel, F., Vandré-Plozzitzka, C., and Kaffernik, H.,** Lipoprotein fractions, lipoprotein lipase and hepatic triglyceride lipase during short time and long-term uptake of ethanol in healthy subjects, *Atherosclerosis,* 57, 281, 1985.

111. **Shapiro, R. H., Scheig, R. L., Drummey, G. D., Mendelson, J. H., and Isselbacher, K. J.,** Effect of prolonged ethanol ingestion on the transport and metabolism of lipids in man, *N. Engl. J. Med.,* 272, 610, 1965.

112. **Kudzma, D. J. and Schonfeld, G.,** Alcoholic hyperlipemia: induction by alcohol but not by carbohydrate, *J. Lab. Clin. Med.,* 77, 384, 1971.

113. **Losowsky, M. S., Jones, D. P., Davidson, C. S., and Lieber, C. S.,** Studies of alcoholic hyperlipemia and its mechanism, *Am. J. Med.,* 35, 794, 1963.

114. **Kallio, V., Saarima, H., and Saarima, A.,** Serum lipids and postprandial lipemia in alcoholics after a drinking bout, *Q. J. Stud. Alcohol,* 30, 565, 1969.

115. **Sirtori, C. R., Agradi, E., and Mariani, C.,** Hyperlipoproteinemia in alcoholic subjects, *Pharmacol. Res. Commun.,* 5, 81, 1973.

116. **Böttiger, L. E., Carlson, L. A., Hultman, E., and Romanus, V.,** Serum lipids in alcoholics, *Acta Med. Scand.,* 199, 357, 1976.
117. **Taskinen, M.-R., Välimäki, M., Nikkilä, E. A., Kuusi, T., Ehnholm, C., and Ylikahri, R.,** High density lipoprotein subfractions and postheparin plasma lipases in alcoholic men before and after ethanol withdrawal, *Metabolism,* 31, 1168, 1982.
118. **Ginsberg, H., Olefsky, J., Farquhar, J. W., and Reaven, G. M.,** Moderate ethanol ingestion and plasma triglyceride levels, *Ann. Intern. Med.,* 80, 143, 1974.
119. **Nestel, P. J., Simons, L. A., and Homma, Y.,** Effects of ethanol on bile acid and cholesterol metabolism, *Am. J. Clin. Nutr.,* 29, 1007, 1976.
120. **Mendelson, J. H. and Mello, N. K.,** Alcohol-induced hyperlipidemia and beta lipoproteins, *Science,* 180, 1372, 1973.
121. **Albrink, M. F. and Klatskin, G.,** Lactescence of serum following episodes of acute alcoholism and its probable relationship to acute pancreatitis, *Am. J. Med.,* 23, 26, 1957.
122. **Zieve, L.,** Jaundice, hyperlipemia and hemolytic anemia: a heretofore unrecognized syndrome associated with alcoholic fatty liver and cirrhosis, *Ann. Intern. Med.,* 48, 471, 1958.
123. **Jones, D. P., Perman, E. S., and Lieber, C. S.,** Free fatty acid turnover and triglyceride metabolism after ethanol ingestion in man, *J. Lab. Clin. Med.,* 66, 804, 1965.
124. **DiLuzio, N. R.,.** An evaluation of plasma triglyceriode formation as a factor in the development of the ethanol-induced fatty liver, *Life Sci.,* 4, 1373, 1965.
125. **Baraona, E., Pirola, R. C., and Lieber, C. S.,** Pathogenesis of postprandial hyperlipemia in rats fed ethanol-containing diets, *J. Clin. Invest.,* 52, 296, 1973.
126. **Baraona, E. and Lieber, C. S.,** Effects of chronic ethanol feeding on serum lipoprotein metabolism in the rat, *J. Clin. Invest.,* 49, 769, 1970.
127. **Ockner, R. K., Hughes, F. B., and Isselbacher, K. J.,** Very low density lipoproteins in intestinal lymph: origin, composition, and role in lipid transport in the fasting state, *J. Clin. Invest.,* 48, 2079, 1969.
128. **Wu, A.-L. and Windmueller, H. G.,** Relative contribution by liver and intestine to individual plasma apolipoproteins in the rat, *J. Biol. Chem.,* 254, 7316, 1979.
129. **Annable, W. and Cooper, C.,** Inhibition of release of hepatic triglyceride by ethanol — a reappraisal, *Biochem. Pharmacol.,* 23, 2063, 1974.
130. **Dajani, R. M. and Kouyoumjian, C.,** A probable direct role of ethanol in the pathogenesis of fat infiltration in the rat, *J. Nutr.,* 91, 535, 1967.
131. **Zakim, D., Alexander, D., and Sleisinger, M. H.,** The effect of ethanol on hepatic secretion of triglyceride into plasma, *J. Clin. Invest.,* 44, 1115, 1965.
132. **Hernell, O. and Johnson, O.,** Effect of ethanol on plasma triglycerides in male and female rats, *Lipids,* 8, 503, 1973.
133. **Madsen, N. P.,** Reduced serum very low density lipoprotein levels after acute ethanol administration, *Biochem. Pharmacol.,* 18, 261, 1969.
134. **Lakshmanan, M. R., Felver, M. E., and Veech, R. L.,** Alcohol and very low density lipoprotein synthesis and secretion by isolated hepatocytes, *Alcoholism Clin. Exp. Res.,* 4, 361, 1980.
135. **Matsudu, Y., Baraona, E., Salaspuro, M., and Lieber, C. S.,** Effects of ethanol on liver microtubules and Golgi apparatus, *Lab. Invest.,* 41, 455, 1979.
136. **Gordon, E.,** Effect of intoxicating dose of ethanol on lipid metabolism in an isolated perfused rat liver, *Biochem. Pharmacol.,* 21, 2991, 1972.
137. **Mørland, J. and Øye, J.,** Effect of acute and chronic ethanol treatment on hepatic triglyceride release from the rat perfused liver, *Eur. J. Pharmacol.,* 27, 238, 1974.
138. **Schapiro, R. H., Drummey, G. D., Shimizu, Y., and Isselbacher, K. J.,** Studies on the pathogenesis of the ethanol-induced fatty liver. II. Effect of ethanol on palmitate-1-C$^{14}$ metabolism by the isolated perfused rat liver, *J. Clin. Invest.,* 43, 1338, 1964.
139. **Carter, E. A., Drummey, G. D., and Isselbacher, K. J. ,** Ethanol stimulates triglyceride synthesis by the intestine, *Science,* 174, 1245, 1971.
140. **Baraona, E. and Lieber, C. S.,** Intestinal lymph formation and fat absorption: stimulation by acute ethanol administration and inhibition by chronic ethanol feeding, *Gastroenterology,* 68, 495, 1975.
141. **Mistilis, S. P. and Ockner, R. K.,** Effect of ethanol on endogenous lipid and lipoprotein metabolism in small intestine, *J. Lab. Clin. Med.,* 80, 34, 1972.
142. **Schlamp, R., Gärtner, U., and Becker, K.,** Studies concerning the hyperlipidemic action of ethanol, *Acta Hepatol Gastroenterol.,* 21, 93, 1974.
143. **Tobias, H. and Dawson, A. M.,** Free fatty acid mobilisation and plasma triglyceride clearance in alcoholic hyperlipemia, *Gastroenterology,* 50, 393, 1966.
144. **Nikkilä, E. A., Taskinen, M.-R., and Huttunen, J. K.,** Effect of acute ethanol load on postheparin plasma lipoprotein lipase and hepatic lipase activities and intravenous fat tolerance, *Horm. Metab. Res.,* 10, 220, 1978.

145. **Schneider, J., Panne, E., Braun, H., Mordasini, R., and Kaffernik, H.,** Ethanol-induced hyperlipo-proteinemia. Crucial role of preceeding ethanol intake in the removal of chylomicrons, *J. Lab. Clin. Med.,* 101, 114, 1983.

146. **Olivencrona, T., Hernell, O., Johnson, O., Fex, G., Wallinder, L., and Sandgren, O.,** Effect of ethanol on some enzymes inducible by fat-free refeeding, *Q. J. Stud. Alcohol,* 33, 1, 1972.

147. **Redgrave, T. G. and Martin, G.,** Effect of chronic ethanol consumption on the catabolism of chylomicron triacylglycerol and cholesteryl ester in the rat, *Atherosclerosis,* 28, 69, 1977.

148. **Lakshmanan, M. R. and Ezekiel, M.,** Relationship of alcoholic hyperlipidemia to the feed-back regulation of hepatic cholesterol synthesis by chylomicron remnants, *Alcoholism Clin. Exp. Res.,* 6, 482, 1982.

149. **Lakshmanan, M. R. and Ezekiel, M.,** Effect of chronic ethanol feeding upon the catabolism of protein and lipid moieties of chylomicrons and very low density lipoproteins in vivo and in the perfused heart system, *Alcoholism Clin. Exp. Res.,* 9, 327, 1985.

150. **Ekman, R., Fex, G., Johansson, B. G., Nilsson-Ehle, P., and Wadstein, J.,** Changes in plasma high density lipoproteins and lipolytic enzymes after long-term heavy ethanol consumption, *Scand. J. Clin. Lab. Invest.,* 41, 709, 1981.

151. **Kessler, J. I., Kniffen, J. C., and Janowitz, H. D.,** Lipoprotein lipase inhibition in the hyperlipemia of acute alcoholic pancreatitis, *N. Engl. J. Med.,* 269, 943, 1963.

152. **Johansson, B. G. and Nilsson-Ehle, P.,** Alcohol consumption and high density lipoproteins, *N. Engl. J. Med.,* 298, 633, 1978.

153. **Hulley, S. B. and Gordon, S.,** Alcohol and high-density lipoprotein cholesterol, *Circulation,* 64 (Suppl. 3), 57, 1981.

154. **Danielsson, B., Eksam, R., and Fex, G.,** Changes in plasma high density lipoprotein in chronic male alcoholics during and after abuse, *Scand. J. Clin. Lab. Invest.,* 38, 113, 1979.

155. **Berg, B. and Johansson, B. G.,** Effects on parameters of liver function, plasma lipid concentrations and lipoprotein patterns, *Acta Med. Scand.,* Suppl. 552, 13, 1973.

156. **Belfrage, P., Berg, B., Hagerstrand, I., Nilsson-Ehle, P., Tornzvist, H., and Wiebe, T.,** Alterations of lipid metabolism in healthy volunteers during long-term ethanol intake, *Eur. J. Clin. Invest.,* 7, 127, 1977.

157. **Fraser, G. E., Anderson, J. T., Foster, N., Goldberg, R., Jacobs, D., and Blackbourn, H.,** The effect of alcohol on serum high density lipoprotein, *Atherosclerosis,* 46, 275, 1983.

158. **LaPorte, R. E., Falco-Gerard, L., and Kuller, L. H.,** The relationship between alcohol consumption, liver enzymes and high density lipoprotein cholesterol, *Circulation,* 64 (Suppl. 3), 67, 1981.

159. **Barboriak, J. J., Alaupovic, P., and Cushman, P.,** Abstinence induced changes in plasma apolipoprotein levels of alcoholics, *Drug Alcohol Depend.,* 8, 337, 1981.

160. **Puchios, P., Fontan, M., Gentillini, J.-L., Gelez, P., and Fruchart, J.-C.,** Serum apolipoprotein A-II, a biochemical indicator of alcohol abuse, *Clin. Chim. Acta,* 185, 185, 1984.

161. **Cushman, P., Barboriak, J. J., and Kalbfleisch, J.,** Alcohol: high density lipoproteins, apolipoproteins, *Alcoholism Clin. Exp. Res.,* 10, 154, 1986.

162. **Dai, W. S., LaPorte, R. E., Hom, D. L., Kuller, L. H., D'Antonio, J. A., Gutai, J. P., Wozniczak, M., and Wohlfahrt, B.,** Alcohol consumption and high density lipoprotein concentration among alcoholics, *Am. J. Epidemiol.,* 122, 620, 1985.

163. **Haskell, W. L., Camargo, C., Williams, P. T., Vranizan, K. M., Kraus, R. M., Lindgren, F. T., and Wood, P. D.,** The effect of cessation and resumption of moderate alcohol intake on serum high-density lipoprotein fractions, *N. Engl. J. Med.,* 310, 805, 1984.

164. **Barboriak, J. J.,** Alcohol, lipids and heart disease, *Alcohol,* 1, 341, 1984.

Chapter 8

# INTERACTIONS OF ETHANOL AND CARBOHYDRATE METABOLISM

## John G. T. Sneyd

## TABLE OF CONTENTS

# I. INTRODUCTION

Most of the literature on the interactions of ethanol and carbohydrate metabolism deals either with effects of ethanol on blood glucose concentrations or on pathways of carbohydrate metabolism in the liver. Accordingly, this review will deal predominantly with these two topics, beginning with the observed effects of ethanol on blood glucose concentrations *in vivo*, then moving on to the causes of these effects, which involves discussion of the interaction of ethanol with pathways of carbohydrate metabolism in the liver. This subject has been reviewed previously by Reitz.[1]

# II. THE EFFECT OF ETHANOL ON THE BLOOD GLUCOSE CONCENTRATION

The effect of ethanol on the blood glucose concentration varies with the nutritional state. In the fed state, when liver glycogen stores are high, the consumption of ethanol either produces hyperglycemia or fails to alter the blood glucose concentration. When glycogen stores are low, ethanol produces hypoglycemia.

## A. Hypoglycemia

Hypoglycemia after drinking alcoholic beverages was described over 40 years ago.[2] The early literature has been reviewed by Madison.[3] The first reported cases were chronic alcoholics who had been drinking denatured ethanol and the hypoglycemia was at first attributed to the effects of adulterants such as methanol or to the effects of chronic alcoholism itself. The occurrence of ethanol-induced hypoglycemia in well-nourished children showed that chronic alcoholism was not necessary for the condition to develop, and careful studies by Field et al.[4] and Freinkel et al.[5] delineated the condition more precisely. Field et al.[4] showed that ethanol did not produce hypoglycemia if subjects were fasted overnight or for 24 h; the administration of ethanol to normal subjects after 44 h of fasting led to significant hypoglycemia. If glucagon was administered during the period of hypoglycemia the blood glucose did not rise, which indicates that liver glycogen stores were largely depleted. Glucagon had its usual hyperglycemic effect on the blood glucose in subjects who were fasted for only 24 h before being given ethanol.

The studies of Freinkel et al.[5,6] produced very similar results. This group studied nine chronic alcoholics admitted in a state of hypoglycemia. After an overnight fast the administration of ethanol produced mild hypoglycemia in these patients, whereas after a 3-d fast the hypoglycemic response to ethanol was immediate and profound.

There seems to be general consensus that in subjects fasted for more than 36 h, ethanol will produce hypoglycemia — sometimes profound hypoglycemia. Some clinicians are convinced that ethanol produces more severe hypoglycemia in children and in patients with diabetes mellitus, but good data are hard to find.

## B. Hyperglycemia

In subjects who have not fasted or have fasted for only short periods of time, ethanol either raises the blood glucose concentration or has very little effect.[4] This effect of ethanol has not been studied in as much detail as the hypoglycemic effect. There is some evidence that the hyperglycemic effect of ethanol is mediated by an increased secretion of catecholamines[7,8] and/or glucagon,[9] leading to phosphorylase activation and glycogenolysis. However, studies with perfused liver preparations and isolated liver cells (see below) have shown that ethanol will, under some circumstances, stimulate gluconeogenesis. It seems likely that the variable glycemic response to ethanol in the fed state is caused by variations in catecholamine secretion and possibly other factors.

## III. THE MECHANISM OF PRODUCTION OF HYPOGLYCEMIA BY ETHANOL

### A. Experiments *In Vivo*

The fall in blood glucose which follows the ingestion of ethanol by fasting man or animals could be caused by a decrease in the output of glucose by the liver or an increase in peripheral glucose utilization. In the studies of human subjects referred to above, both Field et al.[4] and Freinkel et al.[5] showed that glucagon did not elevate the blood glucose during hypoglycemia, and they concluded that the hypoglycemia was caused by an inhibition of gluconeogenesis by ethanol. Since the conversion of fructose to glucose was not impaired, the inhibition was presumably restricted to gluconeogenesis from substrates such as lactate, pyruvate, or alanine.

Wilson et al.[10] produced hypoglycemia by infusing ethanol (6.8 g/h) into normal medical students who had fasted for 36 h. The measurement of glucose turnover using [6-³H]glucose showed that ethanol decreased both the appearance rate and disappearance rate of glucose. The plasma glucose concentration did not rise in response to glucagon administration. They concluded that ethanol had caused hypoglycemia by inhibiting gluconeogenesis. The fall in glucose utilization was presumably due to declining availability of substrate and to low insulin levels. Their findings differ from those of Searle et al.[11] who measured glucose turnover during ethanol infusion by injecting a single bolus of [6-¹⁴C]glucose. Their methodology has been criticized.[10]

Careful studies of both glucose output and peripheral glucose utilization were carried out by Lochner et al.[12] They used dogs which had been starved for 2 to 3 d when the hepatic glucose output is derived almost entirely from gluconeogenesis from alanine, lactate, and glycerol. The administration of ethanol not only decreased the output of glucose by the liver, but also inhibited peripheral glucose utilization. Hypoglycemia resulted when the fall in glucose output exceeded the inhibition of peripheral glucose utilization. In some experiments the fall in peripheral utilization was enough to lead to hyperglycemia.

There is no convincing evidence that ethanol-induced hypoglycemia is caused by increased insulin secretion. The studies of Field et al.[4] and Freinkel et al.[5] in human subjects showed low rather than high concentrations of plasma insulin, as did the experiments of Bagdade et al.[13] One exception is the study by O'Keefe and Marks[14] which suggested that lunchtime gin and tonic may lead to hyperglycemia followed by a reactive hypoglycemia, but this suggestion does not appear to have been substantiated.

### B. Work with Isolated Tissues and Organs

Studies of the effect of ethanol on gluconeogenesis *in vivo* throw only limited light on the mechanisms involved, and most work on the mechanism by which ethanol affects gluconeogenesis and/or glycolysis has been carried out with isolated tissues or organs. Little work has been done with isolated human tissues for obvious reasons, and this review will therefore concentrate on work carried out on experimental animals — mainly, rats and dogs.

#### 1. The Role of the Liver in the Control of Blood Glucose

The liver plays an important role in the control of blood glucose concentrations. It may either remove glucose from the bloodstream or release glucose to the bloodstream, depending upon the hormonal state of the animal and the concentration of glucose in the blood perfusing it. Immediately after a meal there is net uptake of glucose, some of which is converted to glycogen. In the fasting state the liver is a net producer of glucose by the breakdown of glycogen or by gluconeogenesis from precursors such as lactate, alanine, pyruvate, and glycerol.

Gluconeogenesis is, thus, important in maintaining the blood glucose concentration during starvation when liver glycogen stores are low, but it also has an important role in removing lactate produced by exercising muscle and by red blood cells. Gluconeogenesis is subject to control by hormones and by the availability of substrates. The details of this regulation are complex and will not be dealt with here. The control of gluconeogenesis has been well reviewed by Hue et al.,[15] and quantitative analysis of the control of the pathway has been carried out by Groen et al.[16]

It is clear that the liver can carry out both gluconeogenesis or glycolysis under the appropriate circumstances. For example, if livers from fed or fasted animals are perfused with high concentrations of glucose, there is net production of lactate from glucose.[17] On the other hand, if livers from fed or fasted animals are perfused with high concentrations of lactate, then there is net production of glucose, that is, gluconeogenesis is operating.[18] The gluconeogenic and glycolytic pathways share a number of enzymes, but there are several key steps where reactions in the glycolytic pathway and the gluconeogenic pathway are catalyzed by different enzymes.[15] The reactions catalyzed by these enzymes are essentially irreversible. If the reactions in the two pathways operate simultaneously, substrate cycling takes place and there is interconversion of two substrates with the expenditure of energy. Substrate cycles exist between glucose and glucose 6-phosphate; fructose 6-phosphate and fructose 1,6-bisphosphate; and between phosphoenolpyruvate and pyruvate. The direction of flux through the gluconeogenic and glycolytic pathways is determined by the activity of these irreversible steps. Clearly, if the liver is carrying out gluconeogenesis, then the steps catalyzed by gluconeogenic enzymes must exceed the rate of the glycolytic enzymes. Thus, gluconeogenesis and glycolysis cannot proceed simultaneously in the liver unless the two pathways are operating independently, either in separate compartments in the same cell or in different cells.

In intact animals or man it is not always clear whether gluconeogenesis or glycolysis is predominant. After a prolonged fast the liver maintains the blood glucose concentration by gluconeogenesis from lactate, alanine, glycerol, and other precursors. Under these circumstances the activity of phosphofructokinase in the liver is very low or undetectable.[19] In the postabsorptive state the liver maintains the blood glucose concentration both by the breakdown of glycogen and by gluconeogenesis. Glycogen breakdown accounts for some two thirds of the net glucose production.[20] The major substrate for gluconeogenesis may well be lactate produced by the gastrointestinal tract. Immediately after a carbohydrate-containing meal or a glucose load the situation is less clear-cut. Under these circumstances the liver shows a net glucose uptake and some glucose is converted to glycogen. There is, however, good evidence that much of the glycogen that is synthesized following a glucose load is not formed directly from glucose, but is derived from three carbon precursors, i.e., it is formed by gluconeogenesis.[20,21]

The three carbon precursors (presumably lactate) are formed from glucose in the peripheral tissues or in the gut. A recent study by Kuwajima et al.[22] has shown that when fasted rats were refed on a sucrose-based diet the fructose 2,6-bisphosphate levels (*vide infra*) in the liver rose 20-fold within the first hour and liver glycogen was laid down rapidly. Despite the high levels of fructose 2,6-bisphosphate, isotopic studies showed that the bulk of the liver glycogen synthesized was gluconeogenic in origin. They point out that carbon flow through fructose 1,6-bisphosphatase in the presence of such high fructose 2,6-bisphosphate levels is a metabolic paradox and suggest that there may be "metabolic zonation" in the liver; that is, some cells would have a high capacity for glycolysis and a low capacity for gluconeogenesis while in others the reverse situation would be obtained. There is good evidence for metabolic zonation in the liver. Jungermann[23] showed that the key enzymes of glycolysis — glucokinase and pyruvate kinase — were located mainly in the perivenous zone, whereas the key enzymes of gluconeogenesis — phosphoenolpyruvate carboxykinase, fructose 1,6-bisphosphatase, and glucose 6-phosphatase — were found mainly in the periportal zone. He suggested that gluconeogenesis would be predominantly catalyzed by periportal hepatocytes, while glycolysis would be preferentially mediated by perivenous cells. Quistorff[24] showed that cells isolated from the periportal region of the liver carried out gluconeogenesis at 1.6 times the rate of perivenous cells. Nevertheless, the important question — "does glycolysis occur in one part of the liver *in vivo* while gluconeogenesis is proceeding in another?" — has still not been answered.

After glucose feeding some of the administered glucose is converted to fatty acid in the liver which would imply that glycolysis is proceeding. These observations, too, are hard to explain without postulating metabolic zonation, unless there are temporal differences between fatty acid synthesis and gluconeogenesis or the substrate for fatty acid synthesis is, in fact, lactate formed in peripheral tissues from the ingested glucose.

### 2. Effects of Ethanol on Glucose Metabolism in the Liver In Vitro

Experiments with slices of rat liver showed direct impairment of glucose formation from precursors such as lactate and alanine when ethanol was added.[6] Since gluconeogenesis from fructose was not impaired under the same circumstances and fructose does not require $NAD^+$ for its conversion to glucose, it was postulated that the effect of ethanol was mediated by a rise in NADH and a fall in $NAD^+$.

Krebs and co-workers[25,26] studied the effect of ethanol on gluconeogenesis in perfused livers from rats starved for 48 h. Ethanol inhibited gluconeogenesis from lactate by 66%; concentrations of ethanol higher than 10 m$M$ produced less inhibition, presumably because such high concentrations also inhibit alcohol dehydrogenase. Gluconeogenesis from glycerol, dihydroxyacetone, proline, serine, alanine, fructose, and galactose was also inhibited by ethanol, but gluconeogenesis from pyruvate was not inhibited, and gluconeogenesis in the kidney cortex where the activity of alcohol dehydrogenase is very low was not inhibited by ethanol. Inhibition of gluconeogenesis from lactate has also been observed in isolated rat hepatocytes[27-30] and in these cells gluconeogenesis from pyruvate is actually increased by ethanol. These findings suggested very strongly that the inhibition of gluconeogenesis by ethanol was caused by an increase in the $NADH/NAD^+$ ratio, which lowered the concentration of pyruvate when lactate, alanine, or serine was the substrate, so that the concentration of pyruvate in the liver became rate limiting for pyruvate carboxylase and, hence, for gluconeogenesis. When exogenous pyruvate (10 m$M$) was the substrate for gluconeogenesis its concentration was not rate limiting. There was no satisfactory explanation for the inhibition of gluconeogenesis from glycerol, fructose, or dihydroxyacetone, although there was some indirect evidence that once again the inhibition was caused by an increase in the $NADH/NAD^+$ ratio.

There is general agreement with these findings from other workers, although Williamson et al.[31] claimed that ethanol increased the rate of gluconeogenesis from alanine in perfused livers from starved rats. The reason for this discrepancy is not obvious.

## IV. THE MECHANISM OF THE HYPERGLYCEMIC EFFECT OF ETHANOL

### A. Direct Effect of Ethanol in the Perfused Liver

Although the hyperglycemic effect of ethanol has been ascribed to effects on the secretion of catecholamines or glucagon,[7-9] ethanol also has direct effects on the liver. When it is added to the perfusion medium of livers from fed rats it stimulates glucose output.[32,33] In these experiments Topping et al. used homologous blood for the perfusion medium. They have pointed out that the use of other perfusion media, particularly those with a low hematocrit, may lead to diminished hormone responsiveness and possibly other abnormalities as well. Accordingly, work carried out on the effects of ethanol on glucose metabolism in liver slices or livers perfused with media with a low hematocrit should be treated with caution. Since liver glycogen concentrations were unaffected and lactate uptake was increased by ethanol infusion, the increase in glucose output must be caused by an increase in the rate of gluconeogenesis. Although a great deal of attention has been focused on the change in the NADH/NAD$^+$ ratio when ethanol is being metabolized, it has been pointed out by Topping[32] that ethanol also increases the concentration of acetyl CoA in the liver and increases the mitochondrial ratio of acetyl CoA to CoA. An increase in this ratio would inactivate pyruvate dehydrogenase and, thus, direct carbon flux through pyruvate carboxylase into gluconeogenesis. Metabolites of alcohol may play some role in stimulating gluconeogenesis. Both acetaldehyde[34] and acetate[35] have been shown to stimulate gluconeogenesis in isolated rat liver cells. Their mechanisms of action are presumably different, since acetaldehyde stimulated gluconeogenesis from pyruvate, whereas acetate stimulated gluconeogenesis from lactate only.

## B. The Effects of Ethanol on Fructose 2,6-Bisphosphate Concentrations

More recent work has concentrated on the substrate cycle between fructose 6-phosphate and fructose 1,6-bisphosphate catalyzed by the enzymes phosphofructo-1-kinase and fructose 1,6-bisphosphatase. Work from the group of Hers[19] has shown that this substrate cycle operates only in fed animals, and substrate cycling can be inhibited by glucagon under these conditions. In livers from starved animals the virtual absence of cycling between fructose 6-phosphate and fructose 1,6-bisphosphate is attributed to very low activity of phosphofructo-1-kinase.[19] The action of glucagon to inhibit phosphofructo-1-kinase and the action of glucose to stimulate it was shown by van Schaftingen et al.[36] in an elegant series of experiments to be caused by changes in the concentration of a new mediator which they identified as fructose 2,6-bisphosphate.

Shortly afterwards, this mediator was identified by Pilkis et al.[37] Glucose increases the concentration of fructose 2,6-bisphosphate in the liver, whereas glucagon decreases it; the mechanism by which glucose and glucagon alter the concentration of fructose 2,6-bisphosphate is described by Hers and van Schaftingen.[38] Ethanol decreases the concentration of fructose 2,6-bisphosphate in the liver of fed rats and hepatocytes from either fed or fasted rats.[39] This effect of ethanol is very striking, the fructose 2,6-bisphosphate concentration *in vivo* falling to less than about 5% of the initial value after 1 h. van Schaftingen et al.[40] have studied the mechanism by which ethanol lowers fructose 2,6-bisphosphate. In isolated hepatocytes, ethanol caused a rapid increase in the concentration of glycerol 3-phosphate and a slower decrease in the concentration of fructose 2,6-bisphosphate without any change in the hexose 6-phosphates. There was also a slow activation of fructose 2,6-bisphosphatase and inactivation of phosphofructo-2-kinase (the enzymes involved in the degradation and formation of fructose 2,6-bisphosphate). Inhibitors of alcohol dehydrogenase such as 4-methyl pyrazole blocked this effect of ethanol, as did high concentrations of glucose or dihydroxyacetone. These effects on fructose 2,6-bisphosphate were attributed to the ability of glycerol 3-phosphate to inhibit the synthesis of fructose 2,6-bisphosphate by phosphofructo-2-kinase. Glycerol 3-phosphate also prevented the inhibition of fructose 2,6-bisphosphatase by fructose 6-phosphate. Thus glycerol 3-phosphate both inhibited the formation of fructose 2,6-bisphosphate and stimulated its removal. Glycerol 3-phosphate also acted indirectly to accelerate the inactivation of phosphofructo-2-kinase by the cyclic AMP-dependent protein kinase. The net effect of these changes is an increase in gluconeogenesis.

## V. THE EFFECTS OF ETHANOL ON GLYCOLYSIS

Since the importance of glycolysis in liver metabolism is far from clear, it is hard to be certain about the effects of ethanol on this pathway. In the postabsorptive state when the liver is carrying out gluconeogenesis, the rate of glycolysis in the liver is not known. Under these circumstances ethanol lowers the concentration of fructose 2,6-bisphosphate (see earlier discussion), inhibits phosphofructo-1-kinase, and activates fructose 1,6-bisphosphatase. This would have the effect of inhibiting glycolysis.

The work of Thurman and Scholz[17] has shown that ethanol inhibits glycolysis in perfused livers from fed rats, and it was suggested that this inhibition caused the increase in oxygen uptake observed in the presence of ethanol. However, more recent work with hepatocytes from fed rats[41] showed that the effect of ethanol on oxygen uptake depended on the presence of other substrates, which suggests that glycolysis may not be inhibited by ethanol under all conditions. It has also been shown[42] that pretreatment of rats with a large dose of ethanol leads to subsequent observation of low rates of glycolysis during liver perfusion. This inhibition of glycolysis has been suggested as a cause on the increased rates of oxygen uptake and ethanol oxidation observed in these livers. The effect is known as the swift increase in alcohol metabolism (SIAM) effect and is described in more detail in Volume II, Chapter 2. However, in another study[41] with isolated hepatocytes, oxygen uptake was not increased by ethanol pretreatment in cells from fed rats, although a small increase was observed in cells from starved animals. The increased oxygen uptake in cells from starved rats was accompanied by an increase in rates of gluconeogenesis under some conditions, but this appeared to be a consequence rather than a cause of the increased oxygen uptake. These results suggest that the SIAM effect may not be mediated simply by inhibition of glycolytic flux, as do other results published by Yuki et al.[43] It is clear that more work is needed to clarify the effects of ethanol on glycolysis.

## VI. SUMMARY OF THE MECHANISMS OF THE HYPERGLYCEMIC AND HYPOGLYCEMIC ACTIONS OF ETHANOL

In fasted men or animals when carbohydrate stores are low, gluconeogenesis is proceeding rapidly and phosphofructokinase is inactive. Under these circumstances the major effect of ethanol appears to diminish the concentration of pyruvate by increases in the NADH/NAD$^+$ ratio and, hence, to decrease gluconeogenesis and glucose output by the liver. It is not clear whether ethanol will lower fructose 2,6-bisphosphate in the livers of fasted animals, as experiments *in vivo* have been carried out with fed rats and experiments *in vitro* on isolated hepatocytes have been carried out in the presence of glucose.

In livers from fed rats the fructose 2,6-bisphosphate concentration is high and phosphofructokinase is active. Ethanol decreases the concentration of fructose 2,6-bisphosphate, stimulates fructose 1,6-bisphosphatase, and inactivates phosphofructo-1-kinase. The net effect is a stimulation of gluconeogenesis and of glucose output by the liver. Other effects of ethannol, e.g., to increase the ratio of acetyl CoA to CoA or to increase acetaldehyde or acetate concentrations, may play a role in stimulating gluconeogenesis, but their quantitative importance is uncertain. It is also clear that ethanol will stimulate glycogenolysis in the livers of fed animals, probably via the release of catecholamines or glucagon, but it is hard to assess the relative importance of this effect vs. the stimulation of gluconeogenesis. The importance of the inhibition of glycolysis by ethanol is even more difficult to quantitate. There is no doubt that under some experimental conditions *in vitro* ethanol will inhibit glycolysis in the liver, but whether this occurs *in vivo* is not known. Much of the difficulty stems from our lack of knowledge of how rapidly glycolysis proceeds in the liver *in vivo*.

## VII. THE LONG-TERM EFFECTS OF ETHANOL ON CARBOHYDRATE METABOLISM

The long-term effects of ethanol on carbohydrate metabolism are indirect, i.e., they are mediated by ethanol-induced tissue damage. Patients with pancreatitis produced by chronic ethanol ingestion show impaired glucose tolerance[44] and may become frankly diabetic. The diabetes is of the insulin-dependent type, but many patients do not develop ketoacidosis, presumably because they have lost the capacity to secrete glucagon as well as insulin.[44] In patients with alcoholic cirrhosis, impaired glucose metabolism is common, although the disturbance is usually mild.[45] The cause of impaired glucose tolerance is uncertain, but peripheral resistance to insulin may play a part as may the high plasma glucagon. By the time destruction of liver tissue is severe enough to produce significant impairment of gluconeogenesis or glycogenolysis, the effects of cirrhosis on glucose metabolism are overshadowed by other aspects of chronic liver failure.

# REFERENCES

1. **Reitz, R. C.,** Effects of ethanol on the intermediary metabolism of liver and brain, in *Biochemistry and Pharmacology of Ethanol,* Majchrowicz, E. and Nobel, E. P., Eds., Plenum Press, New York, 1979, 353.
2. **Brown, T. M. and Harvey, A. M.,** Spontaneous hypoglycemia in "smoke" drinkers, *J.A.M.A.,* 117, 12, 1941.
3. **Madison, L. L.,** Ethanol-induced hypoglycemia, in *Advances in Metabolic Disorders,* Vol. 3, Levine, R. and Luft, R., Eds., Academic Press, London, 1968, 85.
4. **Field, J. B., Williams, H. F., and Mortimore, G. E.,** Studies on the mechanism of ethanol-induced hypoglycemia, *J. Clin. Invest.,* 42, 497, 1963.
5. **Freinkel, N., Singer, D. L., Arky, R. A., Bleicher, S. J., Anderson, J. B., and Silbert, C. K.,** Ethanol hypoglycemia. I. Carbohydrate metabolism of patients with clinical ethanol hypoglycemia and the experimental reproduction of the syndrome with pure ethanol, *J. Clin. Invest.,* 42, 1112, 1963.
6. **Freinkel, N., Cohen, A. K., Arky, R. A., and Foster, A. E.,** Ethanol hypoglycemia. II. A postulated mechanism of action based on experiments with rat liver slices, *J. Clin. Endocrinol.,* 25, 76, 1965.
7. **Perman, E. S.,** Effect of ethanol and hydration on the urinary excretion of adrenaline and noradrenaline and on the blood sugar of rats, *Acta Physiol. Scand.,* 51, 68, 1961.
8. **Erwin, G. V. and Towell, J. F.,** Ethanol-induced hyperglycaemia mediated by the central nervous system, *Pharmacol. Biochem. Behav.,* 18 (Suppl. 1), 559, 1983.
9. **Potter, D. E. and Morris, J. W.,** Ethanol-induced changes in plasma glucose, insulin, and glucagon in fed and fasted rats, *Experimentia,* 36, 1003, 1980.
10. **Wilson, N. N., Brown, P. M., Juul, S. M., Prestwich, S. A., and Sonsken, P. H.,** Glucose turnover and metabolic and hormonal changes in ethanol-induced hypoglycemia, *Br. Med. J.,* 282, 849, 1981.
11. **Searle, C. L., Shames, D., Cavalieri, R. R., Bagdade, J. D., and Porte, D. J.,** Evaluation of ethanol hypoglycemia in man. Turnover studies with $^{14}$C-6-glucose, *Metabolism,* 23, 1023, 1974.
12. **Lochner, A., Wulff, J., and Madison, L. L.,** Ethanol-induced hypoglycemia. I. The acute effects of ethanol on hepatic glucose output and peripheral glucose utilization in fasted dogs, *Metabolism,* 16, 1, 1967.
13. **Bagdade, J. D., Bierman, E. L., and Porte, D.,** Counter-regulation of basal insulin secretion during ethanol hypoglycemia in diabetic and normal subjects, *Diabetes,* 21, 65, 1972.
14. **O'Keefe, S. J. D. and Marks, V.,** Lunchtime gin and tonic: a cause of reactive hypoglycemia, *Lancet,* 1, 1286, 1977.
15. **Hue, L., Felix, J. E., and van Schaftingen, E.,** Hormonal regulation of gluconeogenesis, in *Short-Term Regulation of Liver Metabolism,* Hue, L. and van de Werve, G., Eds., Elsevier/North-Holland, Amsterdam, 1981, 141.
16. **Groen, A. K., van Roermund, C. W. T., Vervoorn, R. C., and Tager, J. M.,** Control of gluconeogenesis in rat liver cells. Flux control coefficients of the enzymes in the gluconeogenic pathway in the absence and presence of glucagon, *Biochem. J.,* 237, 379, 1986.
17. **Thurman, R. G. and Scholz, R.,** Interaction of glycolysis and respiration in perfused liver, *Eur. J. Biochem.,* 75, 13, 1977.
18. **Exton, J. H.,** Gluconeogenesis, *Metabolism,* 21, 945, 1972.
19. **van Schaftingen, E., Hue, L., and Hers, H.-G.,** Study of the fructose 6-phosphate/fructose 1,6-bisphosphate cycle in the liver *in vivo, Biochem. J.,* 192, 263, 1980.
20. **Radziuk, J.,** Sources of carbon in hepatic glycogen synthesis during absorption of an oral glucose load in humans, *Fed. Proc.,* 41, 110, 1982.
21. **Katz, J. and McGarry, J. D.,** The glucose paradox. Is glucose a substrate for liver metabolism?, *J. Clin. Invest.,* 74, 1901, 1984.

22. **Kuwajima, M., Goldens, S., Katz, J., Unger, R. H., Foster, D. W., and McGarry, J. D.,** Active hepatic glycogen synthesis from gluconeogenic precursors despite high tissue levels of fructose 2,6-bis-phosphate, *J. Biol. Chem.,* 261, 2632, 1986.
23. **Jungermann, K.,** Functional significance of hepatocyte heterogeneity for glycolysis and gluconeogenesis, *Pharmacol. Biochem. Behav.,* 18 (Suppl. 1), 409, 1983.
24. **Quistorff, B.,** Gluconeogenesis in periportal and perivenous hepatocytes of rat liver, isolated by a new high-yield digitonin/collagenase perfusion technique, *Biochem. J.,* 229, 221, 1985.
25. **Krebs, H. A., Freedland, R. A., Hems, R., and Stubbs, M.,** Inhibition of hepatic gluconeogenesis by ethanol, *Biochem. J.,* 112, 117, 1969.
26. **Krebs, H. A.,** The effects of ethanol on the metabolic activities of the liver, *Adv. Enzyme Regul.,* 6, 467, 1968.
27. **Stubbs, M. and Krebs, H. A.,** The accumulation of aspartate in the presence of ethanol in rat liver, *Biochem. J.,* 150, 41, 1975.
28. **Cornell, M. W., Lund, P., and Krebs, H. A.,** The effect of lysine on gluconeogenesis from lactate in rat hepatocytes, *Biochem. J.,* 142, 327, 1974.
29. **Crow, K. E., Cornell, N. W., and Veech, R. L.,** Lactate-stimulated ethanol oxidation in isolated rat hepatocytes, *Biochem. J.,* 172, 29, 1978.
30. **Phillips, J. W., Berry, M. N., Grivell, A. R., and Wallace, P. G.,** Some unexplained features of hepatic ethanol oxidation, *Alcohol,* 2, 57, 1985.
31. **Williamson, J. R., Scholz, R., Browning, E. T., Thurman, R. G., and Fukami, M. H.,** Metabolic effects of ethanol in perfused rat liver, *J. Biol. Chem.,* 244, 5044, 1969.
32. **Topping, D. L., Clark, D. G., Illman, R. J., and Trimble, R. P.,** Inhibition by insulin of ethanol-induced hyperglycemia in perfused livers from fed rats, *Horm. Metab. Res.,* 14, 361, 1982.
33. **Topping, D. L., Clark, D. G., Storer, G. B., Trimble, R. P., and Illman, R. J.,** Acute effects of ethanol on the perfused rat liver, *Biochem. J.,* 184, 97, 1979.
34. **Cederbaum, A. I. and Dicker, E.,** Evaluation of the role of acetaldehyde in the actions of ethanol on gluconeogenesis by comparison with the effects of crotonol and crotonaldehyde, *Alcoholism Clin. Exp. Res.,* 6, 100, 1982.
35. **Whitton, P. D., Rodrigues, L. M., and Hems, D. A.,** Stimulation by acetate of gluconeogenesis in hepatocyte suspensions, *FEBS Lett.,* 98, 85, 1979.
36. **van Schaftingen, E., Hue, L., and Hers, H.-G.,** Fructose 2,6-bisphosphate, the probable structure of the glucose- and glucagon-sensitive stimulator of phosphofructokinase, *Biochem. J.,* 192, 897, 1980.
37. **Pilkis, S. J., El-Maghrabi, M. R., Pilkis, J., Claus, T., and Cumming, D. A.,** Fructose 2,6-bisphosphate. A new activator of phosphofructokinase, *J. Biol. Chem.,* 256, 3171, 1981.
38. **Hers, H.-G. and van Schaftingen, E.,** Fructose 2,6-bisphosphate 2 years after its discovery, *Biochem. J.,* 206, 1, 1982.
39. **Claus, T. H., Schlumpf, J. R., El-Maghrabi, M. R., and Pilkis, S. J.,** Regulation of the phosphorylation and activity of 6-phosphofructo 1-kinase in isolated hepatocytes by α-glycerolphosphate and fructose 2,6-bisphosphate, *J. Biol. Chem.,* 257, 7541, 1982.
40. **van Schaftingen, E., Bartrons, R., and Hers, H.-G.,** The mechanism by which ethanol decreases the concentration of fructose 2,6-bisphosphate in the liver, *Biochem. J.,* 222, 511, 1984.
41. **Stowell, K. M. and Crow, K. E.,** The effect of acute ethanol treatment on rates of oxygen uptake, ethanol oxidation and gluconeogenesis in isolated rat hepatocytes, *Biochem. J.,* 230, 595, 1985.
42. **Yuki, T. and Thurman, R. G.,** The swift increase in alcohol metabolism. Time course for the increase in hepatic oxygen uptake and the involvement of glycolysis, *Biochem. J.,* 986, 119, 1980.
43. **Yuki, T., Israel, Y., and Thurman, R. G.,** The swift increase in alcohol metabolism. Inhibition by propylthiouracil, *Biochem. Pharmacol.,* 31, 2403, 1982.
44. **Sarles, H., Sahel, J., Staub, J. L., Bourry, J., and Langier, R.,** Chronic pancreatitis, in *The Exocrine Pancreas,* Howat, A. T. and Sarles, H., Eds., W. B., Saunders, London, 1979, 409.
45. **Conn, H. O.,** *Cirrhosis in Diseases of the Liver,* 5th ed., Schiff, L. and Schiff, E. R., Eds., Lippincott, Philadelphia, 1982, 887.

Chapter 9

# HEPATIC GLUTATHIONE METABOLISM: ALTERATIONS INDUCED BY ALCOHOL CONSUMPTION

**Hernan Speisky, Hector Orrego, and Yedy Israel**

## TABLE OF CONTENTS

# I. INTRODUCTION

In this chapter we address the acute and chronic effects of ethanol on hepatic glutathione (GSH), including its possible pathogenic role in alcoholic liver injury. In order to provide the reader with a framework on the biology of hepatic GSH, we have devoted the initial section to examining the functions and biological significance of this tripeptide in the liver. Subsequently, we discuss the available information on the effects of ethanol on liver GSH in humans, so that the studies in experimental animals discussed later can be viewed in this context.

# II. FUNCTIONS AND BIOLOGICAL SIGNIFICANCE OF HEPATIC GLUTATHIONE

GSH (L-gamma-glutamyl-L-cysteinyl-glycine) is found in virtually all mammalian cells in concentrations ranging from 0.1 to 10 m*M*. Two salient structural features characterize the molecule of GSH (Figure 1): first, a gamma-glutamyl peptide linkage which makes GSH resistant to normal peptidase activity; and, second, the thiol moiety of the cysteinyl residue, which endows the tripeptide with the ability to function as an electron donor and as a nucleophilic compound.[1-3]

In the liver, the concentration of GSH ranges between 5 and 7 m*M*, levels which are about 1000-fold higher than those in plasma.[3] Over 90% of intracellular GSH occurs in the reduced or thiol form, although glutathione disulfide (GSSG), thio-ethers, and, to a lesser extent, GSH-mixed disulfides (less than 1%) also contribute to the total pool of GSH. Although the bulk of intracellular GSH is found in the cytosol, the existence of a minor (0.2 to 0.45 m*M*) mitochondrial pool of GSH in hepatocytes has also been described.[4-6] The biological half-life of GSH in different tissues varies from less than an hour in the kidney to several days in erythrocytes, nervous tissue, lung, and spleen. In rat liver, GSH has a half-life which ranges from 2 to 4 h.[4,7,8] Research conducted in recent years has revealed that in the liver most of the utilization of GSH and, thus, of its turnover is accounted for by the efflux of GSH molecules into the extracellular space (i.e., blood and bile).[9-11] While the biological function of extracellular GSH is not yet well understood, the intracellular presence of the tripeptide is known to be intimately associated with a large number of important protective, metabolic, and coenzymatic functions (for comprehensive reviews see Orrenius et al.,[12] Reed and Fariss,[13] and Meister[14]).

Among its protective functions, GSH is of major importance as part of a cellular mechanism for the reductive removal of highly toxic peroxides such as hydrogen and lipid peroxides.[12,15] GSH is also actively involved in protecting cell components against free radical-mediated attack by acting as a major free-radical scavenger. A number of studies attest to the importance of GSH as a cellular reductant. For instance, an inverse linear relationship exists between the intracellular GSH concentration and the susceptibility of endothelial cells to hydrogen peroxide-induced lysis.[16,17] Similarly, manipulations leading to a decreased GSH synthesis in human lymphoid cells have been shown to result in an increased cell sensitivity (diminished viability) to the effects of irradiation.[18]

Another major role of GSH in cell protection refers to its ability to interact with a large number of toxic electrophiles, including carcinogens and drug metabolites, to form GSH conjugates. In fact, considerable evidence indicates that the intrahepatic availability of GSH is a key determinant in cell protection against liver injury induced by toxic electrophiles.[13,19,20] This is best exemplified by the inverse correlation between the intrahepatic GSH concentration, the degree of binding of reactive metabolites to cell components, and the extent of liver cell damage induced by the administration of a large dose of acetaminophen.[21,22] The amount of GSH present in the liver has been shown to be of importance in determining the

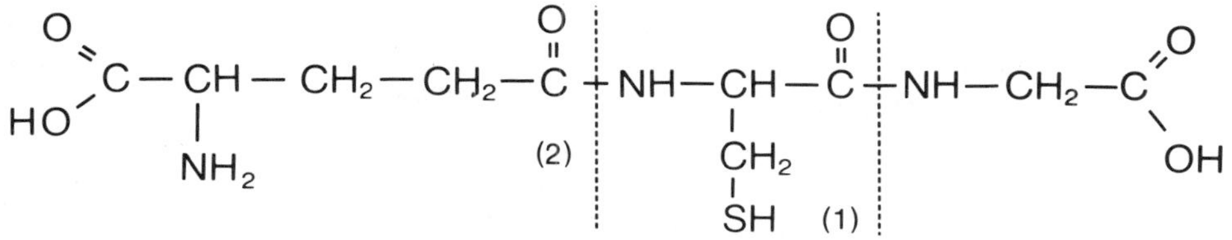

FIGURE 1. The structure of glutathione: γ-glutamyl-cysteinyl-glycine. (1) Cysteinyl thiol; (2) γ-glutamyl peptide linkage.

susceptibility of the liver cell to the toxic effects of compounds such as bromobenzene,[23,24] iodobenzene,[22] acetaminophen,[21] formaldehyde,[25] dichloro-ethylene,[26] carbon tetrachloride,[27] chloroform,[28] and thioacetamide.[23] In general, pharmacological manipulations or biological conditions which deplete liver GSH are associated with an enhancement of the hepatotoxicity induced by the above-mentioned compounds, while treatments which increase liver GSH availability have been shown to ameliorate or prevent their toxicity.[13]

In addition to its protective functions, GSH also plays a significant role regulating a number of metabolic and transport processes.[1-3] Among others, GSH is important in the regulation of the thiol/disulfide redox state, the synthesis of proteins and DNA, the maintenance of structural integrity of organelles and cell membranes, and the activity of many enzymes. In most of these cell functions, the regulatory role of GSH is exerted (or determined) through changes in the intracellular ratio of GSH to GSSG.

An additional function of GSH relates to its participation in the gamma-glutamyl cycle, a series of six enzymatically catalyzed reactions which account for the synthesis and degradative utilization of the tripeptide. Meister and Tate[29] have postulated that in the gamma-glutamyl cycle GSH, by serving as a donor of its gamma-glutamyl moiety, participates in the translocation of some amino acids across cell membranes. GSH is also believed to serve as an intracellular form of storage of cysteine,[7,30] and its release from the liver has been suggested to constitute a form of interorgan transport of its sulfur-containing amino acid (see Kaplowitz et al.[3]). In addition, GSH also participates as a coenzyme in several metabolic and detoxification reactions.[31]

## A. Biosynthesis of Glutathione

GSH is synthesized in the cytosol from its constituent amino acids in two consecutive reactions,[32] each requiring one molecule of ATP (Reactions 1 and 2). The first reaction involves the formation of a gamma-glutamyl linkage between glutamate and cysteine and is catalyzed by gamma-glutamyl-cysteine synthetase:

L-glutamate + L-cysteine + ATP $\rightleftharpoons$ L-gamma-glutamyl-cysteine + ADP + Pi (1)

In the second reaction, GSH synthetase catalyzes the addition of glycine to gamma-glutamyl-cysteine to form the tripeptide GSH:

L-gamma-glutamyl-cysteine + glycine + ATP $\rightleftharpoons$ glutathione + ADP + Pi (2)

GSH regulates its own biosynthesis by feedback inhibition of the rate-limiting step, the formation of gamma-glutamyl-cysteine. This inhibition, which is competitive and nonallosteric, is exerted on the enzyme gamma-glutamyl synthetase at the gamma-glutamyl site, with GSH having an apparent $K_i$ value of approximately 2.3 m$M$.[33,34] Thus, with normal GSH concentrations (5 to 7 m$M$), the activity of gamma-glutamyl-cysteine synthetase is less than maximal. Manipulations *in vivo*, which lead to a decrease in hepatic GSH, are associated with an increase in the rate of GSH synthesis, indicating that the feedback regulatory mechanism is operative *in vivo*.[8,35] In addition to this mechanism, the rate of GSH biosynthesis appears to be strongly limited by the intracellular availability of its precursor amino acid L-cysteine.[7,30,36] In fact, this amino acid seems to play a rate-limiting role as its intrahepatic concentration (0.2 to 0.4 m$M$) approximates that of its $K_m$ (0.35 m$M$) for gamma-glutamyl-cysteine synthetase.[34,37] The concentration of free cysteine within the liver is at least one order of magnitude lower than that of GSH. As free cysteine can sustain GSH synthesis for only a short time, the synthesis of the tripeptide should depend heavily on the biosynthesis and/or mobilization of cysteine from other pools. The dependence of normal GSH synthesis on extrahepatic sources of cysteine has been shown in isolated hepatocytes in which the intracellular concentration of GSH gradually decreases unless the incubation medium contains this precursor amino acid.[37] Under physiological conditions, hepatic cysteine is derived mainly from the diet and from protein breakdown.[38] In plasma, cysteine occurs predominantly in the oxidized form as cystine. However, since cystine is poorly taken up by rat hepatocytes, it does not constitute an important source of cysteine for the liver.[12,38] Rat kidney cells, in turn, actively utilize cystine as a major source of cysteine for the intracellular synthesis of GSH.[12] As an alternative, the liver has the ability to utilize methionine as a source of intracellular cysteine. Conversion of methionine into cysteine occurs in the hepatocytes via the cystathionine, or transulfuration pathway.[37,38] Direct administration of cysteine does not constitute an ideal way of increasing intracellular GSH, since this amino acid is rapidly metabolized, and when given in high doses it is toxic.[39] In turn, administration of *N*-acetyl-cysteine, extensively used in the treatment of acute intoxication with acetaminophen,[40] appears to serve as a rapid and effective source of cysteine for the intracellular replenishment of GSH.[41] Recently, a synthetic derivative of cysteine, L-2-oxo-thiazolidine-4-carboxylate, has been proposed as a useful alternative source of cysteine.[42] This compound, which readily enters the liver cell, has been shown to rapidly undergo enzymatic conversion into free cysteine and to serve as a precursor in the synthesis of GSH. Intrahepatic GSH levels may also be increased by supplying the tripeptide to hepatocytes in the ester form.[43] Administration of the monomethyl or monoethyl ester of GSH to mice has been shown to substantially increase both liver and kidney GSH.[43] In contrast to GSH, these compounds are believed to be readily taken up by the hepatocytes, after which they are hydrolyzed releasing free GSH.

## B. Glutathione Utilization

### 1. Role of Glutathione in Conjugation Reactions

The most widely known biological role of GSH is that of serving as a nucleophile for the conjugation of a number of foreign compounds or their metabolites. The resulting GSH conjugates are the precursors of a generic group of excretion products known as mercapturic acids (see Boyland and Chasseaud[44]). Formation of GSH conjugates can proceed spontaneously with some electrophilic compounds, but, in general, is greatly facilitated by the action of GSH transferases.[45] The list of compound types known to function as electrophiles in reactions with GSH is long and includes halogenated nitroaromatic compounds, organophosphorus compounds, α-, and β-unsaturated carbonyl compounds, sulfates, quinones, and epoxides. Although GSH conjugation is an important mechanism in the detoxification of foreign compounds, many endogenous compounds (e.g., steroid hormones, leukotrienes, prostaglandins) are also substrates for conjugation.

### 2. Reactions of Oxidation and Reduction

As mentioned earlier, the roles of GSH and the liver are not limited to the conjugation of electrophiles, but also include its function as the major cellular reductant in the removal of both hydrogen and organic peroxides. This latter function of GSH is accomplished in a reaction catalyzed by GSH peroxidases of which two types have been described: a selenium-dependent one, which is active in the metabolism of both $H_2O_2$ and organic peroxides, and a selenium-independent enzyme, which exhibits peroxidase activity towards organic peroxides but not towards hydrogen peroxide.

Peroxides are continuously formed in the cell as a consequence of aerobic metabolism.[12,46] Under normal conditions, a major group of organic peroxides found intracellularly is derived from the free-radical-mediated attack of polyunsaturated fatty acids. Since these highly reactive peroxides can become very toxic for the cell if they accumulate, their reductive removal, via the GSH peroxidase system, becomes an important detoxification reaction. Although $H_2O_2$ is largely metabolized by catalase located in peroxisomes, GSH peroxidase is believed to be responsible for the reduction of $H_2O_2$ formed in other cellular compartments.[12] The reductive metabolism of lipid peroxides and $H_2O_2$ leads to the formation of less reactive lipid alcohols (hydroxy fatty acids) and water, respectively. In such a reaction stoichiometric amounts of GSH are oxidized into GSSG. Most of the GSSG thus formed is subsequently reduced back to GSH in a reaction catalyzed by the NADPH-dependent GSH reductase.[47] In addition to this mechanism for removing cellular GSSG, a small fraction of GSSG is continuously excreted into the bile under normal conditions.[9] However, under conditions of cellular oxidative stress the formation of GSSG may exceed the capacity of the reductase, leading to an intracellular accumulation of GSSG and to a subsequent increase in the biliary excretion of the disulfide. In fact, Akerboom et al.,[48] using the perfused rat liver, demonstrated a linear relationship between the intracellular content of GSSG and its rate of efflux into the bile. Oxidation of GSH to GSSG during drug biotransformation may result not only from an increased GSH peroxidase activity, but also directly from the interaction between GSH and oxygen-free radical intermediates formed during the metabolism of xenobiotics.[47]

### 3. Efflux of Hepatic Glutathione into the Circulation

Studies conducted in recent years indicate that the efflux of GSH molecules from the hepatocyte into the extracellular space constitutes the major component of the hepatic turnover of GSH.[2,3,9-11] Under normal conditions, about 90 to 95% of the total utilization of GSH in the liver can be accounted for by the efflux process. The remaining 5 to 10% may be accounted for by its consumption in reactions of conjugation, oxidation, or by its enzymatic breakdown. Translocation of GSH molecules takes place across both the sinusoidal and the canalicular cell membrane. The sinusoidal efflux of GSH accounts for about 80 to 85% of the total efflux (see Kaplowitz et al.[3]). While this efflux appears to constitute the major (90%) source of circulating GSH,[11] arterial plasma GSH is believed to largely reflect the hepatic content of the tripeptide.[49] Depletion of liver GSH by fasting or by the administration of diethylmaleate or acetaminophen has been shown to result in proportional decreases in arterial plasma GSH concentration.[49]

Sinusoidal efflux of GSH occurs in the direction of a large (over 100-fold) concentration gradient (5 to 7 m$M$ intracellular compared to 20 to 40 $\mu M$ in the hepatic vein plasma). Studies *in vitro*[50,51] indicate that GSH efflux is a carrier-mediated process and that at GSH concentrations close to those present in the liver of fed and fasted animals, efflux of GSH occurs at rather uniform rates.[50] Recent studies by Sies and Graf[52] conducted in the perfused liver suggest the existence of a hormone dependence for the efflux of GSH. These investigators found that vasopressin, phenylephrine, and adrenaline stimulate GSH efflux, whereas glucagon or dibutyryl cyclic AMP have no effect. The exact physiological significance of these effects, however, remains to be established. The transport of GSH across liver cell membranes appears to be a unidirectional phenomenon, as livers perfused with radiolabeled GSH are essentially impermeable to the tripeptide.[53]

*4. Glutathione Degradation: Role of Gamma-Glutamyl Transferase*

Gamma-glutamyl transferase (GGT), normally bound to the external surface of the membrane of a variety of cells, is the only known enzyme capable of initiating the degradative utilization of GSH.[2,29,31] GGT catalyzes the hydrolytic cleavage of the gamma-glutamyl moiety of GSH in a reaction which leads to the formation of glutamate and cysteinyl-glycine. In the presence of amino acids or peptides, GGT also acts as a transpeptidase catalyzing the formation of cysteinyl-glycine and gamma-glutamyl peptides. In the kidney, GGT-mediated hydrolysis of GSH seems to be of major importance as a mechanism to recover GSH, in the form of its constituent amino acids, from the extracellular space back into the cell. The reaction of transpeptidation, on the other hand, is believed to serve as an alternative mechanism for the transport of some amino acids in the form of gamma-glutamyl amino acids into the cell.[2,29] Both the hydrolysis of GSH and the formation and transport of gamma-glutamyl amino acids been have demonstrated to occur physiologically in the kidney of experimental animals.[2,54,55] The transpeptidation reaction of GSH with extracellular cystine leads to the formation of gamma-glutamyl-cystine. It has been demonstrated[56] that administration of gamma-glutamyl cystine markedly increases the intracellular levels of GSH. Elegant studies by Anderson and Meister[56] indicate that this occurs because gamma-glutamyl-cystine enters the cell and is subsequently reduced to gamma-glutamyl-cysteine. The latter compound bypasses the rate-limiting step in the synthesis of GSH and serves as a direct substrate for the GSH synthetase. It has been postulated[2] that tissues containing substantial GGT activity can form gamma-glutamyl cystine and that the existence of such a reaction can constitute an alternative pathway for the synthesis of GSH.

## III. LIVER GLUTATHIONE LEVELS IN ALCOHOLICS WITH DIFFERENT DEGREES OF LIVER DAMAGE

### A. General Considerations in the Study of Liver GSH in an Alcoholic Population

Given the many essential roles of GSH in metabolic and detoxification processes (see Section II), the study of the effects of alcohol consumption on liver GSH may prove to be of special relevance for the understanding of alcoholic liver disease. In contrast to a large number of studies conducted in experimental animals (see Section IV), only limited data are available on the effects of alcohol on liver GSH in humans. Since a number of clinical and nonclinical factors can influence hepatic GSH concentrations, studies in humans are often frought with problems of interpretation.[57] Among others, factors likely to interfere with a correct interpretation of the data are those related to the procedure of obtaining the liver samples (generally percutaneous liver biopsies) and to the characteristics of the alcoholic and control populations. With regard to liver biopsies, a general assumption in the assessment of biochemical parameters is that the sampled tissue is representative of the liver as a whole. This to a large extent appears to be true in normal livers. Nevertheless, in pathologic livers there is always the possibility of sampling error, specifically in conditions such as alcoholic liver cirrhosis. Furthermore, the liver in abnormal states often has components that can dilute the contribution of the hepatocytes to the values of the biochemical parameters studied, when expressed per unit of tissue weight. For instance, the accumulation of fat, fibrous tissue, or inflammatory cells, and an actual decrease in the number of metabolically viable hepatocytes (affected by liver cell necrosis) are all factors which are likely to confound the information which is sought. In some cases of liver cirrhosis, for example, most of the weight of a liver biopsy can be accounted for by collagen, fat, and infiltrating cells. Since during alcoholic liver damage retention of export proteins[58] (e.g., albumin, transferrin) can occur, expression of data per unit of protein should take into account the possible confounding effects of protein accumulation. With regard to population sampling, the selection and comparison of alcoholics with nonalcoholic controls is also a difficult task. Alcoholics frequently present with a number of nonalcohol-related factors which are capable of influencing liver GSH. Excessive smoking, heavy coffee drinking, and concomitant use of drugs capable of depleting GSH, such as acetaminophen,[19,20] aspirin,[59] cocaine,[60] and morphine,[61] are all common features in the alcoholic. Furthermore, alcoholics frequently show malnutrition and/or mineral (e.g., selenium) and vitamin deficiencies (vitamin E) which are likely to influence the levels of GSH in the liver.

## B. Hepatic Glutathione Content in Alcoholics

In a detailed study, Videla et al.[62] reported that shortly after withdrawal (1 to 5 d), GSH levels in the liver of alcoholics were markedly lower (by about 50%) than those found in the liver of alcoholics studied after more prolonged periods of time (up to 23 d). The recovery of hepatic GSH was significantly correlated ($r = 0.71$ $p < 0.001$) with the length of abstinence between 1 and 25 d. The investigators indicate that the recovery was observed only in patients presenting no hepatocellular necrosis. While the recognition that upon abstinence a necrotic liver loses its ability to recover GSH would constitute a most important observation, it is unfortunate that in patients with necrosis GSH recovery was studied for only 13 d, rather than for 23 d as for the group without necrosis. When GSH recovery is studied for short periods of time, a marked interindividual variation can mask any trend towards recovery. However, the data of Videla et al. clearly reveal that GSH levels are significantly decreased in the liver of alcoholics without active liver disease. These investigators further demonstrated that GSH levels did not correlate with hepatocyte size, thus, suggesting that dilution due to hepatomegaly did not account for the decreased levels of GSH. Further, GSH levels did not correlate with fat accumulation measured by the degree of histologically observed steatosis in the biopsies. Studies by Shaw et al.[63] have also shown that the livers of alcoholics have a decreased content of GSH. In this study, in which the majority of patients had fatty liver, the mean duration of abstinence was 7.5 d; alcoholics had liver GSH levels 30 to 35% lower than those of control patients with nonalcoholic liver disease. Interestingly, the depression in liver GSH was of similar magnitude in the well-nourished patients as compared to those with history of inadequate nutrition. However, as indicated by these authors,[63] subtle nutritional deficiencies (e.g., vitamins, minerals) could not be discarded. The same considerations apply to the patients studied by Videla et al.[62] in whom nutritional deficiencies are more likely to have existed in recently admitted patients. Woodhouse et al.[64] have compared hepatic GSH levels in patients with alcoholic cirrhosis vs. those of individuals with normal liver biopsies. In the former, GSH levels were significantly reduced (33%) and did not appear to be related to the degree of hepatic fibrosis. Unfortunately, there is no indication in this study as to the length of abstinence in these patients. Preliminary studies by Lauterburg et al.[65] indicate that patients with alcoholic hepatitis present markedly decreased (66%) hepatic GSH levels as compared to patients presenting chronic persistent hepatitis of nonalcoholic etiology. Chronic alcoholics drinking heavily up to 36 h prior to testing were also found to have markedly reduced (45%) circulating plasma GSH levels, which, according to the investigators,[65] closely reflects the hepatic content of the tripeptide. Studies by Poulsen et al.[66] have shown that hepatic GSH levels are not significantly decreased in patients with moderate hepatic impairment of nonalcoholic etiology. In a more recent study, Jewell et al.[67] reported a significantly lower (40%) GSH concentration in the liver of alcoholic patients compared with that of a selected group of nonalcoholics with normal liver function tests and histology. The reduction in GSH levels in the alcoholics did not correlate with the severity of liver damage.[67] Since alcoholics with minimal liver disease also presented depressed GSH levels, the authors suggest that the decrease in the liver GSH constitutes an "early event relative to other diagnostic and prognostic indicators of alcoholic liver disease."[67]

From the foregoing it seems apparent that (1) alcoholics present low levels of hepatic

GSH, with reductions ranging from 30 to 50%; (2) low liver GSH levels are not related to the degree of histological damage or hepatic dysfunction; and (3) the levels of GSH appear to be depressed even after 1 week of abstinence, but they tend to return to normal upon a more prolonged abstinence period of the order of 3 to 4 weeks. The exact mechanism(s) which lead to a decreased GSH content in the liver of alcoholics has, at present, not been established. The reader is referred to Section IV for the discussion of relevant data available in humans and in experimental animals.

### C. Possible Implications of a Decreased Hepatic GSH Content in Alcoholics

The exact implications of a decreased GSH availability in the liver of human alcoholics are not clear. However, based on the key role of GSH in cell protection, a decreased GSH availability is likely to render the liver of alcoholics more susceptible to cell damage induced by compounds such as free radicals, hydrogen and lipid peroxides, and toxic reactive electrophiles. A clear example of a possible pathogenic link between low hepatic GSH levels in alcoholics and susceptibility to chemically induced liver damage is found in the observation that an increased susceptibility to hepatic injury and mortality due to acetaminophen occurs in alcoholics at doses which are generally nontoxic or not lethal in normal individuals.[68,69] As shown by early studies in experimental animals,[19-21] acetaminophen metabolites bind to GSH causing its cellular depletion. When GSH stores are depleted by 60 to 70%, substantial binding of acetaminophen-derived electrophiles to cell macromolecules takes place, leading to a marked impairment in essential cell functions and ultimately to cell death.[24] Decreased hepatotoxicity of acetaminophen has been shown in humans after administration of substances that allow the rapid replenishment of cellular GSH.[40] It should be noted that in addition to a decreased capacity to conjugate electrophiles, due to lower GSH levels, the livers of alcoholics also present an increased drug-metabolizing capacity,[70] such that the combination of both an increased electrophile load and a decreased conjugation capacity is likely to underlie their enhanced susceptibility to acetaminophen hepatotoxicity. In experimental animals given ethanol, an increased generation of toxic electrophiles seems to be a major mechanism for their increased susceptibility to acetaminophen.[71] It should be further noted that the presence of alcohol itself can inhibit the activity of microsomal systems, so that acute ethanol administration can lead to a reduction in acetaminophen hepatotoxicity.[72] Conceivably, a greater susceptibility to acetaminophen toxicity could occur in alcoholics shortly after ethanol withdrawal, a condition in which GSH stores still remain low and the drug-metabolizing capacity is increased. Such a consideration is likely to apply not only to acetaminophen, but also to other xenobiotics which are activated by the microsomal system and require GSH for their detoxification.

As indicated previously, GSH plays an active role in protecting the cell against lipid peroxides by either preventing their formation or by facilitating their removal. Thus, a decreased GSH availability in the liver of alcoholics is likely to result in a decreased capacity of the liver cell to counteract lipid peroxidative processes. Alternatively, decreased intrahepatic GSH levels could be the consequence of an increased lipid peroxidation. Shaw et al.[63] reported the simultaneous occurrence of increased lipid peroxidative indices and decreased GSH levels in the liver of alcoholics. Unfortunately, no correlation between lipid peroxides and GSH levels was presented. In a work by Suematsu et al.,[73] alcoholics were found to have increased serum and liver lipid peroxides on admission, but GSH levels were not measured. Heavy drinkers with the highest levels of lipid peroxides also exhibited the highest incidence of liver cell necrosis as compared to individuals presenting low lipid peroxide levels.[73] Contrary to the view that the decreased GSH levels are a consequence of an increased lipid peroxidation, studies by Jewell et al.[67] have shown, in alcoholics, the occurrence of markedly decreased hepatic GSH levels in the absence of an increased level of GSSG. In addition to a decreased availability of hepatic GSH, alcoholics also exhibit diminished blood selenium levels,[74,75] which are likely to result in a decreased GSH peroxidase activity and thereby further compromise their capacity to detoxify lipid peroxides.

## IV. EFFECTS OF ETHANOL ADMINISTRATION ON HEPATIC GSH IN EXPERIMENTAL ANIMALS

### A. Acute Effects

Estler and Ammon,[76] in 1966, were first to report the effect of an acute dose of ethanol on liver GSH. In their study, the administration of 4.1 g of ethanol per kilogram to fed mice elicited a 22% decrease in liver GSH in 2 h. During the last decade a number of studies, both *in vivo*[77-87] and *in vitro*,[88,89] have confirmed the ability of ethanol to deplete liver GSH in rodents.

While most studies have focused on changes in liver GSH, ethanol administration has also been shown to decrease, although to a smaller extent, the content of GSH in kidney.[81,86] In the liver, the magnitude of the depletion has been reported to range from 20 to over 50%, depending on the dose of ethanol and the time elapsed after its administration.[78,79] While the depletion of GSH does not depend on the sex of the animals,[79] its magnitude appears to depend upon the nutritional status, as fasted animals are more susceptible to this effect than fed ones.[79] This observation, confirmed in our laboratory,[90] suggests that in alcoholics malnutrition and fasting might render them more susceptible to the depleting effects of ethanol on liver GSH.

Recent studies[84,85] addressing the mechanism(s) of the depletion of hepatic GSH by ethanol have examined the kinetic changes which underlie the perturbations of the steady-state GSH levels. Results from turnover studies indicate that the ethanol-induced GSH depletion occurs in association with an increase in the rate of utilization of hepatic GSH.[85] Ethanol given acutely to rats markedly increases (47%) the concentration of GSH in blood from the posthepatic inferior vena cava, but not in its prehepatic segment, thus indicating that ethanol enhances the output of hepatic GSH into the circulation.[85] Such an interpretation is consistent with the report by Videla et al.[82] that acute ethanol administration increases arterial plasma GSH in rats given ethanol acutely. Although the mechanism(s) by which ethanol increases liver GSH efflux is not clear, studies in perfused livers[50,90] have shown that GSH output can be increased by increasing the liver perfusion rate. Since acute ethanol administration increases liver blood flow[91] by over 50%, an increased flow might result in an increased hepatic output (plasma GSH concentration times liver blood flow) of GSH. While an enhanced GSH output would not necessarily result in a higher GSH concentration in hepatic vein blood,[84] increased GSH levels should be expected in posthepatic inferior vena cava blood.[85] The latter will be enriched in GSH as a result of an enhanced hepatic output of GSH secondary to a faster blood flow. While a possible effect of ethanol on the GSH carrier mechanism has not been investigated, perfusion of livers with 60 m$M$ ethanol had no direct enhancing effect on GSH output into the hepatic vein.[90]

In addition to an alcohol-induced increased output of hepatic GSH, the GSH depletion has also been postulated to result from a direct, nonenzymatic binding of GSH to ethanol-derived acetaldehyde.[88] Aldehydes avidly bind to thiols to form hemimercaptals and thiazolidine carboxylic derivatives. However, when studied under conditions which resemble those prevailing in alcoholics *in vivo* (4 to 8 m$M$ GSH, 70 $\mu M$ acetaldehyde, pH 7.4, 37°C), binding of acetaldehyde with GSH was found to occur at rates which could account for only 6% of the rate at which GSH is lost *in vivo* from the liver of animals given ethanol.

Furthermore, no evidence has been presented to indicate the *in vivo* occurrence of a condensation product between acetaldehyde and GSH. It therefore seems apparent that binding of acetaldehyde to GSH, which has been generally assumed to be quantitatively important,[88,92] would play only a minor role in the depletion of GSH induced by ethanol *in vivo*.

Another suggested mechanism, to which much attention has been devoted, postulates that the depletion of hepatic GSH results from its increased oxidation into GSSG during the removal of lipid peroxides formed during the metabolism of ethanol.[79,93] Much of the controversy associated with this postulated mechanism derives from the fact that while some investigators have reported increases in lipid peroxidation after acute ethanol administration,[77,79,82,83,86,94-98] others have failed to observe such an effect.[84,85,87,99,100] At present, there is no satisfactory explanation for this discrepancy. Studies by Videla and co-workers[79,83] have demonstrated the concomitant occurrence of a severe GSH depletion (over 50%) with increases in lipid peroxides and GSSG levels in the livers of animals given ethanol acutely. The same authors, however, have reported that a depletion in liver GSH of a smaller extent (35% or less) occurred in complete absence of increases in lipid peroxides.[79,83] While both findings are consistent with the postulate that an increase lipid peroxidation would occur only in conditions of maximal GSH depletion,[79] they further imply the existence of an ethanol-mediated nonoxidative mechanism capable of depleting liver GSH.[85] In fact, a number of laboratories,[81,84,94] including ours,[85,87] have demonstrated the occurrence of depletions in liver GSH in the absence of increases in either hepatic or biliary GSSG, the latter being an excellent indicator of intracellular oxidative stress.[48]

In addition to an increased rate of GSH utilization, two groups of investigators have reported a reduction in the rate of synthesis of GSH after ethanol administration.[84,85] The mechanism by which this effect occurs is not clear. In recent studies, however, we have found that after ethanol administration the concentration of hepatic cysteine (the limiting precursor amino acid) was not decreased at times preceding those at which GSH levels were significantly depressed.[85] Furthermore, a condensation product between cysteine and acetaldehyde, which could have reduced the availability of cysteine for GSH synthesis, was not detected in the liver of animals administered ethanol acutely.[85] The study of the mechanism by which ethanol impairs GSH synthesis *in vivo* is of interest and warrants its further study. In fact, under conditions in which reactive electrophiles have depleted the initial GSH content, protection against these compounds wll largely depend on the ability of the liver cell to rapidly resynthesize the tripeptide.

## B. Chronic Effects

Unlike the acute effects of ethanol, the effects of chronic ethanol consumption on liver GSH have been more controversial. Although most studies have reported increased[101-108] levels of GSH in the liver of rodents given ethanol chronically, some investigators have found the levels to be either unchanged[97,109,110] or decreased.[80,81,111] A comparison of the different experimental conditions employed in the above-mentioned studies suggests that neither the length of ethanol administration, the technique of pair-feeding employed, the sex, nor the age of the animals are factors likely to underlie the observed discrepancies.

As discussed earlier (Section III), alcoholics present a decreased content of hepatic GSH. Similarly, baboons fed alcohol chronically also show significantly lower (40%) hepatic GSH levels.[97] Acute infusion of ethanol (1.6 g/kg) to baboons fed alcohol chronically further decreased hepatic GSH to levels 30% of those seen in control baboons fed an isocaloric carbohydrate diet.[97] Unlike baboons, rodents, in general, appear to develop resistance to the GSH-depleting effects of ethanol,[101-108] but do not develop liver damage upon administration of ethanol in liquid diets. This species difference is of interest in that, in baboons, chronic alcohol consumption can lead to the development of liver cirrhosis.[92] Thus, the mechanisms responsible for the compensatory response to the GSH-depleting effects of ethanol in rodents may acquire special relevance in relation to alcoholic liver damage.

Recent studies indicate that prolonged ethanol administration would result in an increased hepatic rate of synthesis and utilization of GSH.[104] These kinetic changes were reported to occur in association with increased hepatic GSH levels. At present, the exact nature of the events leading to increased GSH levels and turnover is unknown. A close association was observed[104] between the increased hepatic GSH turnover and the activity of GGT in the liver of alcohol-fed animals. In recent studies addressing the biological significance of the increases in hepatic GGT which follow chronic alcohol consumption, we observed that in perfused livers of alcohol-treated rats, GGT as an ecto-enzyme is substantially increased in basolateral plasma membranes.[53] In the liver of these animals, the degradative removal of GSH from the circulation was markedly enhanced (over 100%).[53] Notably, a strong correlation (r = 0.96) was observed between total liver GGT activity and GGT-mediated GSH removal from the circulation.[53] Based on the biological role of GGT in the kidney, an increased removal of circulating GSH could conceivably contribute to an increased recovery of GSH precursors into the liver. For example, GGT-mediated transpeptidation of GSH with cystine can lead to the formation of gamma-glutamyl cystine, which has been shown to lead to increased intracellular levels of GSH.[56] Such a mechanism could prove to be of special importance in the economy of GSH in the liver of ethanol-fed animals, as these have been recently shown to lose GSH into the circulation at rates substantially higher than those of controls.[106]

Another compensatory mechanism to recover intrahepatic GSH could result from an increased transformation of methione into cysteine leading to an enhanced synthesis of the tripeptide. Finkelstein et al.[113] showed that prolonged ethanol administration to rats is associated with increases in the activity of enzymes of the cystathionine pathway. Consistently, chronic alcohol consumption has been shown in rats to increase dietary methionine requirements.[92] Furthermore, plasma and hepatic $\alpha$-amino-$n$-butyric acid (a secondary metabolite generated in the cystathionine pathway) is also markedly increased in experimental animals and in humans after prolonged alcohol consumption.[114] In line with this, the incorporation of $^{35}$S-methionine into GSH has been reported to be increased in the liver of alcohol-fed rats.[115] Interestingly, in alcoholic cirrhotics, the route of conversion of methionine into cysteine is significantly impaired.[116] Conceivably, this condition could contribute to the decreased GSH levels seen in the liver of these subjects.

## ACKNOWLEDGMENTS

Studies by the authors discussed in this review were supported in part by grants from the Medical Research Council of Canada (MT-8928) and from the U.S. National Institute of Alcoholism and Alcohol Abuse (AA-06573).

## REFERENCES

1. **Kosower, E. M. and Kosower, N.,** The glutathione status of cells, *Int. Rev. Cytol.*, 54, 109, 1978.
2. **Meister, A. and Anderson, M. E.,** Glutathione, *Annu. Rev. Biochem.*, 52, 711, 1983.
3. **Kaplowitz, N., Aw, T. Y., and Ookhtens, M.,** The regulation of hepatic glutathione, *Annu. Rev. Pharmacol. Toxicol.*, 25, 715, 1985.
4. **Meredith, M. and Reed, D.,** Status of the mitochondrial pool of glutathione in the isolated hepatocyte, *J. Biol. Chem.*, 257, 3747, 1982.
5. **Romero, F. J. and Sies, H.,** Subcellular glutathione contents in isolated hepatocytes treated with L-Buthionine Sulfoximine, *Biochem. Biophys. Res. Commun.*, 123, 1116, 1984.
6. **Griffith, O. W. and Meister, A.,** Origin and turnover of mitochondrial glutathione, *Proc. Natl. Acad. Sci. U.S.A.*, 82, 4668, 1985.
7. **Higashi, T., Tateishi, N., Naruse, A., and Sakamoto, Y.,** A novel physiological role of liver glutathione as a reservoir of L-cysteine, *J. Biochem. (Tokyo)*, 83, 117, 1977.
8. **Lauterburg, B. H. and Mitchell, J. R.,** Regulation of hepatic glutathione turnover in rats *in vivo* and evidence for kinetic homogeneity of the hepatic glutathione pool, *J. Clin. Invest.*, 67, 1415, 1981.

9. **Kaplowitz, N., Eberle, D. E., Petrini, J., Touloukian, J., Corvasce, M. C., and Kuhlenkamp, J.**, Factors influencing the efflux of hepatic glutathione into bile in rats, *J. Pharmacol. Exp. Ther.*, 224, 141, 1983.

10. **Lauterburg, B. H., Smith, C. V., Hughes, H., and Mitchell, J. R.**, Biliary excretion of glutathione and glutathione disulfide in the rat: regulation and response to oxidative stress, *J. Clin. Invest.*, 73, 124, 1984.

11. **Lauterburg, B. H., Adams, J. D., and Mitchell, J. R.**, Hepatic glutathione homeostatis in the rat: efflux accounts for glutathione turnover, *Hepatology*, 4, 586, 1984.

12. **Orrenius, S., Ormstad, K., Thor, H., and Jewell, S. A.**, Turnover and functions of glutathione studied in isolated hepatic and renal cells, *Fed. Proc.*, 42, 3177, 1983.

13. **Reed, D. J. and Fariss, M. W.**, Glutathione depletion and susceptibility, *Pharmacol. Rev.*, 36, 25S, 1983.

14. **Meister, A.**, New aspects of glutathone biochemistry and transport: selective alteration of glutathione metabolism, *Fed. Proc.*, 43, 3031, 1984.

15. **Tsan, M. F., Danis, E. H., de Vecchio, P. J., and Rosano, C. L.**, Enhancement of intracellular glutathione protects endothelial cells against oxidant damage, *Biochem. Biophys. Res. Commun.*, 127, 270, 1985.

16. **Harlan, J. M., Levine, J. D., Callahan, K. S., and Schwartz, B. R.**, Glutathione redox cycle protects cultured endothelial cells against lysis by extracellularly generated hydrogen peroxide, *J. Clin. Invest.*, 73, 106, 1984.

17. **Dethmers, J. K. and Meister, A.**, Glutathione export by human lymphoid cells: depletion of glutathione by inhibition of its synthesis decreases export and increases sensitivity to radiation, *Proc. Natl. Acad. Sci. U.S.A.*, 78, 7492, 1981.

18. **Chance, B., Sies, H., and Boveris, A.**, Hydroperoxide metabolism in mammalian organ, *Physiol. Rev.*, 59, 605, 1979.

19. **Mitchell, J. R., Jollow, D. J., Potter, W. Z., David, D. C., Gillette, J. R., and Brodie, B. B.**, Acetaminophen-induced hepatic necrosis. I. Role of drug metabolism, *J. Pharmacol. Exp. Ther.*, 187, 185, 1973.

20. **Mitchell, J. R., Jollow, D. J., Potter, W. Z., Gillette, J. R., and Brodie, B. B.**, Acetaminophen hepatic necrosis. IV. Protective role of glutathione, *J. Pharmacol. Exp. Ther.*, 187, 211, 1973.

21. **Pessayre, D., Dolder, A., Artigou, J. Y., Wandscheer, J. C., Descatoire, V., Degott, C., and Benhamou, J. P.**, Effect of fasting on metabolite-mediated hepatotoxicity in the rat, *Gastroenterology*, 77, 264, 1979.

22. **Casini, A. F., Pompella, A., and Comporti, M.**, Liver glutathione depletion induced by bromobenzene, iodobenzene, and diethylmaleate poisoning and its relation to lipid peroxidation and necrosis, *Am. J. Pathol.*, 118, 225, 1985.

23. **Strubelt, O., Dost-Kempf, E., Siegers, C. P., Younes, M., Volpel, M., Preuss, U., and Dreckmann, J. G.**, The influence of fasting on the susceptibility of mice to hepatotoxic injury, *Toxicol. Appl. Pharmacol.*, 60, 66, 1981.

24. **Mitchell, J. R., Hughes, H., Lauterburg, B. H., and Smith, C. V.**, Chemical nature of reactive intermediates as determinant of toxicologic responses, *Drug Metab. Rev.*, 13, 527, 1982.

25. **Ku, R. H. and Billings, R. E.**, Relationships between formaldehyde metabolism and toxicity and glutathione concentrations in isolated rat hepatocytes, *Chem. Biol. Interact.*, 51, 25, 1985.

26. **Jaeger, R. J., Conolly, R. B., and Murphy, S. D.**, Effect of 18 hr fast and glutathione depletion on 1,1-dichloro-ethylene-induced hepatotoxicity and lethality in rats, *Exp. Mol. Biol.*, 20, 187, 1974.

27. **Lindstrom, T. D., Anders, M. W., and Remmer, H.**, Effect of phenobarbital and diethyl maleate on carbon tetrachloride toxicity in isolated rat hepatocytes, *Exp. Mol. Pathol.*, 28, 18, 1978.

28. **Ekstrom, T. and Hogberg, J.**, Chloroform-induced glutathione depletion and toxicity in freshly isolated hepatocytes, *Biochem. Pharmacol.*, 29, 3059, 1980.

29. **Meister, A. and Tate, S. S.**, Glutathione and related gamma-glutamyl-compounds; biosynthesis and utilization, *Annu. Rev. Biochem.*, 45, 559, 1976.

30. **Sho, E. S., Sahyoun, N., and Stegink, L. D.**, Tissue glutathione as a cysteine reservoir during fasting and refeeding of rats, *J. Nutr.*, 111, 914, 1981.

31. **Meister, A.**, Glutathione, in *The Liver Biology and Pathobiology*, Arias, I., Popper, H., Schachter, D., and Shafritz, P., Eds., Raven Press, New York, 1982, 297.

32. **Snoke, J. E. and Bloch, K.**, Biosynthesis of glutathione, in *Glutathione*, Colowick, S., Lazarow, A., Racker, E., Schware, D. R., Stadtman, E., and Waelsch, H., Eds., Academic Press, New York, 1954, 129.

33. **Davis, J. S., Balinski, J. B., Harington, J. S., et al.**, Assay, purification, properties and mechanism of action of gamma-glutamylcysteine synthetase from the liver of the rat and xenopus laevis, *Biochem. J.*, 133, 667, 1973.

34. **Richman, P. G. and Meister, A.**, Regulation of gamma-glutamyl cysteine synthetase by nonallosteric feedback inhibition by glutathione, *J. Biol. Chem.*, 250, 1422, 1975.

35. **Lauterburg, B. H., Vaishnar, Y., Stillwell, W. G., and Mitchell, J. R.,** The effects of age and glutathione depletion on hepatic glutathione turnover *in vivo* determined by acetaminophen probe analysis, *J. Pharmacol. Exp. Ther.,* 213, 54, 1980.

36. **Vina, J., Reginald, H., and Krebs, H. A.,** Maintenance of glutathione content in isolated hepatocytes, *Biochem. J.,* 170, 627, 1978.

37. **Beatty, P. and Reed, D. J.,** Influence of cysteine upon the glutathione status of isolated rat hepatocytes, *Biochem. Pharmacol.,* 30, 1227, 1981.

38. **Reed, D. J. and Orrenius, S.,** The role of methionine in glutathione biosynthesis in isolated hepatocytes, *Biochem. Biophys. Res. Commun.,* 77, 1258, 1977.

39. **Meister, A.,** Selective modifications of glutathione metabolism, *Science,* 220, 471, 1983.

40. **Peterson, R. B. and Rumack, B. H.,** Treating acute acetaminophen poisoning with acetyl cysteine, *J.A.M.A.,* 237, 2406, 1977.

41. **Lauterburg, B. H., Corcoran, G. B., and Mitchell, J. R.,** Mechanism of action of N-acetylcysteine in the protection against the hepatotoxicity of acetaminophen in rats *in vivo, J. Clin. Invest.,* 71, 980, 1983.

42. **Williamson, J. M. and Meister, A.,** Stimulation of hepatic glutathione formation by administration of L-2-oxothiazolidine-4-carboxylate, a 5-oxoprolinase substrate, *Proc. Natl. Acad. Sci. U.S.A.,* 78, 936, 1982.

43. **Puri, R. and Meister, A.,** Transport of glutathione, as gamma-glutamyl cysteinyl-glycyl ester, into the liver and kidney, *Proc. Natl. Acad. Sci. U.S.A.,* 80, 5258, 1983.

44. **Boyland, E. and Chasseaud, L. F.,** The role of glutathione and glutathione S-transferases, in mercapturic acid biosynthesis, *Adv. Enzymol.,* 32, 173, 1969.

45. **Jacoby, W. B.,** The glutathione-S-transferases: a group of multifunctional detoxification proteins, *Adv. Enzymol.,* 46, 383, 1978.

46. **Bast, A. and Haenen, G. R. M. M.,** Cytochrome P-450 and glutathione: what is the significance of their interrelationship in lipid peroxidation?, *Trends Biochem. Sci.,* 9, 510, 1984.

47. **Orrenius, S. and Moldeus, P.,** The multiple roles of glutathione in drug metabolism, *Trends Pharmacol. Sci.,* 5, 432, 1984.

48. **Akerboom, T. P. M., Bilzer, M., and Sies, H.,** The relationship of biliary glutathione disulfide and intracellular glutathione disulfide content in perfused rat liver, *J. Biol. Chem.,* 257, 4248, 1982.

49. **Adams, J. D., Lauterburg, B. H., and Mitchell, J. R.,** Plasma glutathione and glutathione disulfide in the rat: regulation and response to oxidative stress, *J. Pharmacol. Exp. Ther.,* 227, 749, 1983.

50. **Ookhtens, M., Hobdy, K., Corvasce, M. C., Aw, T. Y., and Kaplowitz, N.,** Sinusoidal efflux of glutathione in the perfused rat liver: evidence for a carrier-mediated process, *J. Clin. Invest.,* 75, 258, 1985.

51. **Inoue, M., Kinne, R., Tran, T., and Arias, I. M.,** Glutathione transport across plasma membranes: analysis using isolated rat liver sinusoidal-membrane vesicles, *Eur. J. Biochem.,* 138, 491, 1983.

52. **Sies, H. and Graf, P.,** Hepatic thiol and glutathione efflux under the influence of vasopressin, phenyl-ephrine, and adrenaline, *Biochem. J.,* 226, 545, 1985.

53. **Speisky, H. Gunasekara, A., Varghese, G., and Israel, Y.,** Basolateral gamma-glutamyl transferase ectoactivity in rat liver: effects of chronic alcohol consumption, *Alcohol Alcoholism (Suppl. 1),* p. 245, 1987.

54. **Abbott, W. A., Bridges, R. J., and Meister, A.,** Extracellular metabolism of glutathione accounts for its disappearance from the basolateral circulation from the kidney, *J. Biol. Chem.,* 259, 15393, 1984.

55. **Inoue, M., Shinozuka, S., and Morino, Y.,** Mechanism of renal peritubular extraction of plasma glutathione, *Eur. J. Biochem.,* 157, 605, 1986.

56. **Anderson, M. E. and Meister, A.,** Transport and direct utilization of gamma-glutamyl cyst(e)ine for glutathione synthesis, *Proc. Natl. Acad. Sci. U.S.A.,* 80, 707, 1983.

57. **Shi, E. C. P., Fisher, R., McEvoy, M., Vantol, R., Rose, M., and Ham, J. M.,** Factors influencing hepatic glutathione concentrations: a study in surgical patients, *Clin. Sci.,* 62, 279, 1982.

58. **Baraona, E. and Lieber, C. S.,** Effects of alcohol on hepatic transport of proteins, *Annu. Rev. Med.,* 33, 281, 1982.

59. **Kaplowitz, N., Kuhlenpamp, J., Goldstein, L., and Reeve, J.,** Effect of salicylates and phenobarbital on hepatic glutathione in the rat, *J. Pharmacol. Exp. Ther.,* 212, 240, 1980.

60. **Kloss, M. W., Rosen, G. M., and Rauckman, E. J.,** Cocaine-mediated hepatotoxicity, *Biochem. Pharmacol.,* 33, 169, 1984.

61. **James, R. C., Goodman, D. R., and Harbison, R. D.,** Hepatic glutathione and hepatotoxicity: changes induced by selected narcotics, *J. Pharmacol. Exp. Ther.,* 221, 708, 1982.

62. **Videla, L. A., Iturriaga, H., Pino, M. E., Bunout, D., Valenzuela, A., and Ugarte, G.,** Content of hepatic reduced glutathione in chronic alcoholic patients: influence of the length of abstinence and liver necrosis, *Clin. Sci.,* 66, 283, 1984.

63. **Shaw, S., Rubin, K. P., and Lieber, C. S.,** Depressed hepatic glutathione and increased diene conjugates in alcoholic liver disease: evidence of lipid peroxidation, *Dig. Dis. Sci.,* 28, 585, 1983.

64. **Woodhouse, K. W., Faith, M. W., Mutch, E., Wright, P., James, O. F. W., and Rawlins, M. D.,** The effect of alcoholic cirrhosis on the activities of microsomal aldrin epoxidase, 7-ethoxycoumarin 0-deethylase and epoxyde-hydrolase, and on the concentrations of reduced glutathione in human liver, *Br. J. Clin. Pharmacol.,* 15, 667, 1983.
65. **Lauterburg, B. H., Velez, M. E., and Mitchell, J. R.,** Plasma glutathione as an index of intrahepatic GSH in man: response to acetaminophen and chronic ethanol abuse, *Hepatology,* 4, 1051(A), 1984.
66. **Poulsen, H. E., Ranek, L., and Andreasen, P. B.,** The hepatic glutathione content in liver diseases, *Scand. J. Clin. Lab. Invest.,* 41, 573, 1981.
67. **Jewell, S. A., Di Monte, D., Gentile, A., Guglielmi, A., Altomare, E., and Albano, O.,** Decreased hepatic glutathione in chronic alcoholic patients, *J. Hepatol.,* 3, 1, 1986.
68. **Seeff, L. B., Cucherini, B. A., Zimmerman, H. J., Adler, E., and Benjamin, S. B.,** Acetaminophen hepatotoxicity in alcoholics: a therapeutic misadventure, *Ann. Intern. Med.,* 104, 399, 1986.
69. **McClain, C. J., Kromhout, J. P., Peterson, F. J., and Holtzman, J. L.,** Potentiation of acetaminophen hepatotoxicity by alcohol, *J.A.M.A.,* 244, 251, 1980.
70. **Rubin, E. and Lieber, C. S.,** Hepatic microsomal enzymes in man and rats: induction and inhibition by ethanol, *Science,* 162, 690, 1968.
71. **Walker, R. M., McElligott, T. F., Power, E. M., Mazey, T. E., and Racz, W. J.,** Increased acetaminophen-induced hepatotoxicity after chronic ethanol consumption in mice, *Toxicology,* 28, 193, 1983.
72. **Sato, C. and Lieber, C. S.,** Mechanism of the preventive effect of ethanol on acetaminophen-induced hepatotoxicity, *J. Pharmacol. Exp. Ther.,* 218, 811, 1981.
73. **Suematsu, T., Matsumura, T., Sato, N., Miyamoto, T., Ooka, T., Kamada, T., and Abe, H.,** Lipid peroxidation in alcoholic liver disease in humans, *Alcoholism Clin. Exp. Res.,* 5, 427, 1981.
74. **Dworkin, B. M., Rosenthal, W. S., Gordon, G. G., and Jankowski, R. H.,** Diminished blood selenium levels in alcoholics, *Alcoholism Clin. Exp. Res.,* 8, 535, 1984.
75. **Dworkin, B., Rosenthal, W. S., Jancowski, R. H., Gordon, G. G., and Haldea, D.,** Low blood selenium levels in alcoholics with and without advanced liver disease, *Dig. Dis. Sci.,* 30, 838, 1985.
76. **Estler, C. J. and Ammon, H. P. T.,** Glutathione und SH-gruppenhaltige enzyme in der leber weiber nach einmahger alkoholgabe, *Med. Pharmacol. Exp.,* 15, 299, 1966.
77. **Comporti, M., Benedetti, A., and Chieli, E.,** Studies on in vitro peroxidation of liver lipids in ethanol treated rats, *Lipids,* 8, 498, 1973.
78. **Macdonald, C. M., Dow, J., and Moore, M. R.,** A possible protective role for sulfhydryl compounds in acute alcoholic liver injury, *Biochem. Pharmacol.,* 26, 1529, 1977.
79. **Videla, L. C., Fernandez, V., Ugarte, G., and Valenzuela, A.,** Effect of acute ethanol intoxication on the content of reduced glutathione of the liver in relation to its lipoperoxidative capacity in the rat, *FEBS Lett.,* 111, 6, 1980.
80. **Fernandez, V. and Videla, L. A.,** Effect of acute and chronic ethanol ingestion on the content of reduced glutathione of various tissues of the rat, *Experientia,* 37, 392, 1981.
81. **Guerri, C. and Grisolia, S.,** Changes in glutathione in acute and chronic alcohol intoxication, *Pharmacol. Biochem. Behav.,* 13 (Suppl. 1), 53, 1980.
82. **Videla, L. A., Fernandez, V., Fernandez, N., and Valenzuela, A.,** On the mechanism of the glutathione depletion induced in the liver by acute ethanol ingestion, *Subst. Alcohol Actions Misuse,* 2, 153, 1981.
83. **Videla, L. A., Fernandez, V., Marinis, A., Fernandez, N., and Valenzuela, A.,** Liver lipidperoxidative pressure and glutathione status following acetaldehyde and aliphatic pretreatments in the rat, *Biochem. Biophys. Res. Commun.,* 104, 965, 1982.
84. **Lauterburg, B. H., Davies, S., and Mitchell, J. R.,** Ethanol supresses hepatic glutathione synthesis in rats in vivo, *J. Pharmacol. Exp. Ther.,* 230, 7, 1984.
85. **Speisky, H., Macdonald, A., Giles, G., Orrego, H., and Israel, Y.,** Increased loss and decreased synthesis of hepatic glutathione after acute ethanol administration: turnover studies, *Biochem. J.,* 225, 565, 1985.
86. **Kera, Y., Komura, S., Ohbora, Y., Kiriyama, T., and Inoue, K.,** Ethanol induced changes in lipid peroxidation and non-protein sulfhydryl content, *Res. Commun. Chem. Pathol. Pharmacol.,* 47, 203, 1985.
87. **Speisky, H., Bunout, D., Orrego, H., Giles, H. G., Gunasekara, A., and Israel, Y.,** Lack of changes in diene conjugate levels following ethanol-induced glutathione depletion or hepatic necrosis, *Res. Commun. Chem. Pathol. Pharmacol.,* 48, 77, 1985.
88. **Vina, J., Estrela, J. M., Guerri, C., and Romero, F.,** Effect of ethanol on glutathione concentration in isolated hepatocytes, *Biochem. J.,* 188, 549, 1980.
89. **Pascale, R., Garcea, R., Daino, L., Frassetto, S., Ruggiu, M., Pirisi, L., Stramentinoli, G., and Feo, F.,** The role of S-adenosylmethionine in the regulation of glutathione pool and acetaldehyde production in acute ethanol intoxication, *Res. Commun. Subst. Abuse,* 5, 321, 1984.
90. **Speisky, H.,** Hepatic Glutathione Content and Gamma-Glutamyl Transferase Activity: Effects of Acute and Chronic Ethanol Administration, Ph.D. thesis, Annals of the University of Toronto, Canada, 1986.

91. **McKaigney, J. P., Carmichael, F. J., Saldivia, V., Israel, Y., and Orrego, H.,** Role of ethanol metabolism in the ethanol-induced increase in splanchnic circulation, *Am. J. Physiol.*, 250, G518, 1986.
92. **Lieber, C. S.,** Alcohol, liver injury and protein metabolism, *Pharmacol. Biochem. Behav.*, 13 (Suppl. 1), 17, 1980.
93. **Videla, L. A. and Valenzuela, A.,** Alcohol ingestion, liver glutathione and lipoperoxidation: metabolic interrelations and pathological implications, *Life Sci.*, 31, 2395, 1982.
94. **Siergers, C. P., Jeb, U., and Younes, M.,** Effects of phenobarbital, GSH-depletors, CCly and ethanol on the biliary efflux of glutathione in rats, *Arch. Int. Pharmacodyn. Ther.*, 266, 315, 1983.
95. **Hashimoto, S. and Recknagel, R. O.,** No chemcial evidence of hepatic lipid peroxidation in acute ethanol toxicity, *Exp. Mol. Pathol.*, 8, 225, 1968.
96. **Pesh-Imam, M. and Recknagel, R. O.,** Lipid peroxidation and the concept of anti-oxygenic potential: vitamin E changes in acute experimental CCly, Brcl$_3$ and ethanol induced liver injury, *Toxicol. Appl. Pharmacol.*, 42, 463, 1977.
97. **Finkelstein, J. D., Cello, J. P., and Kyle, W. E.,** Ethanol-induced changes in methionine metabolism in rat liver, *Biochem. Biophys. Res. Commun.*, 61, 525, 1974.
98. **Litov, S., Green, D. L., Downey, J. E., and Tappel, A. L.,** The role of lipidperoxidation during chronic and acute exposure to ethanol as determined by pentane expiration in the rat, *Lipids,* 16, 52, 1981.
99. **Dianzani, M. U.,** Lipid peroxidation in ethanol poisoning: a critical reconsideration, *Alcohol Alcoholism,* 20, 161, 1985.
100. **DiLuzio, N. R. and Stege, T. E.,** The role of ethanol metabolites in hepatic lipid peroxidation, in *Alcohol and the Liver,* Fisher, M. M. and Rankin, J. G., Eds., Plenum Press, New York, 1977, 45.
101. **Shaw, S., Jayatilleke, E., Ross, W. A., Gordon, E. R., and Lieber, C. S.,** Ethanol-induced lipid peroxidation: potentiation by long-term alcohol feeding and attenuation by methionine, *J. Lab. Clin. Med.,* 98, 417, 1981.
102. **Hassing, J. M., Hupka, A. L., Stohs, S. J., and Yoon, P. C.,** Hepatic glutathione levels in D-penicillamine-fed ethanol-dependent rats, *Res. Commun. Chem. Pathol. Pharmacol.*, 25, 389, 1979.
103. **Nishimura, M., Stein, H., Berges, W., and Teschke, R.,** Gamma-glutamyl-transferase activity of liver plasma membrane: induction following chronic alcohol consumption, *Biochem. Biophys. Res. Commun.,* 99, 142, 1981.
104. **Hetu, C., Yelle, L., and Joly, J. G.,** Influence of ethanol on hepatic glutathione content and on the activity of glutathione S-transferase and epoxide hydrase in the rat, *Drug Metab. Dispos.*, 10, 246, 1982.
105. **Morton, S. and Mitchell, M. C.,** Effects of chronic ethanol on glutathione turnover in the rat, *Biochem. Pharmacol.*, 34, 1559, 1985.
106. **Pierson, J. L. and Mitchell, M. C.,** Increased hepatic efflux of glutathione after chronic ethanol feeding, *Biochem. Pharmacol.*, 35, 1533, 1986.
107. **Harata, J., Nagata, M., Sasaki, E., Ishiguro, I., Ohta, Y., and Murakami, Y.,** Effect of prolonged alcohol administration on activities of various enzymes scavenging activated oxygen radicals and lipidperoxide level in the liver of rats, *Biochem. Pharmacol.*, 32, 1795, 1983.
108. **Hetu, C., Dumont, A., and Joly, J. G.,** Effect of chronic ethanol administration on bromobenzene liver toxicity in the rat, *Toxicol. Appl. Pharmacol.*, 67, 166, 1983.
109. **Aykac, G., UySal, M., Yalcim, A. S., Kocak-Toker, N., Sivas, A., and Oz, H.,** The effect of chronic ethanol ingestion on hepatic lipidperoxide, glutathione, glutathione peroxidase and glutathione transferase, *Toxicology,* 36, 71, 1985.
110. **Kaplan, E., DeMaster, E. G., and Nagasawa, H. T.,** Effect of pargyline on hepatic glutathione levels in rats treated acutely and chronically with ethanol, *Res. Commun. Chem. Pathol. Pharmacol.*, 30, 577, 1980.
111. **Moldeus, P., Andersson, B., Norlig, A., and Ormstad, K.,** Effect of chronic ethanol administration on drug metabolism on isolated hepatocytes with emphasis on paracetamol activation, *Biochem. Pharmacol.,* 29, 1741, 1980.
112. **Misslbeck, N. G., Campbell, T. C., and Roe, D. A.,** Increased in hepatic gamma-glutamyl transferase (GGT) activity following chronic ethanol intake in combination with a high fat diet, *Biochem. Pharmacol.,* 35, 399, 1986.
113. **Finkelstein, J. D., Cello, J. P., and Kyle, W. E.,** Ethanol- induced changes in methionine metabolism in rat liver, *Biochem. Biophys. Res. Commun.*, 61, 525, 1974.
114. **Vendemiale, G., Jayatilleke, E., Shaw, S., and Lieber, C. S.,** Depression of biliary glutathione excretion by chronic ethanol feeding in the rat, *Life Sci.,* 34, 1065, 1984.
115. **Shaw, S. and Lieber, C. S.,** Increased hepatic production of alpha-amino-n-butyric acid in rats and baboons, *Gastroenterology,* 78, 108, 1980.
116. **Horowitz, J. H., Rypins, E. B., Henderson, J. M., Heymsfield, S. B., Moffitt, S. D., Bain, R. P., Chawla, R. K., Bleier, J. C., and Rudman, D.,** Evidence for impairment of trans-sulfuration pathway in cirrhosis, *Gastroenterology,* 81, 668, 1981.

Chapter 10

INTERACTIONS OF ETHANOL AND DRUG METABOLISM

**Jorg Mørland, Egil Bodd, and Gaut Gadeholt**

TABLE OF CONTENTS

# I. INTRODUCTION

The present chapter deals with the interaction of ethanol and drug metabolism. We have organized the text according to the various metabolic reactions to which a drug might be subject. Under each heading (reaction type) we have reviewed reports on those drugs for which effects of ethanol intake have been given. This approach may make it easier to see whether ethanol exerts a general effect on that particular type of reaction.

By "drug" in this review we mean all exogenous compounds affecting living processes, taken for therapeutic or other reasons, or introduced into the body as a consequence of environmental contamination. It should be stressed that only effects of ethanol, after single intake or due to long-term consumption, on drug metabolism will be considered, and that effects of drug intake on ethanol and acetaldehyde metabolism will be covered by other chapters in this book. We have not discussed decreased drug metabolism caused by loss of functional liver tissue due to alcoholic cirrhosis, as this, in our opinion, is a rather remote interaction between ethanol and drug metabolism. The effects of congeners are not discussed.

We have mainly reviewed studies in humans. However, demonstration of a metabolic ethanol-drug interaction in a human system is often difficult, since determination of ethanol, drug, and metabolite concentrations is performed in a compartment (usually blood or urine) distinct from the site of interaction (usually the liver). Many studies have measured disappearance of a parent drug, but have made no measurements of its metabolites in blood, urine, bile, or feces. In such studies conclusions about ethanol effects on drug metabolism are based only on indirect experimental evidence from blood concentration vs. time curves. Studies on humans are also often limited by ethical considerations. We have included some animal data to elucidate possible mechanisms with more direct evidence.

The interaction of ethanol with drug metabolism may manifest itself through changed clearance, drug bioavailability, or metabolite patterns. If a drug is eliminated mainly by metabolism, total body clearance will increase if the rate of drug metabolism is increased. Inhibition of drug metabolism will decrease drug clearance. If the drug has substantial first-pass metabolism (presystemic clearance), inhibition of drug metabolism will increase bioavailability, while increased drug metabolism will reduce bioavailability. Inhibition of a metabolic reaction will decrease the production of metabolites and, if these are active, decrease the effects (therapeutic/toxic) mediated by such metabolites. Increased drug metabolism would, on the contrary, enhance such effects.

Ethanol may therefore change the clinical response to a drug by either altered drug metabolism through changed clearance, bioavailability, and metabolite levels, or by combinations of such effects. The main emphasis has been put on presenting documented effects on clearance and bioavailability. From a practical point of view it should be noticed that some of the clinically most important metabolic interactions will result from increased drug concentrations during the absorption phase rather than from delayed elimination.

# II. ETHANOL INTERACTION WITH DRUG OXIDATION

Drug oxidation (hydroxylation, O- and N-dealkylation, and oxidation of nitrogen and sulfur) is mainly a function of the hepatic microsomal cytochromes P-450, but also of the mixed-function flavoprotein oxidase (FAD monooxygenase). During activity, these enzymes react with NADPH and molecular oxygen and depend on the availability of these substances to function. The products are water or hydrogen peroxide and $NADP^+$. In humans a number of different cytochrome P-450 isoenzymes have been isolated and purified. This multiplicity allows a large number of combinations of isoenzymes with resulting highly individual isoenzyme patterns, which leads to interindividual variations in metabolic patterns of drugs. This is most obvious for drugs with several possible metabolic pathways.[1-3] Several drugs have a number of different metabolic pathways. We have discussed each drug according to what appeared to be the main metabolic pathway.

## A. Effects of Acute Ethanol Intake

In principle, there are several possible sites of acute interaction between the cytochrome P-450 system and ethanol: the interested reader is referred to specialized literature for details about the mechanisms of P-450 action[4-7] and the interaction with ethanol.[8]

### 1. N-Dealkylation

Acute ethanol intake (blood ethanol concentration [BAC] between 17 and 22 m$M$) decreased the formation of N-desmethylpropoxyphene from propoxyphene expressed as a 36% decrease in the average plasma ratio for metabolite and parent drug.[9] Neither the $C_{max}$ nor apparent half-life was changed for either parent drug or metabolite.

Sellers et al.[10] investigated the effect of ethanol (blood concentration between 17 and 22 m$M$) on the disposition of i.v. diazepam and found decreased elimination of diazepam during the 8-h observation period and decreased formation of N-desmethyldiazepam during the same period. This indicated inhibition by ethanol of the formation of the major metabolite of diazepam in man. These authors circumvented the problem of ethanol interference with drug absorption by using the i.v. route. Other studies using oral administration of diazepam have given conflicting results on ethanol effects, probably due to differences in experimental design. Details that differed were the type of diazepam preparation used and sequence of ethanol and diazepam administration (for example, as powder suspended in water in the control situation and as a solution in 50% ethanol in the experimental situation,[11] or ethanol, 11 mmol/kg followed by diazepam administration).[12] The type and strength of the alcoholic beverage differed between the studies.[13] The reader is referred to the review by Lane et al.[14] for further details.

Chlordiazepoxide peak concentration after oral intake of the drug was increased following an ethanol dose of 17 mmol/kg.[15] This could be a result of alteration in the N-demethylation of the drug, but it could also be an effect of a change in distribution pattern in the body. A study using i.v. chlordiazepoxide[16] and ethanol dosing sufficient to maintain blood ethanol levels of 11 to 33 m$M$ for 32 h showed that ethanol decreased chlordiazepoxide clearance by 37% (44% of unbound drug) and increased the elimination half-life by 66%.

Taeuber et al.[17] found that a BAC of 22 m$M$ increased the area under the curve (AUC) of the benzodiazepine clobazam by about 50%. This drug has N-demethylation as a major transformation pathway.

This ethanol effect is not shared by all benzodiazepines. For example, clotiazepam elimination (mainly hydroxylation and N-demethylation) was not influenced by acute ethanol administration.[18] This may have been due to the low dose of ethanol (417 mmol per person weighing from 43 to 77 kg) with no maintenance doses during the 36-h observation period.

Dorian et al.[19] showed that the first-pass metabolism of amitriptyline was dramatically decreased by alcohol. Alcohol increased plasma concentrations of amitriptyline in the absorption phase from 38 to 350%. N-demethylation is the main metabolic pathway for this drug. This ethanol-drug interaction was thought to be clinically relevant. A several-fold increase in bioavailability may greatly increase the dangers associated with concomitant ethanol use and overdosage of amitriptyline. At lower dose levels, the central nervous system (CNS) effects of the combination may increase the risk of exaggerated intoxication from relatively low doses of ethanol and amitriptyline. The same group[20] found important impairment in psychomotor functions after 25 mg amitriptyline with 17 to 22 m$M$ ethanol, compared to either substance alone. Much of the pharmacodynamic ethanol effect could be attributed to ethanol-induced changes in free drug concentrations.

Warrington et al.[21] studied the effect of a single lower dose of ethanol (9 mmol/kg) on the pharmacokinetics of amitriptyline. They found a tendency to increased bioavailability of amitriptyline in the absorption phase, but no dramatic change in overall kinetics. Ethanol (blood concentration between 17 and 22 m$M$) decreased the rate of N-demethylation of zimelidine by 46%.[22] The overall change of drug elimination induced by ethanol was considerably less. The antiarrhythmic drug, disopyramide, showed no change in elimination kinetics in the presence of ethanol (given as 17 mmol/kg loading dose and 2.8 mmol/kg and h maintenance doses), but the ratio of *N*-desalkyldisopyramide to disopyramide in urine was decreased, suggesting some impairment of the oxidative N-dealkylation.[23]

Cushman et al.[24] studied the kinetic interaction between ethanol (concentration in blood up to 17 m$M$) and methadone, which has N-demethylation as a major pathway.[25] They found no discernible effect of a single dose of ethanol on the methadone concentrations at steady state. Ethanol was, however, present for relatively short periods compared to the long half-life of methadone (24 to 36 h). With this experimental approach, effects on methadone metabolism would be difficult to detect.

### 2. Aliphatic and Nonaromatic Ring Hydroxylations

Dorian et al.[19] showed in their study of first-pass metabolism of amitriptyline that there was evidence of decreased nortriptyline hydroxylation in the presence of 17 to 22 m$M$ ethanol. Ethanol (BAC between 17 and 22 m$M$) increased the bioavailability of triazolam by 21% expressed as area under the time/concentration curve.[26] Triazolam is mainly metabolized by hydroxylation in the 4-position of the benzodiazepine ring and the methyl side chain of the triazolo-ring.[27]

Pentobarbital is a barbiturate sedative/hypnotic drug, now rarely used, which is mainly eliminated by hydroxylation. Rubin et al.[28] gave a loading dose of the $^{14}$C-labeled barbiturate, measured the rate of elimination of radioactivity after establishment of the β phase, then gave ethanol as a single dose of 22 mmol/kg, followed by maintenance doses of ethanol in the observation period. They found an approximate doubling of the elimination half-life of radioactivity. This method is not completely satisfactory, as it does not exclude interference from radioactive metabolites in the assay. Meprobamate is another sedative drug that is metabolized by aliphatic oxidation. In an experiment with unlabeled meprobamate, Rubin et al.[28] found a two- to fourfold higher elimination half-life in the presence of ethanol compared to the control situation. Sellers and Holloway[29] have criticized both these studies for using short sampling periods.

In inhalation studies with *m*-xylene,[30] 17 mmol/kg of ethanol by mouth increased the blood xylene concentration 1.5- to 2.0-fold and decreased the urinary excretion of methylhippuric acid by about 50%. The various metabolic pathways of *m*-xylene were not equally affected, as there was no difference in the urinary excretion of 2,4-xylenol, a possible precursor metabolite of methylhippuric acid. Another organic solvent, toluene, is biotransformed by oxidations similar to those for xylene. Wallén et al.[31] found decreased metabolic clearance of toluene after a single dose of ethanol (15 mmol/kg body weight) and considerable increase in the maximal blood toluene concentrations (69%). This acute effect of ethanol was confirmed by Waldron et al.[32] Trichloroethylene is also metabolized by hydroxylation, and trichloroethanol is a major metabolite. The presence of ethanol (average 13 m$M$ in blood) inhibited the conversion by 40%[33] and resulted in a 2.5-fold increase in the trichloroethylene concentration in blood compared to the control situation.

Tolbutamide, a hypoglycemic drug, has a major biotransformation pathway very similar to xylene. Carulli et al.[34] infused about 4 mmol ethanol per kilogram per hour for 6 h and found about 40% prolongation of the elimination half-life of tolbutamide after i.v. injection.

Antipyrine has several metabolic pathways, including N-demethylation and 4-hydroxylation of the nonaromatic ring.[35] Disappearance was not influenced by ethanol (16 mmol/kg as loading dose and 3.5 mmol/kg/h for 12 h).[36] N-Desmethyldiazepam is a metabolite of diazepam and the active metabolite of the prodrug di-potassium chlorazepate. It is metabolized by hydroxylation in the 3-position of the nonaromatic ring to oxazepam. Staak and Moosmayer[37] investigated the 3-hydroxylation of N-desmethyldiazepam and found that the urinary excretion of oxazepam was almost halved by an acute ethanol dose of 22 mmol/kg. In a recent study, intake of a single large dose of ethanol (approximately 4 mol) by seven volunteers, together with sparteine, was followed by a significant reduction of the excretion of 5-dehydroxysparteine in the urine.[163]

### 3. N- and S-Oxidations

Chlormethiazole is a sedative drug often used to treat delirium tremens. It has a large number of oxidized metabolites, some of which are unlikely to be formed via cytochrome P-450, but instead by the microsomal flavoprotein oxidase system.[38] Ethanol (17 mmol/kg loading dose and 2 mmol/kg/h as a maintenance dose) increased the oral bioavailability by 62% from 8.5 to 13.8% and the AUC from 0 to 8 h by 82%.[39] This interaction is likely to be of importance as a cause of death in situations of combined abuse of ethanol and chlormethiazole because of the dramatic increase in bioavailability. The marked metabolic interaction could not be confirmed by Bury et al.,[40] who found slightly decreased rates of elimination in three of four subjects (statistically nonsignificant) after i.v. chlormethiazole. After oral chlormethiazole they found decreased rates in three and increased rates in two of five subjects. They used BACs of about 22 m$M$.

### 4. Aromatic Hydroxylation and Other Aromatic Oxidations

Warfarin has a number of different hydroxylation sites on the aromatic rings.[2] Breckenridge and Orme[41] found that coadminstration of ethanol and warfarin resulted in increased blood levels of warfarin and increased anticoagulation.

Propranolol has hydroxylation in the 4-position as a main metabolic pathway.[42] Another pathway is dealkylation to yield reactive aldehydes, which may damage drug-metabolizing enzymes of the liver[43] and, in addition, give rise to other less reactive metabolites. Sotaniemi et al.[44] studied the metabolism of propranolol given 12 h after commencement of ethanol drinking. The results indicated that the plasma clearance rate of propranolol was enhanced by ethanol. On the other hand, the drug level 1 h after drug intake was positively correlated to the BAC. In this study there appeared to be both enhancement and inhibition of propranolol metabolism by acute ethanol. The dose of ethanol taken in this study was very variable, expressed as large intersubject variations in BACs after the study (from 0 to 33 m$M$). Thus, the experimental situation appears likely to be similar to the normal drinking party.

Under more controlled conditions,[45] BACs of 17 mmol/l increased the area under the concentration vs. time curve for oral propranolol doses by 17% in responsive subjects, but a decrease was observed in one nonresponsive person. Later work by the same group confirmed the slight effects of ethanol on the disposition and effects of this drug.[46] Metabolites were not measured in the study, so there are no clues as to which pathways were affected.

Paracetamol (acetaminophen) has one minor toxic metabolite which is formed by oxidation (in contrast to the major pathways, which are conjugation pathways). The reactive intermediate is believed to be *N*-acetyl *P*-benzoquinone imine,[47] but the excreted metabolite is an aromatic oxidation product, and for this reason the drug is discussed under this heading. Acute ethanol intake (13 mmol/kg and 2 or 3.5 mmol/kg/h as maintenance doses) gave a 67% decrease in the urinary excretion of the glutathione-derived sulfur-containing conjugates (cysteine and mercapturate) of oxidized metabolites, originating from detoxication of the oxidized metabolite.[48] This indicates that the inhibitory effect of ethanol on drug oxidations may also be beneficial and of possible therapeutic application. Banda and Quart[49] have made similar findings with a less complete characterization of urinary metabolites.

The sedative-hypnotic drug methaqualone (not in medical use in many countries, but of considerable interest as a drug of abuse) has aromatic hydroxylation in the 4′-position as a major metabolic pathway.[50] Roden et al.[51] found no effect of evening drinks (11 mmol of ethanol per kilogram body weight in a single dose) on the residual plasma concentrations of methaqualone 24 and 48 h after a single 250-mg tablet, but a significantly higher concentration after 72 h, all in comparison with placebo. This study is difficult to interpret and appears to be of slight relevance to the risk of acute interactions.

## 5. Conclusions

The effect of acute ethanol intake is mainly inhibitory. Not all of the cited studies are conclusive owing to shortcomings in experimental design or analytical methods. Very few of the studies have used chemical determination of metabolites as a measure of biotransformation. Most have followed the disappearance of the parent drug. In some cases the doses of ethanol used may have been too low to produce a detectable interaction.

There is evidence that ethanol causes impaired oxidation of the following drugs: diazepam and other benzodiazepines, chlormethiazole, meprobamate, pentobarbital, amitriptyline, zimelidine, propranolol, disopyramide, *m*-xylene, toluene, trichloroethylene, and possibly methaqualone. The mechanisms responsible may be direct effects on the drug-oxidizing enzymes[8] and possibly an indirect effect of ethanol mediated by elevated corticosteroid levels.[52] Cofactor availability (mainly, NADPH) to cytochrome P-450 associated monooxygenases may be rate limiting for drug oxidation under circumstances of poor availability of carbohydrates. It is uncertain whether this mechanism operates in man *in vivo*. The reader is referred to other review articles[8,53] for more complete discussions of mechanisms.

## B. Effects of Chronic Ethanol Intake

This subject has been reviewed previously from a general point of view[54] and with a focus on human data.[55]

Pelkonen and Sotaniemi[55] found that alcoholics with normal liver histology had normal or elevated levels of hepatic cytochrome P-450 in biopsy material, whereas those with fatty liver had normal values and those with hepatitis or cirrhosis had subnormal values expressed per gram of tissue.

### 1. N-Dealkylation

Sellman et al.[56] studied the single-dose kinetics of 10 mg diazepam given orally in ethanol-free alcoholics. In these subjects there was clear evidence of decreased biological availability of diazepam. Although metabolites were not measured, it may be inferred that the N-demethylation pathway could be affected. Pond et al.[57] found a tendency towards enhanced bioavailability of diazepam after withdrawal for a week in chronic alcoholics, which suggested a reversal of enzyme induction on abstinence. Sellers et al.[58] found no significant effects of ethanol withdrawal on diazepam kinetics.

Theophylline is eliminated by N-demethylation with only a minor fraction (<20%) eliminated as unchanged drug. By investigating a large group of lung patients, Jusko et al.[59]

found 30% higher metabolic clearance in social users of ethanol than in abstainers, but a markedly lower metabolic clearance in cirrhotic patients.

## 2. O-Dealkylation

In liver microsomes from a person consuming 1.1 mol (50 g) ethanol per day, Hoensch et al.[60] found a lower value of ethoxycoumarin O-deethylase and no detectable inhibitory effect of cimetidine or ranitidine compared with controls. Woodhouse et al.[61] studied hepatic biopsy material *in vitro*. They found impairment of both high- and low-affinity *in vitro* microsomal ethoxycoumarin O-deethylation in samples from alcoholic cirrhotics. These patie[nts still had a considerable alcohol intake and clinical chemistry showed evidence of hepatic impairment. Pelkonen and Sotaniemi[55] found normal or elevated ethoxycoumarin O-deethylase activity in liver biopsy material from alcoholics with normal liver histology and decreased values in patients with histological evidence of alcoholic liver disease.

## 3. Aliphatic and Nonaromatic Ring Hydroxylations

Antipyrine has several metabolites in man (see effects of acute ethanol intake). The metabolic clearance of antipyrine in alcoholic patients was found to be elevated prior to treatment with cyanidanol and decreased after treatment (six of seven patients).[62] No untreated alcoholics were studied for comparison, but the treatment brought the metabolic clearance down to the control level. There was no information on the ethanol consumption during the cyanidanol treatment period. Vesell et al.[63] gave ethanol (22 mmol/kg/d) to volunteers for 3 weeks. In all these subjects, antipyrine half-life was decreased, ranging from 4 to 37%.

Tolbutamide is metabolized by oxidation of the *p*-methyl group. Kater et al.[64] (1969) found a decreased elimination half-life of tolbutamide in alcoholic subjects in the absence of ethanol, suggestive of enzyme induction. There was little overlap between alcoholic subjects and controls with respect to tolbutamide half-life. Carulli et al.[34] studied the elimination of tolbutamide in chronic alcoholics and found a 40% shorter elimination half-life in these subjects compared to controls. Kostelnik and Iber[65] found that ethanol withdrawal in alcoholics more than doubled the elimination half-life of tolbutamide after 6 weeks of abstinence compared to 1 week of abstinence. The findings were confirmed in a later study.[66]

Rubin and Lieber[67] gave ethanol (42% of the dietary energy) to volunteers for 12 d and measured pentobarbital hydroxylase activity in liver biopsy material *in vitro* before and after the ethanol feeding period. The hydroxylase activity was doubled by the treatment. Misra et al.[68] found that chronic ethanol consumption for 1 month decreased pentobarbital elimination half-life by 25% and meprobamate elimination half-life by 50%.

## 4. Aromatic Hydroxylation

Kater et al.[64] found a decreased elimination half-life of phenytoin in alcoholic subjects in the absence of ethanol. There was considerable overlap between alcoholic subjects and controls with respect to phenytoin disappearance. Iber[66] also found accelerated disappearance of this drug after ethanol withdrawal in chronic alcoholic patients. Sandor et al.[69] found some evidence of acceleration of disappearance of this drug by ethanol, but could not exclude autoinduction by phenytoin. Kater et al.[64] found decreased elimination half-life of warfarin in alcoholic subjects in the absence of ethanol. There was considerable overlap between alcoholic subjects and controls with respect to warfarin disappearance. Pelkonen and Sotaniemi[55] found normal or elevated values of coumarin 7-hydroxylase activity and benzo[a]pyrene 3- and 9-hydroxylase ("aryl hydrocarbon hydroxylase") activity in alcoholics with normal liver histology, and decreased values in patients with pathological biopsies.

## 5. Conclusions

Chronic ethanol intake appears to increase the metabolism of drugs that are being metabolized by a variety of pathways. Most of the changes are consistent with enzyme induction. Such induction could explain the kinetic changes found for diazepam, meprobamate, pentobarbital, phenytoin, antipyrine, theophylline, tolbutamide, and warfarin. In clinical situations it is likely that enzyme induction and acute inhibition by ethanol may operate together so that the reproducibility of drug concentration and drug effect from dose to dose may be different. Under these circumstances the treatment effect may be very unpredictable.

Chronic consumption of ethanol may induce permanent changes in the isoenzyme composition of cytochrome P-450. In rats, ethanol and some other low molecular weight xenobiotics have been shown to induce similar P-450 changes,[70,71] later identified as formation of a specialized cytochrome P-450, namely, isoenzyme j.[71] The overall effect of such induction on drug metabolism *in vivo* will be highly dependent on the turnover number of the isoenzyme for each particular drug tested, and these are not known yet for human ethanol-induced cytochrome P-450.

The mechanism(s) operating in human beings *in vivo* are unknown, but it can be expected that in a few years the progress in purification and characterization of human P-450 isoenzymes will make ethanol-drug interactions more predictable and easier to identify. Work along such lines has already been carried out on the interaction between quinidine and other drugs.[72] It can be expected that similar work with ethanol interactions may be carried out in the near future.

## III. ETHANOL INTERACTION WITH DRUG CONJUGATION

### A. Ethanol Interaction with Glucuronidation

A drug may conjugate either directly or after phase I metabolism with glucuronic acid in order to make the substance more polar and more readily excreted from the body. The reaction requires energy, and UDP-glucuronic acid is an obligate cofactor. Glucuronidation is the most important conjugation reaction in man, and the liver is considered to be the main reaction site. Examples of drugs which are glucuronidated are propylthiouracil, trichloroethanol, oxazepam, lorazepam, paracetamol, and opiates.

### 1. Effects of Acute Ethanol Intake

Studies *in vitro* have shown that ethanol may inhibit the hepatic glucuronidation of several compounds. The glucuronidation of $p$-nitrophenol, harmol, 2-naphtolol, 4-methylumbelliferone, phenolphthaleine, and morphine was inhibited by the addition of ethanol ($>10$ m$M$) to isolated hepatocytes.[73-75] The mechanism behind the inhibition seen *in vitro* seems to be decreased availability of UDP-glucuronic acid as a result of inhibition at the level of the UDP-glucose dehydrogenase reaction, resulting from the oxidation of ethanol by alcohol dehydrogenase (ADH).[76,77]

There are few experimental data on the acute effects of ethanol on the glucuronidation of drugs in humans. Some of the interaction data which refer only to the disappearance of parent compounds from the body may reflect ethanol interaction with glucuronidation. The actual effect on glucuronidation, however, is impossible to estimate when the conjugate has not been measured. The suggestions from such limited experiments seem to be that acute ethanol has little influence on drug glucuronidation in man.[78,79]

After propylthiouracil administration (300 mg orally), maximum plasma concentration, time to maximum plasma concentration, and the areas under the concentration vs. time curve for total or free drug were not altered in healthy subjects by a low ethanol dose (approximately 6.5 mmol/kg) given with the drug or 2 h after drug intake.[80] Pretreatment with ethanol (20 mmol/kg followed by 2 mmol/kg/30 min to keep the BAC between 17 and 22 m$M$) did not alter propylthiouracil disposition either.

Trichloroethanol, a metabolite of chloral hydrate and trichloroethylene, is mainly excreted after glucuronidation. Trichloroethanol is possibly oxidized to trichloroacetic acid. Sellers et al.[81] found that the plasma concentrations and urinary excretion of trichloroethanol glucuronide were decreased during ethanol exposure, although the AUC for trichloroethanol in plasma was increased by approximately 25%. In this experiment ethanol (11 mmol/kg) was given 30 min after the administration (cross over) of chloral hydrate (15 mg/kg) to healthy subjects. In a crossover study including the measurement of both trichloroethanol and its conjugate, volunteers inhaled 100 ppm trichloroethylene. This was done in the absence and presence of ethanol exposure (about 13 m$M$ BAC).[33] The trichloroethanol glucuronide calculated as percentage of trichloroethanol was approximately doubled during the ethanol exposure. The authors concluded that they have ruled out the possibility of substantial inhibition of trichloroethanol glucuronidation in the presence of ethanol.

In a crossover study with healthy volunteers, there was no evidence of altered glucuronidation of codeine or morphine after the administration of codeine (1 mg/kg) in the presence of approximately 18 m$M$ ethanol.[164]

The investigations described above[33,80,81,164] and[82] below represent the only studies in humans where both the free and glucuronidated drug fractions have been estimated. In the other studies with humans described below the glucuronides were not measured.

Oxazepam is a benzodiazepine which has mainly glucuronides as metabolites. In a crossover study with healthy subjects,[83] oxazepam (30 mg) was administered orally in the absence and presence of ethanol (11 mmol/kg). It was found that the disposition of oxazepam was not altered in the presence of ethanol. A similar result was obtained by administering ethanol (16 mmol/kg) 30 min prior to or simultaneously with oxazepam (30 mg).[82] In this study it was also found that the rate of conjugation was similar during ethanol exposure.

Another benzodiazepine which is inactivated mainly by conjugation is lorazepam. Ethanol (17 mmol/kg, followed by 11 mmol/kg every 5 h) reduced intravenously administered lorazepam (2 mg) clearance by 18% without affecting the elimination half-life, plasma protein binding, or distribution volume.[84] In the same study the effects of ethanol on the pharmacokinetics of lorazepam were investigated in dogs. In dogs, contrary to humans, lorazepam is a high-clearance and high-extraction drug. In dogs the systemic availability of lorazepam increased as a consequence of a 52% reduction in presystemic elimination when ethanol was administered acutely. The total systemic clearance was, however, not inhibited by ethanol in dogs.

Of a therapeutic paracetamol dose, 40 to 60% is excreted as glucuronides in the urine.[85] Paracetamol disposition (1 g administered) was investigated in a crossover design in healthy men given (estimated dose) a 20% (12 mmol/kg) or a 40% (24 mmol/kg) ethanol solution.[86] Significant alterations in paracetamol pharmacokinetics were observed only with the high ethanol dose. AUC was increased by 23% and the elimination rate constant during the β-phase, decreased by 80% indicating inhibition of paracetamol metabolism.

In one study with rats, increased recovery of conjugate in the urine was possibly due to shunting from an oxidative pathway to glucuronidation in the presence of ethanol.[87] When 4-hydroxy-antipyrine was given orally (20 mg/kg) with ethanol (52 mmol/kg), the total urinary excretion of 4-hydroxy-antipyrine glucuronide was slightly increased. The total urinary excretion of norantipyrine glucuronide was not altered after oral norantipyrine (20 mg/kg) with ethanol (52 mmol/kg) compared to controls receiving norantipyrine only.

### 2. Effects of Chronic Ethanol Intake

In a study with male alcoholic cirrhotic patients given oxazepam (30 mg), the half-life of the drug did not significantly differ from the half-life observed with healthy controls.[88] In a study with perfused rat livers after chronic ethanol administration, diminished rates of glucuronidation and UDP-glucuronic acid levels were observed.[89] These findings could not be explained by either alterations in NAD$^+$/NADH ratio, UDP-glucose, or activity of UDP-glucose dehydrogenase.

The biliary excretion of paracetamol and its glucuronide conjugate was decreased in chronically ethanol-fed rats,[90] whereas the paracetamol levels in the livers were similar, suggesting that paracetamol was metabolized to products not measured in the study. The urinary excretion of paracetamol and its conjugates was similar in the ethanol-fed rats and controls.

### 3. Conclusions

The acute ethanol inihibition of glucuronidation observed *in vitro* is not obvious in humans *in vivo*. The diminished rates of glucuronidation observed *in vitro* after chronic ethanol administration to rats have not been observed in man or animals *in vivo*.

## B. Ethanol Interaction with Sulfation

Sulfation is a conjugation reaction in liver cytosol which transfers sulfate from 3-phosphoadenosine-5-phosphosulfate to the alcoholic or phenolic groups of a drug. Examples of drugs which are sulfated are opiates, paracetamol, and 7-hydroxycoumarin.

### 1. Effects of Acute Ethanol Intake

We are not aware of human data concerning acute ethanol interaction with sulfation. The rate of harmol sulfation (200 $\mu M$) was unaffected by the presence of ethanol (10 m$M$) in isolated rat hepatocytes.[91]

### 2. Effects of Chronic Ethanol Intake

Human studies of drug sulfation after chronic ethanol exposure seem to be lacking. In rats fed ethanol chronically the biliary excretion of paracetamol sulfate was reduced, but the urinary excretion of paracetamol sulfate showed no significant difference compared to pair-fed control rats.[90] Using 7-hydroxycoumarin as substrate, perfused livers from rats treated chronically with ethanol showed decreased rates of sulfation and concentrations of 3-phosphoadenosine-5-phosphosulfate compared to controls.[89]

### 3. Conclusions

To our knowledge there are no data available on ethanol effects on drug sulfation in man.

## C. Ethanol Interaction with Acetylation

Several drugs (hydralazine, isoniazid, procainamide, certain sulfonamides, dapsone) are conjugated with acetate in the liver.[92] Acetate is derived from acetyl CoA and the reaction is catalyzed by the enzyme *N*-acetyl transferase, located in the cytosol. It has been shown, at least in rat liver homogenate, that the reaction rate is dependent on the concentration of the cosubstrate acetate, when varied within physiological limits.[93,94] The acetylated products are generally considered to be less pharmacologically active than the parent drugs, although some exceptions are known. Some side effects of drugs known to be acetylated have been linked to their acetylated metabolites.[92,95,96]

### 1. Effects of Acute Ethanol Intake

Lester[97] demonstrated that the half-life of isoniazid was decreased (30%) in the presence of ethanol (blood alcohol concentrations of 5 to 10 m$M$) in two test subjects. A similar observation was made by Evans et al.[98] in one subject. No clear conclusions with regard to involvement of the N-acetylation rate could, however, be drawn, since other pathways of isoniazid metabolism exist and the amount of acetylated products was not measured.

In a study on 16 healthy volunteers using a crossover design, it was found that the presence of ethanol (approximate blood concentration 20 m$M$) for 10 h decreased sulfadimidine half-life by about 20%.[99] It was further shown that the clearance of the parent drug most probably increased. The percentage of acetylated sulfadimidine also increased after ethanol intake. The mechanism underlying this effect was found to probably be an increase of the hepatic acetyl-CoA concentration,[93,94] brought about by ethanol oxidation to acetate.[100-102] Procainamide pharmacokinetics was also found to be influenced by ethanol intake in healthy volunteers.[103] When ethanol was given either 1.5 h after oral intake of procainamide, or 2 h before (in both cases followed by additional ethanol intake for some hours to keep the BAC close to 20 m$M$), total procainamide clearance increased, most markedly in slow acetylators (by 52%). The mean rate of the metabolic clearance increased almost fourfold after ethanol intake, while renal clearance, responsible for the main fraction of the total clearance, was unchanged. The bioavailability of procainamide determined from the area under the concentration vs. time curve was somewhat reduced after ethanol. The corresponding area for the acetylated metabolite, $N$-acetylprocainamide (NAPA), increased after ethanol, as did the serum concentration of NAPA and the percentage of NAPA measured in blood and urine. Ethanol intake did not influence drug distribution or renal clearance and it was concluded that the underlying mechanism was an increased rate of acetylation, as discussed above.[103]

The ratio of monoacetyl dapsone to dapsone in a single plasma sample taken 3 h after an oral dose was increased in ten (five slow and five fast acetylators) and reduced in two (one slow and one fast acetylator) volunteers after ethanol intake (17 mmol/kg followed at hourly intervals by 2.5 mmol/kg for 3 h). The mean BAC was 18 m$M$.[104] From the figures given by the authors it can be calculated that the mean relative increase of the acetylation ratio was 19% ($P < 0.01$) after ethanol when data from fast and slow acetylators were combined.

## 2. Effects of Chronic Ethanol Intake

There appear so far to be no published studies on the effects of long-term ethanol intake on drug acetylation.

## 3. Conclusions

There is evidence that acute ethanol intake increases the rate of acetylation of at least three different drugs in man.

## D. Ethanol Interaction with Glutathione Conjugation

Glutathione conjugation with the formation of mercapturic acid is of minor importance in man, but appears to be more important as a route of detoxification in rats and some other species.[105] Hepatic glutathione production has been found to be very active even in the absence of stress on the glutathione pool,[106] but when the available pool of cellular glutathione has been depleted, as in the case of a paracetamol overdosage, cellular damage may occur by several mechanisms, the details of which are mainly unknown.[107]

To our knowledge, direct effects of ethanol on the enzymic reaction of glutathione conjugation have not been described. It is difficult to distinguish between effects on glutathione conjugation, oxidative reactions leading to formation of reactive metabolites and glutathione depletion, and effects on glutathione metabolism. For more details on glutathione metabolism and ethanol-glutathione interactions see this volume, Chapter 9.

## IV. ETHANOL INTERACTION WITH DRUG REDUCTION

Reductions in drug biotransformation are less common than oxidations. Only microsomal reductions will be discussed in this paragraph (NADPH-dependent in contrast to the NADH-dependent reductions mediated by alcohol dehydrogenase).

The most common microsomal reductions are nitro reductions as exemplified by $p$-nitrobenzoic acid and nitrazepam; azo reductions as seen with, for instance, salazosulfamide; and reductive dehalogenations as exemplified by carbon tetrachloride and halothane. To our knowledge, only few studies with humans concerning the acute and chronic effects of ethanol on reductions have been performed.

## A. Effects of Acute Ethanol Intake
### 1. Nitro Reductions
Similar to the effect caused by anoxia, ethanol increased the reductive metabolism of *p*-nitrobenzoate (2 mm) and nitrazepam (1 m$M$) in perfused rat livers.[108] Ethanol oxidation appears to be a prerequisite for the enhanced reduction rate, since maximum inhibition was reached at 10 m$M$ ethanol and the effect was abolished with pyrazole (4 m$M$).

### 2. Azo Reductions
The reductive splitting of the azo bond of salazosulfamide (1 m$M$) was only slightly increased in the presence of ethanol (38 m$M$) in perfused rat liver.[108]

### 3. Reductive Dehalogenations
There have been several investigations reporting that short-term ethanol intake may increase the hepatotoxicity of carbon tetrachloride in man and rodents.[109-111] It has been shown that the hepatotoxicity of carbon tetrachloride (0.1 to 0.5 ml/kg i.p.) in rats was enhanced by ethanol (130 mmol/kg p.o.) administered 6 h before the halocarbon.[112] The metabolism of carbon tetrachloride in microsomal preparations was increased following the pretreatment with ethanol (87 mmol/kg p.o.) 18 h previously.[113] Enhanced hepatotoxicity of carbon tetrachloride was also observed when ethanol was given in the same manner. While intragastrically administered ethanol (43 mmol/kg) increased the early (3 to 24 h) blood concentrations of carbon tetrachloride in rats, i.p. administration decreased the early blood concentrations of carbon tetrachloride when administered together with the halocarbon.[114] Both routes of ethanol administration led to increased carbon tetrachloride toxicity, indicating that mechanisms other than the actual level of carbon tetrachloride mediate the potentiation of carbon tetrachloride hepatotoxicity by ethanol. These mechanisms are thought to be the generation of free radicals by a cytochrome P-450-dependent step and a direct solvent injury which is not prevented by inhibitors of carbon tetrachloride metabolism.[115]

## B. Effects of Chronic Ethanol Intake
Chronic ethanol in low doses (91 to 180 mmol/kg/24 h) and high doses (approximately 260 mmol/kg/24 h) increased the hepatotoxicity of carbon tetrachloride in rats.[116,117] Cytochrome P-450 from rats pretreated with ethanol (260 mmol/kg/24 h) was more susceptible to destruction by carbon tetrachloride in microsomal preparations than phenobarbital-inducible and β-naphthoflavone-inducible cytochrome P-450.[118] The possible cause of the enhanced hepatotoxicity of carbon tetrachloride, when combined with ethanol chronically or acutely, is the increased formation of toxic metabolites resulting from ethanol induction of cytochrome P-450.

It has been found that chronic ethanol administration to rats enhances halothane toxicity.[119] Recent investigations suggested that the anaerobic pathway of halothane metabolism was exaggerated by microsomal induction with ethanol.[120]

## C. Conclusions
No systematic studies on xenobiotic reduction in man have been performed.

# V. ETHANOL INTERACTION WITH DRUGS METABOLIZED VIA ALCOHOL DEHYDROGENASE

Ethanol is metabolized to acetaldehyde by a reaction catalyzed by ADH. ADH in the liver is localized in the cytosol and the coenzyme in the oxidative reaction is NAD$^+$ which is reduced to NADH. Alcohols apart from ethanol may be metabolized via ADH,[121,122] together with some other exogenous compounds.[123,124] Chloral hydrate may be converted to trichloroethanol by means of ADH,[125] although other enzymes might catalyze this reaction.[126,127] ADH is also found outside the liver, specially in the stomach, intestine,[128] and kidney.[129]

## A. Effects of Acute Ethanol Intake

The intake of ethanol, 2.3 mol (100 g)/d or higher, inhibited the development of acidosis following metabolism of consumed methanol.[130] Other studies in humans[131-133] more directly demonstrated impaired disappearance of methanol present in the blood at concentrations of 10 to 30 m$M$ in subjects in whom ethanol was present at concentrations above 15 to 20 m$M$. At lower ethanol concentrations the elimination rate of methanol increased.[133] Jacobsen et al.[134] reported that the acid metabolite of methanol, formic acid, did not accumulate in a patient having a blood methanol concentration of 15 m$M$ when he was given ethanol. The BAC was 12 m$M$ or higher during the treatment period of 72 h, during which the blood methanol concentration declined to 6 m$M$.

Since methanol metabolites are responsible for the life-threatening methanol intoxication, ethanol treatment has been included in the therapy.[135] The mechanism underlying this effect is thought to be competitive inhibition by ethanol of ADH-mediated methanol oxidation.[136] The conversion of methanol to formaldehyde can be accomplished by ADH *in vitro*.[137,138] Other enzymes might also catalyze the reaction, for example, the microsomal alcohol-oxidizing system[139] and catalase.[140] Metabolism via ADH is probably most important *in vivo*, as indicated by studies with 4-methyl pyrazole which blocked methanol metabolism.[141]

Infusion of ethanol to two patients appeared to inhibit oxaluria resulting from ethylene glycol metabolism.[142] The half-life of ethylene glycol was prolonged from approximately 3 to 17 h when ethanol was given (BACs ranging from 28 to 44 m$M$).[143] Since ethylene glycol metabolites are responsible for the serious effects in this poisoning, ethanol treatment constitutes an important part of the therapy.[142,143] Animal studies[144] have also demonstrated marked protective effects of ethanol on ethylene glycol toxicity in monkeys and rats. In the same studies the recoveries of unmetabolized ethylene glycol from the urine increased after ethanol treatment. The mechanism underlying these effects is considered to be inhibition by ethanol of ethylene glycol metabolism by ADH. ADH might oxidize ethylene glycol to the corresponding aldehyde, glycolaldehyde.[145] Studies *in vitro* have shown that ethanol inhibits the oxidation of ethylene glycol by horse liver ADH, beef liver catalase, and crude rat liver homogenates.[146] Studies in mice and rats have demonstrated that pyrazole and 4-methyl-pyrazole reduced ethylene glycol toxicity in mice and rats, pointing to a critical role of ADH-mediated metabolism in the toxic effects of ethylene glycol.[147,148]

Kaplan et al.[149] found increased blood levels of trichloroethanol after oral chloral hydrate administration as a result of ethanol administration to humans (peak BACs ranging from 17 to 35 m$M$). Even with a rather sensitive gas chromatographic method, chloral hydrate could not be detected in any blood sample indicating rapid conversion to trichloroethanol. Ingestion of ethanol (11 mmol/kg) 30 min after chloral hydrate (peak blood concentration about 17 m$M$) resulted in a significantly increased AUC of trichloroethanol in human volunteers.[81] The concentration of trichloroethanol glucuronide was decreased both in plasma and urine, as were the concentrations of trichloroacetic acid in plasma. Chloral hydrate simultaneously increased the peak blood levels of ethanol by about 15%. Since trichloroethanol is considered to be as effective as chloral hydrate as a sedative and since it has a longer half-life than chloral hydrate, the metabolic interaction with ethanol might at least partly explain the potent combination effect of chloral hydrate and ethanol, "Mickey Finn". The mechanisms contributing to the increased concentrations of trichloroethanol have been considered to be enhanced formation from chloral hydrate via ADH,[81] as well as reduced metabolism of trichloroethanol.[81,150] The first of these mechanisms would involve increased reduction of chloral hydrate to trichloroethanol by the ADH-NADH complex which would be formed as a consequence of ethanol metabolism.

Trichloroethylene is metabolized to chloral hydrate, and interactions with ethanol via ADH might thus be expected also for this drug. Müller et al.[33] found that ethanol intake (BACs about 10 to 15 m$M$) increased trichloroethylene plasma levels after inhalation, while the concentrations of trichloroethanol and trichloroacetic acid decreased. Accordingly, it can be assumed that possible interactions at the ADH step would be of less importance in this interaction.

Based on *in vitro* and animal experiments, a few studies in man, as well as theoretical considerations, it has been suggested that ethanol might also inhibit the metabolism of other drugs by a competitive mechanism involving ADH. Such reactions are the oxidation of isopropanol,[151] the conversion of the tetrahydrocannabinol 11-hydroxy metabolite to the corresponding tetrahydrocannabinol acid,[152] the first oxidation step in the inactivation of digitalis glycosides,[153] transfer of xylene and toluene benzyl alcohol metabolites to the corresponding aldehydes,[32,154] the metabolism of ADH-inhibiting pyrazoles,[155] and the oxidation of retinol to retinaldehyde.[156]

## B. Effects of Chronic Ethanol Intake

The effects of chronic ethanol intake on human liver ADH are uncertain. In some cases increased enzyme activity has been reported.[157] The importance of this to other drugs metabolized via ADH in liver has not been settled. Of possible toxicological importance is the induction of ADH in distal colon and rectum by ethanol feeding,[158] an observation which has been linked to increased incidence of colorectal cancer in daily drinkers.[159]

## C. Conclusions

Acute intake of ethanol may impair the metabolism of methanol and ethylene glycol in man by competition for ADH. On the other hand, acute ethanol intake may increase the reduction of chloral hydrate to trichloroethanol by a mechanism also linked to ADH.

## VI. FINAL COMMENTS

The purpose of our presentation has been to review metabolic interactions between ethanol and drugs. When reviewing the literature on metabolic interactions of ethanol and drugs, we obtained the impression that this field has not been systematically investigated. The studies have been initiated by groups with special interests, as a consequence of anecdotal evidence or for some other coincidental reason. Many unreported interactions might, thus, exist.

This overview has, only to a very limited extent, dealt with ethanol-drug interactions in the CNS. From a clinical point of view such interactions might, however, be the most important ones. Ethanol and drug metabolism in the CNS is usually not considered to be of importance to its central nervous effects, although some observations might indicate important local CNS metabolism.[129,160-162] One could speculate that metabolic interactions between ethanol and drugs at this location could be possible and, thus, be of importance to some of the known CNS interactions between ethanol and drugs. Further research might clarify whether such interactions exist.

# REFERENCES

1. **Vesell, E. S.,** Pharmacogenetic perspectives: genes, drugs and diseases, *Hepatology,* 4, 959, 1984.
2. **Kaminsky, L. S., Dunbar, D. A., Wang, P. P., Beaune, P., Larrey, D., Guengerich, F. P., Schnellmann, R. G., and Sipes, I. G.,** Human cytochrome P-450 composition as probed by in vitro microsomal metabolism of warfarin, *Drug Metab. Dispos.,* 12, 470, 1984.
3. **Beaune, P. H., Kremers, P. G., Kaminsky, L. S., de Graeve, J., Albert, A., and Guengerich, F. P.,** Comparison of monooxygenase activities and cytochrome P-450 concentrations in human liver microsomes, *Drug Metab. Dispos.,* 14, 437, 1986.
4. **White, R. E. and Coon, M. J.,** Oxygen activation by cytochrome P-450, *Annu. Rev. Biochem.,* 49, 315, 1980.
5. **White, R. E. and Coon, M. J.,** Heme ligand replacement reactions of cytochrome P-450. Characterization of the bonding atom of the axial ligand trans to thiolate as oxygen, *J. Biol. Chem.,* 257, 3073, 1982.
6. **Schenkman, J. B., Sligar, S. G., and Cinti, D. L.,** Substrate interaction with cytochrome P-450, *Pharmacol. Ther.,* 12, 43, 1981.
7. **Schenkman, J. B. and Gibson, G. G.,** Status of the cytochrome P-450 cycle, *Trends Pharmacol. Sci.,* 2, 150, 1981.
8. **Muhoberac, B. B., Roberts, R. K., Hoyumpa, A. M., and Schenker, S.,** Mechanism(s) of ethanol-drug interaction, *Alcoholism Clin. Exp. Res.,* 8, 583, 1984.
9. **Sellers, E. M., Hamilton, C. A., Kaplan, H. L., Degani, N. C., and Folz, R. L.,** Pharmacokinetic interaction of propoxyphene with ethanol, *Br. J. Clin. Pharmacol.,* 19, 398, 1985.
10. **Sellers, E. M., Naranjo, C. A., Giles, H. G., Frecker, R. C., and Beeching, M.,** Intravenous diazepam and oral ethanol interaction, *Clin. Pharmacol. Ther.,* 28, 638, 1980.
11. **Hayes, S. L., Pablo, G., Radomski, T., and Palmer, R. F.,** Ethanol and oral diazepam absorption, *N. Engl. J. Med.,* 296, 186, 1977.
12. **MacLeod, S. M., Giles, H. G., Patzalek, G., Thiessen, J. J., and Sellers, E. M.,** Diazepam actions and plasma concentrations following ethanol ingestion, *Eur. J. Clin. Pharmacol.,* 11, 345, 1977.
13. **Laisi, U., Linnoila, M., Seppala, T., Himberg, J. J., and Mattila, M. J.,** Pharmacokinetic and pharmacodynamic interactions of diazepam with different alcoholic beverages, *Eur. J. Clin. Pharmacol.,* 16, 263, 1979.
14. **Lane, E. A., Guthrie, S., and Linnoila, M.,** Effects of ethanol on drug metabolite pharmacokinetics, *Clin. Pharmacokinet.,* 10, 228, 1985.
15. **Linnoila, M., Otterström, S., and Anttila, M.,** Serum chlordiazepoxide, diazepam and thioridazine concentrations after the simultaneous ingestion of alcohol or placebo drink, *Ann. Clin. Res.,* 6, 4, 1974.
16. **Desmond, P. V., Patwardan, R. V., Schenker, S., and Hoyumpa, A. M.,** Short-term ethanol administration impairs the elimination of chlordiazepoxide (Librium®) in man, *Eur. J. Clin. Pharmacol.,* 18, 275, 1980.
17. **Taeuber, K., Badian, M., Brettel, H. F., Royen, T. H., Rupp, W., Sittig, W., and Uihlein, M.,** Kinetic and dynamic interaction of clobazam and alcohol, *Br. J. Clin. Pharmacol.,* 7, 91S, 1979.
18. **Ochs, H. R., Greenblatt, D. J., Verburg-Ochs, B., Harmatz, J. S., and Grehl, H.,** Disposition of clotiazepam: influence of age, sex, oral contraceptives, cimetidine, isoniazid and ethanol, *Eur. J. Clin. Pharmacol.,* 26, 55, 1984.
19. **Dorian, P., Sellers, E. M., Warsh, J. J., Reed, K. L., Hamilton, C., and Fan, T.,** Decreased hepatic first-pass extraction of oral drugs: mechanism of ethanol-amitriptyline interaction, *Clin. Pharmacol Ther.,* 31, 219, 1982.
20. **Dorian, P., Sellers, E. M., Carruthers, G., Hamilton, B., Kaplan, H. L., and Fan, T.,** Amitriptyline and ethanol: pharmacokinetic and pharmacodynamic interaction, *Eur. J. Clin. Pharmacol.,* 25, 325, 1983.
21. **Warrington, S. F. J., Ankier, S. I., and Turner, P.,** An evaluation of possible interactions between ethanol and trazodone and amitriptyline, *Br. J. Clin. Pharmacol.,* 18, 549, 1984.
22. **Naranjo, C. A., Sellers, E. M., Kaplan, H., Hamilton, C., and Khouw, V.,** Acute kinetic and dynamic interactions of zimelidine with ethanol, *Clin. Pharmacol. Ther.,* 36, 654, 1984.
23. **Olsen, H., Bredesen, J. E., and Lunde, P. K. M.,** Effect of ethanol intake on disopyramide elimination by healthy volunteers, *Eur. J. Clin. Pharmacol.,* 25, 103, 1983.
24. **Cushman, P., Kreek, M. J., and Gordis, E.,** Ethanol and methadone in man: a possible drug interaction, *Drug Alcohol Dependence,* 3, 35, 1978.
25. **Jaffe, J. H. and Martin, W. R.,** Opioid analgesics and antagonists, in *Goodman and Gilman's The Pharmacological Basis of Therapeutics,* 7th ed., Goodman Gilman, A., Goodman, L. S., Rall, T., and Murad, F., Eds., Macmillan, New York, 1985, chap. 22.
26. **Dorian, P., Sellers, E. M., Kaplan, H. L., Hamilton, C., Greenblatt, D. J., and Abernethy, D.,** Triazolam and ethanol interaction: kinetic and dynamic consequences, *Clin. Pharmacol. Ther.,* 37, 558, 1985.

27. **Eberts, F. S., Philopoulos, Y., Reineke, L. M., and Vliek, R. W.,** Triazolam disposition, *Clin. Pharmacol. Ther.,* 29, 81, 1981.

28. **Rubin, E., Gang, H., Misra, P. S., and Lieber, C. S.,** Inhibition of drug metabolism by acute ethanol intoxication. A hepatic microsomal mechanism, *Am. J. Med.,* 49, 801, 1970.

29. **Sellers, E. M. and Holloway, M. R.,** Drug kinetics and alcohol ingestion, *Clin. Pharmacokinet.,* 3, 440, 1978.

30. **Riihimäki, V., Savolainen, K., Pfäffli, P., Pekari, K., Sippel, H. W., and Laine, A.,** Metabolic interaction between m-xylene and ethanol, *Arch. Toxicol.,* 49, 253, 1982.

31. **Wallén, M., Näslund, P. H., and Byfält Nordquist, M.,** The effects of ethanol on the kinetics of toluene in man, *Toxicol. Appl. Pharmacol.,* 76, 414, 1984.

32. **Waldron, H. A., Cherry, N., and Johnston, J. D.,** The effects of ethanol on blood toluene concentrations, *Int. Arch. Occup. Environ. Health,* 51, 365, 1983.

33. **Müller, G., Spassowski, M., and Henschler, D.,** Metabolism of trichloroethylene in man, *Arch. Toxicol.,* 33, 173, 1975.

34. **Carulli, N., Manenti, F., Gallo, M., and Salvioli, G. F.,** Alcohol-drug interactions in man: alcohol and tolbutamide, *Eur. J. Clin. Invest.,* 1, 421, 1971.

35. **Woodbury, D. M.,** Analgesic-antipyretics, antiinflammatory agents, and inhibitors of uric acid synthesis, in *The Pharmacological Basis of Therapeutics,* 4th ed., Goodman, L. S. and Gilman, A., Eds., Collier Macmillan, London, 1970, chap. 17.

36. **Dössing, M. and Buch Andreasen, P.,** Ethanol and antipyrine clearance, *Clin. Pharmacol. Ther.,* 30, 101, 1981.

37. **Staak, M. and Moosmayer, A.,** Pharmakokinetishe Untersuchungen über Wechselwirkungen zwischen Dikaliumchlorazepat and Alkohol nach oraler Applikation, *Arzneim. Forsch.,* 28 (2), 1187, 1978.

38. **Offen, C. P., Frearson, M. J., Wilson, K., and Burnett, D.,** 4,5-Dimethylthiazole-N-oxide-S-oxide: a metabolite of chlormethiazole in man, *Xenobiotica,* 15, 503, 1985.

39. **Neuvonen, P. J., Penttilä, O., Roos, M., and Tirkkonen, J.,** Effect of ethanol on the pharmacokinetics of chlormethiazole in humans, *Int. J. Clin. Pharmacol. Ther. Toxicol.,* 19, 552, 1981.

40. **Bury, R. W., Desmond, P. V., Mashford, M. L., Westwood, B., Shaw, G., and Breen, K. J.,** The effect of ethanol administration on the disposition and elimination of chlormethiazole, *Eur. J. Clin. Pharmacol.,* 24, 383, 1983.

41. **Breckenridge, A. and Orme, M.,** Clinical implications of enzyme induction, *Ann. N.Y. Acad. Sci.,* 179, 421, 1971.

42. **Weiner, N.,** Drugs that inhibit adrenergic nerves and block adrenergic receptors, in *Goodman and Gilman's The Pharmacological Basis of Therapeutics,* 7th ed., Goodman Gilman, A., Goodman, L. S., Rall, T. W., and Murad, F., Eds., Macmillan, New York, 1985, chap. 9.

43. **Pritchard, J. F., Schenck, D. W., and Hayes, A. H.,** The inhibition of rat hepatic microsomal propranolol metabolism by a covalently bound reactive metabolite, *Res. Commun. Chem. Pathol. Pharmacol.,* 27, 211, 1980.

44. **Sotaniemi, E. A., Anttila, M., Rautio, A., Stengård, J., Saukko, P., and Järvensivu, P.,** Propranolol and sotalol metabolism after a drinking party, *Clin. Pharmacol. Ther.,* 29, 705, 1981.

45. **Dorian, P., Sellers, E. M., Carruthers, G., Hamilton, C., and Fan, T.,** Propranolol-ethanol pharmacokinetic interaction, *Clin. Pharmacol. Ther.,* 31, 219, 1982.

46. **Dorian, P., Sellers, E. M., Kaplan, G., Carruthers, G., Hamilton, C., and Khouw, V.,** Ethanol-induced inhibition of hepatic uptake of propranolol in perfused rat liver and man, *Eur. J. Clin. Pharmacol.,* 27, 209, 1984.

47. **van de Straat, R., de Vries, J., Kulkens, T., Debets, A. J. J., and Vermeulen, N. P. E.,** Paracetamol, 3-monoalkyl- and 3,5-dialkyl derivatives. Comparison of their microsomal cytochrome P-450 dependent oxidation and toxicity in freshly isolated hepatocytes, *Biochem., Pharmacol.,* 35, 3693, 1986.

48. **Critchley, J. A. J. H., Dyson, E. H., Scott, A. W., Jarvie, D. R., and Prescott, L. F.,** Is there a place for cimetidine or ethanol in the treatment of paracetamol poisoning?, *Lancet,* 1, 1375, 1983.

49. **Banda, P. W. and Quart, B.,** The effect of mild alcohol consumption on the metabolism of acetaminophen in man, *Res. Commun. Chem. Pathol. Pharmacol.,* 38, 57, 1982.

50. **Harvey, S. C.,** Hypnotics and sedatives, in *Goodman and Gilman's The Pharmacological Basis of Therapeutics,* 6th ed., Goodman Gilman, A., Goodman, L. S., and Gilman, A., Eds., Collier Macmillan, London, 1980, chap. 17.

51. **Roden, S., Harvey, P., and Mitchard, M.,** The effect of ethanol on residual plasma methaqualone concentrations and behaviour in volunteers who have taken Mandrax(®), *Br. J. Clin. Pharmacol.,* 4, 245, 1977.

52. **Chung, H. and Brown, D. R.,** Mechanism of the effect of acute ethanol on hexobarbital metabolism, *Biochem. Pharmacol.,* 25, 1613, 1976.

53. **Thurman, R. G. and Kauffman, F. S.,** Factors regulating drug metabolism in intact hepatocytes, *Pharmacol. Rev.,* 31, 229, 1980.

54. **Lieber, C. S.,** Medical disorders of alcoholism. Pathogenesis and treatment, in *Major Problems in Internal Medicine,* Vol. 12, Smith, L. H., Jr., Ed., W. B. Saunders, Philadelphia, 1982, chap. 7.

55. **Pelkonen, O. and Sotaniemi, E.,** Drug metabolism in alcoholics, *Pharmacol. Ther.,* 16, 261, 1982.

56. **Sellman, R., Pekkarinen, A., Kangas, L., and Rajiola, E.,** Reduced concentrations of plasma diazepam in chronic alcoholic patients following an oral administration of diazepam, *Acta Pharmacol. Toxicol.,* 36, 25, 1975.

57. **Pond, S. M., Phillips, M., Benowitz, N. I., Galinski, R. E., Tong, T. G., and Becker, C. E.,** Diazepam kinetics in acute alcohol withdrawal, *Clin. Pharmacol. Ther.,* 25, 832, 1979.

58. **Sellers, E. M., Sandor, P., Giles, H. G., Khouw, V., and Greenblatt, D. J.,** Diazepam pharmacokinetics after intravenous administration in alcohol withdrawal, *Br. J. Clin. Pharmacol.,* 15, 125, 1983.

59. **Jusko, W. J., Gardner, M. J., Mangione, A., Schentag, J. J., Kopu, J. R., and Vance, J. W.,** Factors affecting theophylline clearances: age, tobacco, marijuana, cirrhosis, congestive heart failure, obesity, oral contraceptives, benzodiazepines, barbiturates, and ethanol, *J. Pharm. Sci.,* 68, 1358, 1979

60. **Hoensch, H. P., Hutzel, H., Kirch, W., and Ohnhaus, E. E.,** Isolation of human hepatic microsomes and their inhibition by cimetidine and ranitidine, *Eur. J. Clin. Pharmacol.,* 29, 199, 1985.

61. **Woodhouse, K. W., Williams, F. M., Mutch, E., Wright, P., James, O. F. W., and Rawlins, M. D.,** The effect of alcoholic cirrhosis on the two kinetic components (high and low affinity) of the microsomal O-deethylation of 7-ethoxycoumarin in human liver, *Eur. J. Clin. Pharmacol.,* 26, 61, 1984.

62. **Páhr, A., Horváth, T., Beró, T., Kádas, I., Pakodi, F., Wittmann, I., and Jávor, T.,** Effect of cyanidanol on hepatic drug biotransformation and elimination in patients with chronic alcoholic hepatitis and cirrhosis, in *Cytochrome P-450, Biochemistry, Biophysics and Induction,* Vereczkey, L. and Magyar, K., Eds., Elsevier, Amsterdam, 1985, 311.

63. **Vesell, E. S., Page, J. G., and Passananti, G. T.,** Genetic and environmental factors affecting antipyrine metabolism in man, *Clin. Pharmacol. Ther.,* 12, 192, 1971.

64. **Kater, R. M. H., Roggin, G., Tobon, F., Zieve, P., and Iber, F. L.,** Increased rate of clearance of drugs from the circulation of alcoholics, *Am. J. Med. Sci.,* 258, 35, 1969.

65. **Kostelnik, M. D. and Iber, F. L.,** Correlation of ethanol and tolbutamide clearance rates with microsomal alcohol-metabolizing enzyme activity, *Am. J. Clin. Nutr.,* 26, 161, 1973.

66. **Iber, F. L.,** Drug metabolism in heavy consumers of ethyl alcohol, *Clin. Pharmacol. Ther.,* 22, 735, 1977.

67. **Rubin, E. and Lieber, C. S.,** Hepatic microsomal enzymes in man and rat: induction and inhibition by ethanol, *Science,* 162, 690, 1968.

68. **Misra, P. S., Lefevre, A., Ishii, H., Rubin, E., and Lieber, C. S.,** Increase of ethanol, meprobamate and pentobarbital metabolism after chronic ethanol administration in man and rats, *Am. J. Med.,* 51, 346, 1971.

69. **Sandor, P., Sellers, E. M., Dumbrell, M., and Klouw, V.,** Effect of short- and long-term alcohol use on phenytoin kinetics in chronic alcoholics, *Clin. Pharmacol. Ther.,* 30, 390, 1981.

70. **Gadeholt, G.,** Ethanol and isoniazid induce a hepatic microsomal cytochrome P-450-dependent activity with similar properties towards substrate and inhibitors and different properties from those induced by classical inducers, *Biochem. Pharmacol.,* 33, 3047, 1984.

71. **Ryan, D. E., Koop, D. R., Thomas, P. E., Coon, M. J., and Levin, W.,** Evidence that isoniazid and ethanol induce the same microsomal cytochrome P-450 in rat liver, an isozyme homologous to rabbit liver cytochrome P-450 isozyme 3a, *Arch. Biochem. Biophys.,* 246, 633, 1986.

72. **Guengerich, F. P., Müller-Enoch, D., and Blair, I. A.,** Oxidation of quinidine by human liver cytochrome P-450, *Mol. Pharmacol.,* 30, 287, 1986.

73. **Moldéus, P., Andersson, B., and Norling, A.,** Interaction of ethanol oxidation with glucuronidation in isolated hepatocytes, *Biochem. Pharmacol.,* 27, 2583, 1978.

74. **Moldéus, P., Vadi, H., and Berggren, M.,** Oxidative and conjugative metabolism of p-nitroanisole and p-nitrophenol in isolated rat liver cells, *Acta Pharmacol. Toxicol.,* 39, 17, 1976.

75. **Bodd, E., Drevon, C. A., Kveseth, N., Olsen, H., and Mørland, J.,** Ethanol inhibition of codeine and morphine metabolism in isolated rat hepatocytes, *J. Pharmacol. Exp. Ther.,* 237, 260, 1986.

76. **Bodd, E., Gadeholt, G., Christensson, P. I., and Mørland, J.,** Mechanisms behind the inhibitory effect of ethanol on the conjugation of morphine in rat hepatocytes, *J. Pharmacol. Exp. Ther.,* 239, 887, 1986.

77. **Aw, T. Y. and Jones, D. P.,** Intracellular inhibition of UDP-glucose dehydrogenase during ethanol oxidation, *Chem. Biol. Interact.,* 43, 283, 1983.

78. **Sellers, E. M. and Busto, U.,** Benzodiazepines and ethanol: assessment of the effects and consequences of psychotropic drug interactions, *J. Clin. Psychopharmacol.,* 2, 249, 1982.

79. **Schüppel, R. V. A.,** Drug metabolism as influenced by ethanol, in *Proc. of the European Society of the Studies of Drug Toxicity,* De Baker, S. B. and Neuhaus, G. A., Eds., Excerpta Medica, Amsterdam, 1972, 13.

80. **Long, J. P., Giles, H. G., Bennett, R. N., Dorian, P., Thiessen, J. J., Orrego, H., and Sellers, E. M.,** Effect of ethanol on propylthiouracil disposition, *Clin. Pharmacol. Ther.,* 33, 663, 1983.

81. **Sellers, E. M., Lang, M., Koch-Weser, J., LeBlanc, E., and Kalant, H.,** Interaction of chloral hydrate and ethanol in man. I. Metabolism, *Clin. Pharmacol. Ther.*, 13, 37, 1972.

82. **Von Mallach, H. J., Moosmayer, A., Gottwald, K., and Staak, M.,** Pharmacokinetic studies on absorption and excretion of oxazepam in combination with alcohol, *Arzneim. Forsch.*, 25, 1840, 1975.

83. **Sellers, E. M., Giles, H. G., Greenblatt, D. J., and Naranjo, C. A.,** Differential effects on benzodiazepine disposition by disulfiram and ethanol, *Arzneim. Forsch.*, 30, 882, 1980.

84. **Hoyumpa, A. M., Patwardhan, R., Maples, M., Desmond, P. V., Johnson, R. F., Sinclair, A. P., and Schenker, S.,** Effect of short-term ethanol administration on lorazepam clearence, *Hepatology*, 1, 47, 1981.

85. **Meredith, T. J. and Goulding, R.,** Paracetamol, *Postgrad. Med. J.*, 56, 459, 1980.

86. **Wójcicki, J., Baśkiewicz, Z., Gawrońska-szklarz, B., Kazimierczyk, J., and Kalucki, K.,** The effect of a single dose of ethanol on pharmacokinetics of paracetamol, *Pol. J. Pharmacol. Pharm.*, 30, 749, 1978.

87. **Schüppel, R. V. A.,** Konjugationsreaktionen im Arzneistoffwechsel der Ratte bei akuter Äthanolbelastung, *Naunyn Schmiedebergs Arch. Pharmacol.*, 265, 233, 1969.

88. **Sellers, E. M., Greenblatt, D. J., Giles, H. G., Narjano, C. A., Kaplan, H., and Macleod, S. M.,** Chlordiazepoxide and oxazepam disposition in cirrhosis, *Clin. Pharmacol. Ther.*, 26, 240, 1979.

89. **Reinke, L. A., Moyer, M. J., and Nothley, K. A.,** Diminished rates of glucuronidation and sulfation in perfused liver after chronic ethanol administration, *Biochem. Pharmacol.*, 35, 439, 1986.

90. **Vendemiale, G., Altomare, E., and Lieber, C. S.,** Altered biliary excretion of acetaminophen in rats fed ethanol chronically, *Drug Metab. Dispos.*, 12, 20, 1984.

91. **Sundheimer, D. W., and Brendel, K.,** Factors influencing sulfation in isolated rat hepatocytes, *Life Sci.*, 34, 23, 1984.

92. **Weber, W. W. and Hein, D. W.,** N-acetylation pharmacogenetics, *Pharmacol. Rev.*, 37, 25, 1985.

93. **Weber, W. W. and Cohen, S. N.,** N-acetylation of drugs: isolation and properties of an N-acetyltransferase from rabbit liver, *Mol. Pharmacol.*, 3, 266, 1967.

94. **Olsen, H.,** Interaction between drug acetylation and ethanol, acetate, pyruvate, citrate, and L(−)carnitine in isolated rat liver parenchymal cells, *Acta Pharmacol. Toxicol.*, 50, 67, 1982.

95. **Drayer, D. E. and Reidenberg, M. M.,** Clinical consequences of polymorphic acetylation of basic drugs, *Clin. Pharmacol. Ther.*, 22, 251, 1977.

96. **Lunde, P. K. M., Frislid, K., and Hansteen, V.,** Disease and acetylation polymorphism, *Clin. Pharmacokinet.*, 2, 182, 1977.

97. **Lester, D.,** The acetylation of isoniazid in alcoholics, *Q. J. Stud. Alcohol*, 25, 541, 1964.

98. **Evans, D. A. P., Manley, K. A., and McKusick, V. A.,** Genetic control of isoniazid metabolism in man, *Br. Med. J.*, 2, 485, 1960.

99. **Olsen, H. and Mørland, J.,** Ethanol-induced increase in drug acetylation in man and isolated rat liver cells, *Br. Med. J.*, 2, 1260, 1978.

100. **Lindros, K. O.,** Interference of ethanol and sorbitol with hepatic ketone body metabolism in normal, hyper- and hypothyroid rats, *Eur. J. Biochem.*, 13, 111, 1970.

101. **Siess, E. A., Brocks, D. G., and Wieland, O. H.,** Subcellular distribution of key metabolites in isolated liver cells from fasted rats, *FEBS Lett.*, 69, 265, 1976.

102. **Lindros, K. O. and Hillbom, M. E.,** Hepatic redox state and ketone body metabolism during oxidation of ethanol and fructose in normal, hyper- and hypothyroid rats, *Ann. Med. Exp. Biol. Fenn.*, 49, 162, 1971.

103. **Olsen, H. and Mørland, J.,** Ethanol-induced increase in procainamide acetylation in man, *Br. J. Clin. Pharmacol.*, 13, 203, 1982.

104. **Hutchings, A., Monie, R. D., Spragg, B., and Routledge, P. A.,** Acetylator phenotyping: the effect of ethanol on the dapsone test, *Br. J. Clin. Pharmacol.*, 18, 98, 1984.

105. **Williams, R. T.,** Detoxication mechanisms in man, *Clin. Pharmacol. Ther.*, 4, 234, 1963.

106. **Lautenberg, B. H., Adams, J. D., and Mitchell, J. R.,** Hepatic glutathione homeostasis in the rat: efflux accounts for glutathione turnover, *Hepatology*, 4, 586, 1984.

107. **Albano, E., Rundgren, M., Harvison, P. J., Nelson, S. D., and Moldéus, P.,** Mechanisms of N-acetyl-p-benzoquinone imine cytotoxicity, *Mol. Pharmacol.*, 28, 306, 1985.

108. **Jonen, H. G.,** Enhancement of reductive metabolism of p-nitrobenzoate and nitrazepam in isolated perfused rat liver by ethanol, *Drug Metab. Dispos.*, 7, 176, 1979.

109. **Stevens, H. and Forster, F. M.,** Effect of carbon tetrachloride on the nervous system, *Arch. Neurol. Psychiatry*, 70, 635, 1953.

110. **Cornish, H. H. and Adefuin, J.,** Potentiation of carbon tetrachloride toxicity by aliphatic alcohols, *Arch. Environ. Health*, 14, 447, 1967.

111. **Traiger, G. J. and Plaa, G. L.,** Differences in the potentiation of carbon tetrachloride in rats by ethanol and isopropanol pretreatment, *Toxicol. Appl. Pharmacol.*, 20, 105, 1971.

112. **Sato, C., Nakano, M., and Lieber, C. S.,** Prevention of acetaminophen-induced hepatotoxicity by acute ethanol administration in the rat: comparison with carbon tetrachloride-induced hepatotoxicity, *J. Pharmacol. Exp. Ther.,* 218, 805, 1981.

113. **Sato, A., Nakajima, T., and Koyama, Y.,** Dose-related effects of a single dose of ethanol on the metabolism in rat liver of some aromatic and chlorinated hydrocarbons, *Toxicol. Appl. Pharmacol.,* 60, 8, 1981.

114. **Teschke, R., Vierke, W., and Gellert, J.,** Effect of ethanol on carbon tetrachloride levels and hepatotoxicity after acute carbon tetrachloride poisoning, *Arch. Toxicol.,* 56, 78, 1984.

115. **Berger, M. L., Bhatt, H., Combes, B., and Estabrook, R. W.,** CCl$_4$-induced toxicity in isolated hepatocytes: the importance of direct solvent injury, *Hepatology,* 6, 36, 1986.

116. **Strubelt, O., Obermeier, F., Siegers, C. P., and Völpel, M.,** Increased carbon tetrachloride hepatotoxicity after low-level ethanol consumption, *Toxicology,* 10, 261, 1978.

117. **Hasumura, Y., Teschke, R., and Lieber, C. S.,** Increased carbon tetrachloride hepatotoxicity, and its mechanism, after chronic ethanol consumption, *Gastroenterology,* 66, 415, 1974.

118. **Gadeholt, G.,** Ethanol-inducible cytochrome P-450 is more susceptible to in vitro carbon tetrachloride-mediated destruction than phenobarbital-inducible and β-naphthoflavone-inducible cytochromes P-450, *Acta Pharmacol. Toxicol.,* 55, 216, 1984.

119. **Takagi, T., Ishii, H., Takahashi, H., Kato, S., Okuno, F., Ebihara, Y., Yamauchi, H., Nagata, S., Tashiro, M., and Tsuchiya, M.,** Potentiation of halothane hepatotoxicity by chronic ethanol administration in rat: an animal model of halothane hepatitis, *Pharmacol. Biochem. Behav.,* 18, (Suppl. 1), 461, 1983.

120. **Takahashi, H., Ishii, H., Takagi, T., Nagata, S., and Tsuchia, M.,** Role of lipid peroxidation in the pathogenesis of halothane hepatitis after chronic ethanol administration in rats, *Alcohol Alcoholism,* 21, A 22, 1986.

121. **Von Wartburg, J. P.,** The metabolism of alcohol in normals and alcoholics: enzymes, in *The Biology of Alcoholism,* Vol. 1, Kissin, B. and Begleiter, H., Eds., Plenum Press, New York, 1971, 63.

122. **Borson, W. F. and Li, T.-K.,** Alcohol dehydrogenase, in *Enzymatic Basis of Detoxication,* Vol. 1, Jakoby, W. B., Ed., Academic Press, New York, 1980, 231.

123. **Okuda, K. and Takigawa, N.,** Rat liver 5β-cholestane-3α, 7α,12α,26-tetrol dehydrogenase as a liver alcohol dehydrogenase, *Biochim. Biophys. Acta,* 22, 141, 1970.

124. **Björkheim, I.,** On the role of alcohol dehydrogenase in ω-oxidation of fatty acids, *Eur. J. Biochem.,* 30, 441, 1972.

125. **Cooper, J. and Friedman, P.,** The role of alcohol dehydrogenase in the metabolism of chloral hydrate, *J. Pharmacol. Exp. Ther.,* 129, 373, 1960.

126. **Tabakoff, B., Vugrinic, C., Anderson, R., and Avisatos, S. G. A.,** Reduction of chloral hydrate to trichloroethanol in brain extracts, *Biochem. Pharmacol.,* 23, 455, 1974.

127. **Schultz, J. W. and Weiner, H.,** Role of ethanol in enhancement of chloral hydrate reduction, in *Currents in Alcoholism,* Vol. 3, Seixas, F., Ed., Grune & Stratton, New York, 1978, 363.

128. **Pestalozzi, D. M., Bühler, R., von Wartburg, J. P., and Hess, M.,** Immunohistochemical localization of alcohol dehydrogenase in the human gastrointestinal tract, *Gastroenterology,* 85, 1011, 1983.

129. **Bühler, R., Pestalozzi, D., Hess, M., and Wartburg, J. P.,** Immunohistochemical localization of alcohol dehydrogenase in human kidney, endocrine organs and brain, *Pharmacol. Biochem. Behav.,* 18 (Suppl. 1), 55, 1983.

130. **Røe, O.,** Clinical investigations of methyl alcohol poisoning with special reference to the pathogenesis and treatment of amblyopia, *Acta Med. Scand.,* 113, 558, 1943.

131. **Agner, K., Höök, O., and von Porat, B.,** The treatment of methanol poisoning with ethanol, *Q. J. Stud. Alcohol,* 9, 515, 1949.

132. **McCoy, H. G., Cipolle, R. J., Ehlers, S. M., Sawchuk, R. J., and Zaske, D. E.,** Severe methanol poisoning. Application of a pharmacokinetic model for ethanol therapy and hemodialysis, *Am. J. Med.,* 67, 804, 1979.

133. **Jacobsen, D., Jansen, H., Wiik-Larsen, E., Bredesen, J. E., and Halvorsen, S.,** Studies on methanol poisoning, *Acta Med. Scand.,* 212, 5, 1982.

134. **Jacobsen, D., Øvrebø, S., Arnesen, E., and Paus, P. N.,** Pulmonary excretion of methanol in man, *Scand. J. Clin. Lab. Invest.,* 43, 377, 1983.

135. **Røe, O.,** The metabolism and toxicity of methanol, *Pharmacol. Rev.,* 7, 399, 1955.

136. **Zatman, L. J.,** The effect of ethanol on the metabolism of methanol in man, *Biochem. J.,* 40, 67, 1946.

137. **Blair, A. H. and Vallee, B. L.,** Some catalytic properties of human liver alcohol dehydrogenase, *Biochemistry,* 5, 2026, 1966.

138. **Makar, A. B., Tephly, T. R., and Mannering, G. J.,** Methanol metabolism in the monkey, *Mol. Pharmacol.,* 4, 471, 1968.

139. **Lieber, C. S. and De Carli, L. M.,** Hepatic microsomal ethanol oxidizing system: in vitro characteristics and adaptive properties in vivo, *J. Biol. Chem.,* 245, 2505, 1970.

140. **Tephly, T. R., Parks, R. E., and Mannering, G. J.,** Methanol metabolism in the rat, *J. Pharmacol. Exp. Ther.,* 143, 292, 1964.

141. **Blomstrand, R., Östling-Wintzell, H., Löf, A., McMartin, K., Tolf, B.-R., and Hedström, K.-G.,** Pyrazoles as inhibitors of alcohol oxidation and as important tools in alcohol research: an approach to therapy against methanol poisoning, *Proc. Natl. Acad. Sci. U.S.A.,* 76, 3499, 1979.

142. **Wacker, W. E. C., Haynes, H., Druyan, R., Fisher, W., and Coleman, J. E.,** Treatment of ethylene glycol poisoning with ethyl alcohol, *J.A.M.A.,* 194, 173, 1965.

143. **Peterson, C. D., Collins, A. J., Himes, J. M., Bullock, M. L., and Keane, W. F.,** Ethylene glycol poisoning. Pharmacokinetics during therapy with ethanol and hemodialysis, *N. Engl. J. Med.,* 304, 21, 1981.

144. **Peterson, D. I., Peterson, J. E., Hardinge, M. G., Linda, L., and Wacker, W. E. C.,** Experimental treatment of ethylene glycol poisoning, *J.A.M.A.,* 186, 169, 1963.

145. **Von Wartburg, J.-P., Bethune, J. L., and Valle, B. L.,** Human liver alcohol dehydrogenase. Kinetic and physiochemical properties, *Biochemistry,* 3, 1775, 1964.

146. **Weiss, B. and Coen, G.,** Effect of ethanol on ethylene glycol oxidation by mammalian liver enzymes, *Enzymol. Biol. Clin.,* 6, 305, 1966.

147. **Mundy, R. L., Hall, L. M., and Teaque, R. S.,** Pyrazole as an antidote for ethylene glycol poisoning, *Toxicol. Appl. Pharmacol.,* 28, 320, 1974.

148. **Chou, J. Y. and Richardson, K. E.,** The effect of pyrazole on ethylene glycol toxicity and metabolism in the rat, *Toxicol. Appl. Pharmacol.,* 43, 33, 1978.

149. **Kaplan, H. L., Forney, R. B., Hughes, F. W., and Jain, N. C.,** Chloral hydrate and alcohol metabolism in human subjects, *J. Forensic Sci.,* 12, 295, 1967.

150. **Deitrich, R. A. and Petersen, D. R.,** Interaction of ethanol with other drugs, in *Medical and Social Aspects of Alcohol Abuse,* Tabakoff, B., Sutker, P. B., and Randau, C. L., Eds., Plenum Press, New York, 1983, 247.

151. **Kelner, M. and Bailey, D. N.,** Isopropanol ingestion: interpretation of blood concentrations and clinical findings, *J. Toxicol. Clin. Toxicol.,* 20, 497, 1983.

152. **Wall, M. E., Sadler, B. M., Brine, D., Taylor, H., and Peres-Reyes, M.,** Metabolism, disposition, and kinetics of delta-9-tetrahydrocannabinol in men and women, *Clin. Pharmacol. Ther.,* 34, 352, 1983.

153. **Frey, W. A. and Vallee, B. L.,** Human liver alcohol dehydrogenase—an enzyme essential to the metabolism of digitalis, *Biochem. Biophys. Res. Commun.,* 91, 1543, 1979.

154. **Patel, J. M., Harper, C., and Drew, R. T.,** The biotransformation of p-xylene to a toxic aldehyde, *Drug Metab. Dispos.,* 6, 368, 1978.

155. **Rydberg, U., Buijten, J., and Neri, A.,** Kinetics of some pyrazole derivatives in the rat, *J. Pharm. Pharmacol.,* 24, 651, 1972.

156. **Raskin, N. H., Sligar, K. P., and Steinberg, R. H.,** A pathophysiologic role for alcohol dehydrogenase: is retinol its "natural" substrate?, *Ann. N.Y. Acad. Sci.,* 273, 317, 1976.

157. **Khanna, J. M. and Israel, Y.,** Ethanol metabolism, in *Liver and Bilary Tract Physiology I. International Review of Physiology,* Vol. 21, Javitt, N. B., Ed., University Park Press, Baltmore, 1980, 275.

158. **Seitz, H. K., Czygan, P., Waldherr, R., Veith, S., and Kommerell, B.,** Ethanol and intestinal carcinogenesis in the rat, *Alcohol,* 2, 491, 1985.

159. **Pollack, E. S., Nomura, A. M. Y., Heilbrun, L. K., Stemmerman, G. N., and Green, S. B.,** Prospective study of alcohol consumption and cancer, *N. Engl. J. Med.,* 310, 617, 1984.

160. **Raskin, N. H. and Solokoff, L.,** Brain alcohol dehydrogenase, *Science,* 162, 131, 1968.

161. **Bridges, J. W.,** The role of the drug-metabolizing enzymes, in *Environmental Chemicals, Enzyme Function and Human Disease,* Ciba Foundation Symp. 76, Excerpta Medica, Amsterdam, 1980, 5.

162. **Hahn, E. F.,** The role of brain metabolism in the action of opiates, *Med. Res. Rev.,* 5, 255, 1985.

163. **Ohnhaus,** personal communication.

164. **Bodd, E., Beylich, K. M., Christophersen, A. S., and Mørland, J.,** Oral administration of codeine in the presence of ethanol: a pharmacokinetic study in man, *Pharmacol. Toxicol.,* 61, 297, 1987.

Chapter 11

## EFFECTS OF ETHANOL ON MEMBRANES AND THEIR ASSOCIATED FUNCTIONS

**John M. Littleton**

### TABLE OF CONTENTS

## I. INTRODUCTION

This book is intended to cover the effects of ethanol on human biochemistry. It would, however, be impossible to cover adequately the effects of ethanol on biological membranes by referring solely to work on human tissues. A large proportion of the work discussed here will refer to eukaryotic cell membranes from species other than human. Occasionally, reference will be to work on prokaryotic or model membranes. This is justified on the grounds that there are many features common to the membranes of all cells and it seems likely that the effects of ethanol on these will be similar across species. It would be unthinkable to omit any consideration of the effects of ethanol on biological membranes from a book of this kind, because our increased knowledge of such effects has contributed enormously to our understanding of mechanisms by which ethanol produces its acute effects and by which cellular adaptation to these is induced. Lastly, to give equal coverage to all cell types would make this chapter unwieldy and repetitive. It has been decided, therefore, to concentrate mainly on neuronal cells, since it is presumably on these that ethanol exerts the effects responsible for intoxication, tolerance, and physical dependence. Most of the comments about neuronal membranes will apply, however, to the plasma membranes of other eukaryotic cell types.

## II. BIOLOGICAL MEMBRANE ORGANIZATION

Thermodynamically, it would be predicted that the lipids present in biological membranes would take up a bilayer conformation with polar headgroups of phospholipids facing the aqueous phase and acyl chains directed toward the hydrophobic interior. This pattern is now generally accepted, as is the Singer-Nicholson "fluid-mosaic" model[1] in which membrane proteins are thought to float on either surface of the bilayer or to span the bilayer. The important characteristic of this model is the fluidity of the lipid bilayer allowing movement of the proteins within the planes of the bilayer and causing little hindrance to changes of membrane protein conformation. In addition to the bilayer pattern of membrane lipid organization, there are other formations which are theoretically likely and for which there is experimental evidence.[2,3] Such alternative patterns include the "hexagonal", or $H_{11}$ phase, and the presence of micellar and inverted micellar phases. The significance of these phases is unknown, but an attractive speculation is that they play a functional role in basic membrane processes such as fusion and exocytosis.[2]

This is, of course, a highly simplified view of membrane lipid organization. It ignores, for example, the effect of cholesterol in the membrane which, by insertion of its planar ring structure, tends to make membrane bilayers more at rigid temperatures above their phase transition. It also ignores the role of glycolipids which probably project far above the surface of the membrane phospholipids, extending the influence of the membrane out into the aqueous phase surrounding it. This simplified picture will suffice to begin a description of the effects of ethanol on biological membrane organization and, in particular, of the ways in which ethanol can influence the physical properties of the membrane.

## III. PARTITIONING OF ETHANOL INTO MEMBRANES

The best correlation between the anesthetic potency of members of an homologous series of alcohols and their chemical or physical properties is supplied by hydrocarbon chain length and lipid solubility.[4] Since the hydrocarbon chain provides the hydrophobic part of the molecule, these parameters are clearly closely related. They suggest that the extent of insertion of the hydrocarbon chain into some hydrophobic region of the cell correlates with anesthetic potency.

Clearly, there are many hydrophobic sites within the cell which could be potential targets for alcohols. Alcohols could, for example, influence the function of specific membrane proteins directly by partitioning into hydrophobic sites within these proteins.[5,6] This is made less likely, but not excluded, by the large variation in chain lengths of alcohols which would need to fit into such a hydrophobic pocket on a protein. Alternatively, alcohols could partition at the lipid/protein interface or into the bulk lipid of the bilayer, influencing protein function indirectly by some physicochemical effect.

Before any discussion of how ethanol affects the membrane can begin, it is necessary to consider how much ethanol actually enters the membrane. First experiments utilized model membranes or lipid or octanol phases[4] and suggest that 10 to 15% of ethanol is likely to enter membranes, whereas 85 to 90% remains in the aqueous phase. Similar values were obtained when biological membranes were examined by centrifugation techniques.[4,7] More recently,[8] doubt has been cast on the validity of such techniques for establishing partition coefficients in biological membranes, and much lower values have been proposed for the entry of lipid-soluble molecules into membranes as assessed by a rapid filtration technique. The true situation is still far from clear for all drugs, but it is particularly difficult to clarify, in the case of short chain alcohols such as ethanol, which, because of their polarity, probably associate with the exterior of the membrane rather than genuinely partitioning into the membrane (see below).

## IV. PHYSICAL EFFECTS OF ETHANOL ON MEMBRANES

Given the small size of ethanol and its polar nature, it seems likely that any ethanol entering the membrane would preferentially seek either the polar regions of the phospholipid head groups or, less likely, the spaces in the hydrophobic interior of the membrane core. Evidence for this has been provided by $^2$H nuclear magnetic resonance (NMR) studies of the orientation of ethanol in biological membranes.[9] Predictions can now be made about the physical consequences of entry of ethanol into membranes in this way, but first it must be emphasized that such consequences, at anesthetic concentrations of ethanol, will be slight. Estimates of the mole ratio of ethanol to lipids under these conditions vary considerably, but 1 molecule of ethanol to every 200 lipid molecules is probably the correct order of magnitude.

All studies agree that when measurements are made of the physical nature of lipids at the center of the bilayer the presence of ethanol increases the degree of disorder of these lipids.[9-11] This would be predicted from a partitioning of ethanol at the membrane surface, since the space-filling characteristics of the molecule would separate the phospholipid molecules from each other, reducing acyl chain packing in the hydrophobic core of the bilayer. The reduction in packing leads to greater freedom of movement of the acyl chains and, consequently, greater fluidity in this region of the bilayer.

An inevitable corollary of this should be that the presence of ethanol molecules will restrict molecular motion towards the surface of the bilayer. This has been observed under some conditions,[9] but not others.[12] One of the problems may be the lack of suitable probes with which to study the physical properties close to the phospholipid head groups, which is where the ethanol would be expected to have its major effect in increasing rigidity. Alcohols of slightly longer chain length can be shown to reduce the motion of physical probes reporting on the nature of the membrane in the proximal region of the acyl chains,[12] suggesting that ethanol may have a similar effect toward the margin of the membrane.

It therefore seems likely that ethanol has a rather complex effect on the physical properties of biological membranes, restricting molecular motion at the surface of the bilayer, but increasing motion in the hydrophobic core. It must be recognized that these effects are quantitatively rather small — for example, a similar increase in "fluidity" of the membrane core might be produced by a rise in temperature of the membrane of 1°C.[13] However, it must also be recognized that the change produced is dissimilar in quality from the overall change in molecular motion induced by temperature. This will be an important factor when discussing the functional consequences of this change and the adaptive processes which it initiates.

As stated previously it is likely that membrane lipids can take up conformations other than bilayers in biological membranes. Most evidence to date suggests that ethanol reduces the likelihood of formation of nonbilayer structures.[2] Thus, in addition to its effects on the motion of molecules within the bilayer, ethanol stabilizes the bilayer structure of membranes. This, too, might have functional consequences for the cell membrane.

## V. THE HOMEOVISCOUS ADAPTATION HYPOTHESIS

### A. Biochemistry

The realization that ethanol affected the physical properties of membrane lipids in a way superficially resembling that of increased temperature led to speculation on the mechanism of cellular adaptation which this might induce.[14] Prokaryotic and eukaryotic cells seem to respond to increased temperature (which increases the fluidity of membrane lipids) in a uniform way. The principle is that more rigid lipids are incorporated into the membrane, restoring the fluidity of the membrane toward normal values even at the new growth temperature. Hill and Bangham[14] reasoned that if ethanol caused a fluidization of cell membranes, it should elicit a similar incorporation of rigid lipids into the membrane, and this would confer tolerance to ethanol on the organism.

The major ways in which eukaryotic cells (including those of homeotherms such as mammals) regulate their membrane fluidity seem to be by alteration of cholesterol content and the degree of saturation/unsaturation of acyl chains of phospholipids.[15] In response to an increase in temperature (or, theoretically, to the presence of ethanol) the appropriate response should, therefore, be to increase the cholesterol/phospholipid mole ratio and to increase the saturation/unsaturation index of phospholipid acyl chains. Both these changes have been reported in membranes from central neurons of laboratory animals after chronic ethanol administration,[16,17] and similar changes have been described in several peripheral tissues.[18,19] These are not, however, universal findings and some studies have reported no change in cell membrane phospholipid acyl chain composition[20] or cholesterol content[21] during the development of ethanol tolerance. In addition, the magnitude of the changes reported is very small, and it is hard to believe that they could overcome the effect of ethanol on membranes (although this, too, may be very small). To date there is no fully convincing evidence that altered membrane biochemical composition can confer resistance to ethanol at the cellular level.

### B. Physical Properties

Despite the lack of convincing changes in membrane lipid composition which could underly homeoviscous adaptation to ethanol, there is a good deal of evidence from the physical properties of membrane lipids that something akin to homeoviscous adaptation must occur. The first direct evidence came with the observation that the "fluidizing" effects of ethanol, as indicated by an electron spin resonance probe reporting from the proximal regions of the acyl chains, were considerably less in brain membrane preparations from ethanol-tolerant animals than they were from controls.[22] Similar findings were also obtained with erythrocyte membranes from the same animals. The findings were not exactly as predicted by the homeoviscous adaptation hypothesis, since no difference in the "intrinsic" fluidity of the control and ethanol-tolerant membranes could be observed in the absence of ethanol. In other words, the adaptation took the form of a decrease in the effect of ethanol rather than an alteration which opposed the effect of ethanol.

These observations were made on intact membranes and did not exclude some contribution of membrane proteins to the presumably adaptive response. When liposomes were produced from similar membranes and their physical properties studied, almost identical results were obtained.[23] These studies used a fluorescent probe of the lipid bilayer and demonstrated that some alteration in membrane lipid was responsible for the resistance to the fluidizing effects of ethanol *in vitro*. This study also suggested strongly that interactions between cholesterol and phospholipids were important, but that cholesterol content itself could not confer ethanol resistance.[23]

Despite this finding there have been many attempts to correlate cholesterol concentration and membrane tolerance to ethanol. One of the most interesting possibilities, which explains the decrease in the effect of ethanol without necessitating an alteration in intrinsic membrane fluidity, is that increased cholesterol content excludes ethanol from the membrane.[24] Thus, the partitioning of [14C]ethanol into synaptosomal and erythrocyte membranes (and, indeed, the partitioning of other lipophilic drugs) was reported to be reduced after chronic ethanol administration to rats. It would now require a higher concentration of ethanol in the aqueous phase to produce an equivalent concentration within the ethanol-tolerant membrane.

This explanation of the decreased effect of ethanol *in vitro* on membranes from ethanol-tolerant animals is attractive. However, the experiments are technically difficult and there are wide variations, particularly in ethanol partitioning, between preparations. As described earlier, there is no universally accepted technique for estimation of partition coefficients in normal membrane[8] and the values reported in this study[24] were rather high.

The possibility of altered penetration of ethanol into the membrane as a potential mechanism for adaptation should not be ruled out. Exactly how such cellular mechanisms of adaptation could have evolved against a pharmacological challenge is difficult to envisage.

Recently, data more in keeping with the homeoviscous adaptation hypothesis have been obtained in membranes from animals treated chronically with ethanol. Using an electron spin resonance probe which reports from areas further into the hydrophobic core of the membrane, it has been reported that synaptosomal membranes from ethanol-tolerant mice are more ordered than those from controls.[25] Thus, at least in this area of the membrane there is evidence of some change in lipid composition which opposes the effect of ethanol. Apart from this observation, the great deal of energy devoted to establishing a mechanism for "membrane tolerance" to the fluidizing effect of ethanol seems to have repeatedly demonstrated that it does exist, but there is no unanimity as to its cause.

## C. Summary

That this hypothesis was exciting and attractive when introduced is evidenced by the enthusiasm with which it was almost immediately pursued. With hindsight it clearly relied on oversimplification, both of the effects of ethanol on the membrane and of the properties of the components of the bilayer itself. It should now be possible, using our more sophisticated ideas of how ethanol differentially effects the physical properties of the bilayer surface and core, to predict which chemical changes should decrease or oppose the effects of ethanol in these areas. The adaptation (if it genuinely is adaptation) which occurs must involve phospholipids, and it seems likely that head group interactions and the position and number of double bonds in the acyl chains of these molecules must both play some part. The nature and mechanism of the biochemical changes which underly membrane tolerance are still a crucial question in alcohol research.

## VI. FUNCTIONS OF BIOLOGICAL MEMBRANES

Biological membranes, in general, have a multiplicity of functions which would be impossible to cover in a review like this. It has been decided to concentrate on some of the functions of neuronal membranes using these as an example of the ways in which ethanol may affect similar processes in other cells. Included will be the barrier and transport activities of membranes, receptor response coupling, and the related generation of second messengers from membranes.

### A. Barrier Properties

The membrane lipid bilayer has the important function of separating the external aqueous compartment from the cytosol. In general, its job is to keep hydrophilic molecules out of the cell (unless they are transported or enter through ion channels) and to allow hydrophobic molecules, such as some hormones, to enter readily. If we examine this very simple property there is some evidence that ethanol can affect it in an important way.

Most studies have shown that the presence of ethanol, albeit in rather high concentrations, reduces the barrier properties (e.g., to $Na^+$) of some membranes[26] (but not neuronal membranes[27]) to hydrophilic molecules. There is no convincing evidence that tolerance to this effect of ethanol is demonstrated in membranes from animals which have received ethanol chronically. The neuronal membrane under normal conditions is "leaky" to $K^+$, and this, combined with the $Na^+/K^+$ ATPase system (see below), sets up the membrane ion gradient which gives the resting membrane potential. The presence of ethanol may well make the membrane more permeable to $K^+$,[28] although in biological membranes this probably is more related to permeability through ion channels than through the lipid phase. The reduction in barrier properties to $K^+$ could lead to the hyperpolarization of neuronal membranes which ethanol is sometimes reported to cause.[28]

As mentioned above, the membrane bilayer is designed not only to keep hydrophilic molecules on separate sides of the bilayer, but it is also designed to allow free passage of lipophilic molecules.

No work is known on the effects of the presence of ethanol on lipophilic molecules, but the chronic administration of ethanol has been reported to exclude both ethanol and other lipophilic molecules from the membrane (see above).[28] Whether this influences the penetration of molecules such as steroids into the cell interior does not seem to have been investigated.

### B. Ion Transport

The proteins responsible for transport of molecules across cell membranes must presumably span the membrane and be capable of mobility within the membrane. As such they seem good candidates for being influenced by a compound such as ethanol which produces physical effects within the bilayer. As far as neuronal membranes are concerned, the most important transport systems are probably those responsible for maintaining the ion gradients between the interior and exterior of the cell. Among these the $Na^+/K^+$ ATPase, the $Ca^{2+}/Mg^{2+}$ ATPase, and the $Na^+/Ca^{2+}$ exchange systems are those which influence neuronal function to the greatest extent.

There have been many studies on the effect of ethanol on $Na^+/K^+$ ATPase activity in neurons and other cells. In general, they show that the presence of ethanol in low concentrations ($<50$ m$M$) causes either no effect or a small potentiation of activity,[29,30] whereas higher concentrations produce inhibition of $Na^+/K^+$ ATPase.[29-31] It has been reported that the major inhibitory effect is on a specific form of the $Na^+/K^+$ ATPase, that with a high affinity to ouabain.[31] Failure to distinguish between the high and low affinity forms of the enzyme and different distributions of these in different cell types may explain some of the variability in the extent of inhibition described in different reports. Another factor which may influence the effect of ethanol is the receptor-mediated activation of $Na^+/K^+$ ATPase, for example, that by noradrenaline.[32] This may not apply to all situations, however, and not all workers have obtained this interaction between ethanol and noradrenaline in the brain.

The effect of chronic administration of ethanol on $Na^+/K^+$ ATPase activity is even more controversial. Several reports have found little change either in the activity of $Na^+/K^+$ ATPase or on the effect of ethanol added *in vitro* on this activity.[33] Some reports, however, suggest greater than normal activity of $Na^+/K^+$ ATPase[34,35] and/or a reduced effect of ethanol *in vitro* on this activity.[36]

Much less work has been published on the transport of $Ca^{2+}/Mg^{2+}$ across neuronal membranes. There is evidence that ethanol *in vitro*, albeit at relatively high concentrations, can inhibit $Ca^{2+}/Mg^{2+}$ ATPase,[37] whereas others have suggested that the chronic administration of ethanol *in vivo* leads to an increased activity of $Ca^{2+}/Mg^{2+}$ ATPase.[38] Whether this increased activity is a consequence of adaptive changes in the enzyme or in the lipid surrounding it, or whether it is a response to some other change, for example, an increased intracellular $Ca^{2+}$ concentration, is unknown.

Similarly, relatively little work exists on the effect of ethanol on the $Na^+/Ca^{2+}$ exchange system. This is surprising, given the importance of this mechanism in maintaining the concentration difference in $Na^+$ and $Ca^{2+}$ ions across the neuronal membrane. Part of the problem is that this exchange system is driven by the relative concentrations of $Na^+$ and $Ca^{2+}$ on each side of the membrane. The activity of the pump system in the presence of ethanol is, therefore, influenced by the effect of the drug on these ion concentrations as well as by direct effects of ethanol on the exchange protein. Most evidence suggests that ethanol *in vitro* inhibits $Na^+/Ca^{2+}$ exchange[39] and that the chronic effect of ethanol is to increase $Na^+/Ca^{2+}$ exchange activity.[40]

In general, therefore, ethanol at high concentrations is inhibitory to all these ion transport systems. After chronic administration of ethanol there is some equivocal evidence for compensatory increases in activity of the systems and for a loss of the inhibitory effects of ethanol.

## C. Ion Channels

There has been an explosion of knowledge in the field of membrane ion channels with the application of sensitive electrophysiological techniques, such as patch clamping, to neurons. Unfortunately, this has not yet been applied to the study of actions of ethanol on ion channels. Some evidence exists for effects of ethanol on voltage-operated $Na^+$ channels, voltage-operated $Ca^{2+}$ channels, and receptor-operated $Cl^-$ channels. In addition, some work which suggests an effect of ethanol on $Ca^{2+}$-activated $K^+$ channels has also been published. An attempt will be made here to summarize some of these findings, but this area awaits a rigorous study with the new techniques available.

### 1. Ethanol and Neuronal Sodium Channels

Most evidence suggests that the presence of ethanol inhibits the movement of $Na^+$ through voltage-operated $Na^+$ channels. Historically, electrophysiological studies on invertebrate neurons have shown inhibition by ethanol of the transient movement of $Na^+$ ions and the height of the action potential.[41-43] In general, these experiments required high concentrations of ethanol, but inhibitory effects of ethanol on $Na^+$ movement in electrically stimulated slices of mammalian brain were obtained at 105 m$M$ ethanol,[44] a concentration compatible with life.

More recently, the ability of site-specific drugs to induce movement of $Na^+$ ions into neuronal preparations has been utilized to study the effects of ethanol. The uptake of $^{22}Na^+$ into synaptosomes stimulated by batrachotoxin or veratridine is reversibly inhibited by the presence of ethanol *in vitro*.[45] Although the $IC_{50}$ values for this inhibition are high (300 to 600 m$M$), significant effects can be obtained in the 50- to 100-m$M$ range, suggesting that some of the signs of severe intoxication may be related to inhibitory effects on $Na^+$ channels. The effects of alcohols of different chain lengths closely parallel their lipid solubilities, suggesting that the site of action is within the lipid bilayer or in a highly hydrophobic area of the channel glycoprotein.

There is evidence that the inhibitory effect of ethanol on $Na^+$ channel function is lost rapidly if ethanol remains present in the vicinity of the channel.[46] Thus, the effect of adding ethanol *in vitro* to synaptosomal preparations is less when these are obtained from animals treated acutely *in vivo* with the drug. The maximum tolerance in this parameter is obtained about 6 h after acute ethanol administration and dissipates over about 24 h.[46] There is no change in the $Na^+$ uptake stimulated by batrachotoxin or veratridine in the absence of added ethanol. In contrast, chronic administration of ethanol can lead to an inhibition of $Na^+$ uptake into synaptosomes *in vitro* even when studied in the absence of ethanol. Again, chronic administration leads to a loss of the inhibitory effect of ethanol *in vitro*, but this tolerance dissipates much less rapidly than that obtained after acute administration of ethanol.[46]

The reported changes in $Na^+$ channel function induced by the presence of ethanol *in vitro* and *in vivo* thus seem capable of providing a partial explanation for severe intoxication and for the development of tolerance. It is difficult to see, however, how the latent hyperexcitability of the central nervous system which must underlie ethanol physical dependence could be explained by changes in $Na^+$ channels. To date there is no evidence of increased excitability of neuronal conduction in ethanol dependence or withdrawal, which might suggest that voltage-operated $Na^+$ channels play any part in these processes.

### 2. Ethanol and Neuronal Calcium Channels

Recent evidence[47] suggests there to be at least three types of voltage-operated $Ca^{2+}$ channels on neuronal membranes. The techniques necessary for establishing these divisions have not yet been applied to the study of the effects of ethanol on $Ca^{2+}$ channels. In consequence, our understanding of the actions of ethanol on $Ca^{2+}$ flux through voltage-operated $Ca^{2+}$ channels is still in a fairly rudimentary state.

Most studies conducted in a way appropriate to demonstrating $^{45}Ca^{2+}$ flux through neuronal $Ca^{2+}$ channels agree that ethanol is a moderately potent inhibitor of depolarization-induced $Ca^{2+}$ entry into neurons.[48,49] There is evidence[48] that the extent of inhibition by ethanol varies between different brain regions, and this, too, suggests heterogeneity in the types of $Ca^{2+}$ channel present. Electrophysiological studies on cultured neurons also support the concept that ethanol is inhibitory to $Ca^{2+}$ flux at relatively low concentrations.[50,51] The mechanism of this inhibitory effect of ethanol in unknown. In contrast to many of the other actions of alcohols the inhibition of depolarization-induced $Ca^{2+}$ flux has been reported not to show a close relationship to lipid solubility.[44] Some direct interactions with the channel protein or with the protein lipid interface, therefore, seem likely possibilities.

Drugs of the dihydropyridine type which affect one subtype of the voltage-operated $Ca^{2+}$ channels existing on neurons[52] are now available. Activators of these channels, e.g., BAY K 8644, have been shown to inhibit ethanol intoxication,[53] whereas inhibitors of these channels, e.g., nitrendipine or nimodipine, potentiate intoxication.[54] It seems likely, therefore, that the interaction of ethanol with neuronal $Ca^{2+}$ channels plays an important role in the central depressant effects of the drug.

As in the case of $Na^+$ flux, the inhibitory effects of ethanol on $Ca^{2+}$ flux seem to be lost in preparations taken from animals which have received ethanol *in vivo*.[50,51] Although the effect of ethanol *in vitro* is lost in such preparations, there is no evidence for an increased $Ca^{2+}$ flux observed in the absence of ethanol. Just as with the $Na^+$ channel, therefore, there is no evidence from flux studies that alterations in number or function of $Ca^{2+}$ channels could produce the neuronal hyperexcitability which is a feature of the ethanol physical withdrawal syndrome. However, other lines of evidence do suggest that the subtype of $Ca^{2+}$ channels which are sensitive to the dihydropyridine drugs may be altered in such a way as to be causally related to ethanol physical dependence and withdrawal.

If, as seems certain, the presence of ethanol does inhibit net $Ca^{2+}$ influx into neurons on depolarization, then appropriate adaptations to oppose this action would be either to increase the number of voltage-operated $Ca^{2+}$ channels or to increase the internal sensitivity of the neuron to the $Ca^{2+}$ entering. Since the net $Ca^{2+}$ flux on depolarization is not increased (but rather decreased) in brain preparations from ethanol-tolerant animals,[50,51] it is unlikely that an increase in the total number of $Ca^{2+}$ channels has occurred. There is, however, evidence for an increased sensitivity to $Ca^{2+}$ of such preparations in that the release of some neurotransmitters becomes more sensitive to external $Ca^{2+}$ concentrations.[55] The $Ca^{2+}$ sensitivity of the excitation/secretion process in many cells seems to be controlled by the inositol lipid signaling system and there is some evidence that this operates in neurons, also.[56] The depolarization-induced breakdown of inositol lipids in brain is sensitive to $Ca^{2+}$ and can be influenced by the dihydropyridine $Ca^{2+}$ channel activators and inhibitors.[57] It is possible, therefore, that this subtype of $Ca^{2+}$ channel may influence the $Ca^{2+}$ sensitivity of neurons without making an important contribution to the net $Ca^{2+}$ flux on depolarization. A tenable hypothesis to explain neuronal adaptation to ethanol might be that an increased number of the dihydropyridine-sensitive subtype of $Ca^{2+}$ channel occurs in response to chronic alcohol administration, and that this, in turn, modulates the $Ca^{2+}$ sensitivity of neurotransmitter release via the inositol lipid signaling system.

In support of this hypothesis it has been shown that the acute and chronic administration of ethanol to animals *in vivo* produces an increase in the number of [³H]dihydropyridine binding sites in brain[58] and that this effect can be mimicked in cell cultures.[59] The effects of the dihydropyridine $Ca^{2+}$ channel activator BAY K 8644 on inositol lipid breakdown[60] and on neurotransmitter release[58] are also greater in preparations obtained from ethanol-dependent animals. All this evidence points to an increased number and functional importance of the dihydropyridine-sensitive $Ca^{2+}$ channels in ethanol physical dependence. Lastly, the hyperexcitability of the central nervous system associated with ethanol withdrawal can be completely prevented by dihydropyridine $Ca^{2+}$ channel inhibitors at doses which have no overt sedative effects in control animals.[61]

The current status of neuronal $Ca^{2+}$ channels in ethanol dependence (and, indeed, in other drug-dependence states[62]) is one of rapid growth of interest. To date it has been very difficult to find a role in the central nervous system for the dihydropyridine-sensitive $Ca^{2+}$ channels, but their plasticity in the face of chronic drug administration suggests an important adaptive and modulatory function. The hypothesis outlined above may not be correct, but, at the present, it offers an attractive explanation for many of the effects of alcohol at the molecular and behavioral levels.

### 3. Ethanol and Neuronal Chloride Channels

Neuronal chloride channels are in many instances operated by the receptors for the inhibitory neurotransmitter γ-amino butyric acid (GABA). Since the central depressant drugs benzodiazepines and barbiturates can modify the effect of GABA on its receptor and on the $Cl^-$ channel, it is not surprising that similar effects have been sought for ethanol. As far as behavioral and electrophysiological studies are concerned, the evidence supports the concept that ethanol potentiates GABAergic transmission.[63,64] Whether this is by an action on the GABA receptor is not clear from radioligand binding studies. Thus, although ethanol *in vitro* has been reported to enhance [³H]diazepam binding to detergent-solubilized benzodiazepine receptors[65] and to decrease binding of a radioligand directed toward the $Cl^-$ channel,[66] the concentrations required for these effects are high. In addition, no effect of ethanol on [³H]diazepam binding to the membrane-associated benzodiazepine receptor[67] or on [³H]muscimol binding to the GABA receptor[67] has been observed. If ethanol does affect the ability of GABA to influence $Cl^-$ flux, then it seems likely that this is due to some direct or indirect effect on the $Cl^-$ channel rather than on ligand binding to any site in the GABA-benzodiazepine receptor-chloride ionophore complex.

In support of this explanation for the electrophysiological and behavioral potentiation of GABA by ethanol, it has recently been reported that ethanol does affect $^{36}Cl^-$ flux into "synaptoneurosomes", a preparation of brain which includes presynaptic and postsynaptic membrane vesicles.[68] Ethanol had a markedly stimulatory effect on $Cl^-$ flux at concentrations as low as 20 to 30 m$M$ and this effect was significantly inhibited by picrotoxin (a ligand specific for the chloride channel). Chloride channels coupled to the GABA receptor were probably involved, because the specific GABA receptor antagonist bicuculline also inhibited the effect of ethanol.[68] There have been reports, however, where biochemical effects of ethanol may be attributed to potentiation of $Cl^-$ flux through chloride channels not directly coupled to GABA receptors.[69] It seems likely that ethanol either by a direct action on the channel protein or by interaction with surrounding lipids prolongs the open state of the chloride channel, allowing a greater flux of $Cl^-$.[68]

Although there is compelling evidence that the presence of ethanol in relatively small concentrations can influence the GABA receptor function by this effect on the chloride channel, there is very little evidence for adaptive alterations in the system after long-term administration of the drug *in vivo*. It seems, therefore, that although effects of ethanol on chloride channels can explain some of the acute central depressant effects of ethanol, any adaptive alterations which occur involve not the channel itself, but perhaps the release or effects of the endogenous ligands which act on the receptors surrounding the channel.

### 4. Ethanol and Neuronal Potassium Channels

Neurons have $Ca^{2+}$-dependent $K^+$ channels in their synaptic membranes which have important effects on membrane potential. In particular, the after-hyperpolarization which follows depolarizing stimuli is probably a consequence of increased $K^+$ flux activated by $Ca^{2+}$ entry. This system could be an important site for the inhibitory action of ethanol if it could be shown that ethanol potentiated flux through such channels. This seems a real possibility, both in human erythrocytes and in rat brain synaptosomes,[70] that ethanol increases flux through $Ca^{2+}$-dependent $K^+$ channels. Concentrations required are rather high, but a contribution of this action to the central depressant effects of ethanol cannot be excluded. There is as yet no evidence for adaptation in this system.

## VII. CONCLUSIONS

Ethanol has a wide variety of effects on biological membrane function. Most of these are explicable by nonspecific actions of ethanol within, or at the surface of, the membrane lipid bilayer. This does not mean that the intoxicating effects of ethanol are necessarily caused by nonspecific actions on membranes, since intoxication can, in general, occur at concentrations considerably below those needed to induce physical changes in biological membranes. Of the systems discussed in this review, the effects of ethanol on the $Ca^{2+}$ and $Cl^-$ channels in neuronal membranes seem to be the most promising areas for explaining the acute effects of ethanol. There may, of course, be membrane proteins and functions as yet undiscovered where ethanol proves to have even greater potency, but, to date, these two important channel proteins seem most susceptible. The influence of ethanol on these two systems would be expected to inhibit neurotransmission both presynaptically and postsynaptically. This combination of effects is likely to be more than additive in its net effect.

The development of tolerance to ethanol may involve a strong nonspecific element involving alteration in membrane lipid composition. In many instances, such an alteration can explain the lack of effect of ethanol added *in vitro* to brain preparations obtained from animals made tolerant to the drug. A possible explanation is that the "adapted" membrane is affected less by ethanol, because drug molecules are no longer able to penetrate the lipid bilayer.

In most instances, membrane alterations produced by the chronic administration of ethanol appear to decrease the effect of ethanol (as above) rather than to oppose it by inducing a change in function antagonistic to that produced by the presence of the drug. In other words, membrane function, when observed in the absence of ethanol, does not exhibit the hyper-excitability which would be necessary to explain physical dependence and the ethanol with-drawal syndrome. The same comment is not true when whole cells or aggregates (for example, brain slices) or, of course, the intact animal is studied. It seems very likely that the mechanisms responsible for physical dependence reside not within membrane lipids, but in adaptive changes in cellular signaling systems such as the adenylate cyclase system and the inositol lipid system. One mechanism by which alterations in the inositol lipid system could be induced has already been mentioned. "Up" or "down regulation" of numbers of receptor proteins as well as numbers of ion channels could provide the mechanisms necessary for the development of hyperexcitability in the central nervous system underlying physical dependence.

The study of the effects of ethanol on biological membranes is beginning to provide a clearer picture. A relatively coherent concept of the way in which ethanol can affect mem-brane function is developing, and a greater awareness of the advantages and limitations of biochemical and biophysical techniques as applied to the study of ethanol is being gained. The actions of ethanol on biological membranes are of undoubted importance for the be-havioral, psychiatric, and pathological effects of the drug. An understanding of these actions would be of immeasurable benefit to the development of strategies for therapeutic intervention in alcoholism.

# REFERENCES

1. **Singer, S. J. and Nicholson, G. L.,** The fluid mosaic model of the structure of cell membranes, *Science,* 175, 720, 1972.
2. **Cullis, P. R. and De Kruijff, B.,** Lipid polymorphism and the functional roles of lipids in biological membranes, *Biochim. Biophys. Acta,* 559, 399, 1979.
3. **Boggs, J. M.,** Intermolecular hydrogen bonding between lipids: influence on organization and function of lipids in membranes, *Can. J. Biochem.,* 58, 755, 1980.
4. **Seeman, P.,** The membrane actions of anesthetics and tranquilizers, *Pharmacol. Rev.,* 24, 583, 1972.
5. **Franks, N. P. and Lieb, W. R.,** Molecular mechanisms of general anaesthesia, *Nature,* 300, 487, 1982.
6. **Franks, N. P. and Lieb, W. R.,** Do general anaesthetics act by competitive binding to specific receptors?, *Nature,* 310, 599, 1984.
7. **Roth, S. and Seeman, P.,** The membrane concentrations of neutral and positive anesthetics (alcohols, chlorpromazine, morphine) fit the Meyer Overton rule of anesthesia. Negative narcotics do not, *Biochim. Biophys. Acta,* 255, 207, 1972.
8. **Conrad, M. J. and Singer, S. J.,** Evidence for a large internal pressure on biological membranes, *Proc. Natl. Acad. Sci. U.S.A.,* 76, 5202, 1979.
9. **Hitzemann, R. J., Schueler, H.E., Graham-Brittain, C., and Kreishman, G. P.,** Ethanol-induced changes in neuronal membrane order. An NMR study, *Biochim. Biophys. Acta,* 859, 189, 1986.
10. **Chin, J. H. and Goldstein, D. B.,** Effects of low concentrations of ethanol on the fluidity of spin labelled erythrocyte and brain membranes, *Mol. Pharmacol.,* 13, 435, 1977.
11. **Lenaz, G., Curatola, G., Mazzanti, L., Bertoli, E., and Pastuszko, A.,** Spin label studies on the effects of anesthetics in synaptic membranes, *J. Neurochem.,* 32, 1689, 1979.
12. **Chin, J. H. and Goldstein, D. B.,** Disordering effect of ethanol at different depths in the bilayer of mouse brain membranes, *Alcoholism (N.Y.),* 5, 256, 1981.
13. **Richards, C. D., Martin, K., Gregory, S., Keightley, C. A., Hesketh, T. R., Smith, G. A., Warren, G. B., and Metcalf, J. C.,** Degenerate perturbations of protein structure as the mechanism of anaesthetic action, *Nature,* 276, 775, 1978.
14. **Hill, M. W. and Bangham, A. D.,** General depressant drug dependency. A biophysical hypothesis, *Adv. Exp. Med. Biol.,* 59, 1, 1975.

15. **Cossins, A. R.,** Adaptation of biological membranes to temperature. The effect of temperature acclimation of goldfish upon the viscosity of synaptosomal membranes, *Biochim. Biophys. Acta,* 470, 395, 1977.

16. **Chin, J. H., Parsons, L. M., and Goldstein, D. B.,** Increased cholesterol content of erythrocyte and brain membranes in ethanol tolerant mice, *Biochim. Biophys. Acta,* 513, 358, 1978.

17. **Littleton, J. M. and John, G. R.,** Synaptosomal membrane lipids of mice during continuous exposure to ethanol, *J. Pharm. Pharmacol.,* 29, 579, 1977.

18. **Littleton, J. M., Grieve, S. J., Griffiths, P. J., and John, G. R.,** Ethanol-induced alteration in membrane phospholipid composition: possible relationship to development of cellular tolerance to ethanol, in *Biologicol Effects of Ethanol,* Begleiter, H., Ed., Plenum Press, New York, 1980, chap. 2.

19. **Smith, T. L., Vickers, A. E., Brendel, K., and Gerhart, M. J.,** Effects of ethanol diets on cholesterol content and phospholipid acyl composition of rat hepatocytes, *Lipids,* 17, 124, 1982.

20. **Smith, T. L. and Gerhart, M. J.,** Alterations in brain lipid composition of mice made physically dependent to ethanol, *Life Sci.,* 31, 1419, 1982.

21. **Wing, D. R., Harvey, D. J., Hughes, J. P., Dunbar, P. G., McPherson, K. A., and Paton, W. D. M.,** Effects of chronic ethanol administration on the composition of membrane lipids in the mouse, *Biochem. Pharmacol.,* 31, 3431, 1982.

22. **Chin, J. H. and Goldstein, D. B.,** Drug tolerance in biomembranes. A spin label study of the effects of ethanol, *Science,* 196, 684, 1977.

23. **Johnson, D. A., Lee, N. M., Cooke, R., and Loh, H. H.,** Ethanol-induced fluidization of brain lipid bilayers: required presence of cholesterol in membranes for the expression of tolerance, *Mol. Pharmacol.,* 15, 739, 1979.

24. **Rottenberg, H., Waring, A., and Rubin, E.,** Tolerance and cross tolerance in chronic alcoholics: reduced membrane binding of ethanol and other drugs, *Science,* 213, 583, 1981.

25. **Lyon, R. C. and Goldstein, D. B.,** Changes in synaptic membrane order associated with chronic ethanol treatment in mice, *Mol. Pharmacol.,* 23, 56, 1983.

26. **Seeman, P., Kwant, W. O., Goldberg, M., and Chan-Wong, M.,** The effects of ethanol and chlorpromazine on the passive membrane permeability to $Na^+$, *Biochim. Biophys. Acta,* 241, 349, 1971.

27. **Wallgren, H., Nikander, P., von Boguslawsky, P., and Linkola, J.,** Effects of ethanol, tert-butanol and chlormethiazole on net movements of sodium and potassium in electrically stimulated cerebral tissue, *Acta Physiol. Scand.,* 91, 83, 1974.

28. **Berry, M. S. and Pentreath, V. W.,** The neurophysiology of alcohol, in *Psychopharmacology of Alcohol,* Sandler, M., Ed., Raven Press, New York, 1980, chap. 4.

29. **Israel, Y., Kalant, H., and Laufer, I.,** Effect of ethanol on $Na^+$, $K^+$, $Mg^{2+}$,-stimulated microsomal ATPase activities, *Biochem. Pharmacol.,* 14, 1803, 1965.

30. **Gordon, L. M., Sauerheber, R. D., Esgate, J. A., Dipple, I., Marchmont, R. J., and Houslay, M. D.,** The increase in bilayer fluidity of rat liver plasma membranes achieved by the local anaesthetic benzyl alcohol affects the activity of intrinsic membrane enzymes, *J. Biol. Chem.,* 255, 4519, 1980.

31. **Nhamburo, P. T., Salafsky, B., Tabakoff, B., and Hoffman, P. L.,** Selective effects of ethanol on two forms of brain $(Na^+, K^+)$ ATPase, *Alcohol Alcoholism,* 21, A80, 1986.

32. **Rangaraj, N. and Kalant, H.,** α-Adrenoreceptor-mediated alteration of ethanol effects on $(Na^+ + K^+)$ ATPase of rat neuronal membranes, *Can. J. Physiol. Pharmacol.,* 57, 1098, 1979.

33. **Goldstein, D. B. and Israel, Y.,** Effects of ethanol on mouse brain $(Na^+ + K^+)$-activated ATPase, *Life Sci.,* 11, 957, 1972.

34. **Israel, Y., Kalant, H., Leblanc, E., Bernstein, J. C., and Salazar, I.,** Changes in cation transport and $(Na^+ + K^+)$-activated ATPase produced by chronic administration of ethanol, *J. Pharmacol. Exp. Ther.,* 174, 330, 1970.

35. **Knox, W., Perrin, R. G., and Sen, A. K.,** Effect of chronic administration of ethanol on $(Na^+ + K^+)$-activated ATPase activity in six areas of the cat brain, *J. Neurochem.,* 19, 2881, 1972.

36. **Levental, M. and Tabakoff, B.,** Sodium-potassium activated ATPase activity as a measure of neuronal membrane characteristics in ethanol-tolerant mice, *J. Pharmacol. Exp. Ther.,* 212, 316, 1980.

37. **Kondo, M. and Kaseri, M.,** The effect of n-alcohols on sarcoplasmic reticulum vesicles, *Biochim. Biophys. Acta,* 311, 391, 1973.

38. **Guerri, C. and Grisolia, S.,** Chronic ethanol treatment affects synaptosomal membrane bound enzymes, *Pharmacol. Biochem. Behav.,* 18 (Suppl. 1), 45, 1983.

39. **Michaelis, M. L. and Michaelis, E. K.,** Alcohol and local anaesthetic effects of $Na^+$-dependent $Ca^{2+}$ fluxes in brain synaptic membrane vesicles, *Biochem. Pharmacol.,* 32, 963, 1983.

40. **Michaelis, M. L., Michaelis, E. K., and Tehan, T.,** Alcohol effects on synaptic membrane $Ca^{2+}$ fluxes, *Pharmacol. Biochem. Behav.,* 18 (Suppl. 1), 19, 1983.

41. **Armstrong, C. M. and Binstock, L.,** The effects of several alcohols on the properties of the squid giant axon, *J. Gen. Physiol.,* 48, 265, 1964.

42. **Moore, J. W., Ulbricht, W., and Takata, M.,** Effect of ethanol on the sodium and potassium conductances of the squid giant axon membrane, *J. Gen. Physiol,* 48, 279, 1964.

43. **Bergmann, M. C., Klee, M. W., and Faber, D. S.,** Different sensitivities to ethanol of early transient voltage clamp currents of Aplysia neurones, *Pfluegers Arch. Gesamte Physiol. Menschen Tiere,* 348, 139, 1974.
44. **Harris, R. A. and Bruno, P.,** Membrane disordering by anesthetic drugs: relationship to synaptosomal sodium and calcium fluxes, *J. Neurochem.,* 44, 1274, 1985.
45. **Mullin, M. J. and Hunt, W. A.,** Actions of ethanol on voltage-sensitive sodium channels: effects of neurotoxin-stimulated sodium uptake in synaptosomes, *J. Pharmacol. Exp. Ther.,* 232, 413, 1985.
46. **Mullin, M. J., Hunt, W. A., Dalton, T. K., and Majchrowicz, E.,** Alterations in neurotoxin-stimulated $^{22}$Na$^+$ influx in synaptosomes after acute and chronic ethanol treatment, *Fed. Proc.,* 44, 4843, 1985.
47. **Nowcky, M. C., Fox, A. P., and Tsien, R. W.,** Three types of neuronal calcium channel with different calcium agonist sensitivity, *Nature,* 316, 440, 1985.
48. **Harris, R. A. and Hood, W. F.,** Inhibition of synaptosomal calcium uptake by ethanol, *J. Pharmacol. Exp. Ther.,* 213, 562, 1980.
49. **Leslie, S. W., Barr, E., Chandler, L. J., and Farrar, R. P.,** Inhibition of fast- and slow-phase depolarization-dependent synaptosomal calcium uptake by ethanol, *J. Pharmacol. Exp. Ther.,* 225, 571, 1983.
50. **Oakes, S. G. and Pozos, R. S.,** Electrophysiological effects of acute ethanol exposure. II. Alterations in the calcium component of action potentials from sensory neurons in dissociated culture, *Dev. Brain Res.,* 5, 251, 1982.
51. **Triestman, S. N., Camacho-Nasi, P., and Wilson, A.,** Alcohol effects on voltage dependent currents in identified cells, *Alcoholism Clin. Exp. Res.,* 9, 201, 1985.
52. **Miller, R. J.,** Multiple calcium channels and neuronal function, *Science,* 235, 46, 1987.
53. **Dolin, S. J. and Little, H. J.,** The effects of BAY K 8644 on the general anaesthetic potencies of ethanol and argon, *Br. J. Pharmacol.,* 89, 622P, 1986.
54. **Isaacson, R. L., Molina, J. C., Draski, L. J., and Johnston, J. E.,** Nimodipine's interactions with other drugs. I. Ethanol, *Life Sci.,* 36, 2195, 1985.
55. **Lynch, M. A. and Littleton, J. M.,** Possible association of alcohol tolerance with increased synaptic Ca$^{2+}$ sensitivity, *Nature,* 303, 175, 1983.
56. **Tanaka, C., Fujiwara, H., and Fujii, Y.,** Acetylcholine release from guinea pig caudate slices evoked by phorbol ester and calcium, *FEBS Lett.,* 195, 129, 1986.
57. **Kendall, D. A. and Nahorski, S. R.,** Dihydropyridine calcium channel activators and antagonists influence depolarization-evoked inositol phospholipid hydrolysis in brain, *Eur. J. Pharmacol.,* 115, 31, 1985.
58. **Dolin, S. J., Little, H. J., Littleton, J. M., and Pagonis, C.,** Dihydropyridine-sensitive calcium channels in rat brain are increased in ethanol physical dependence, *Br. J. Pharmacol.,* 90, 210P, 1987.
59. **Messing, R. D., Carpenter, C. L., Diamond, I., and Greenberg, D. A.,** Ethanol regulates calcium channels in clonal neuronal cells, *Proc. Natl. Acad. Sci. U.S.A.,* 83, 6213, 1986.
60. **Hudspith, M. J. and Littleton, J. M.,** Enhanced effect of BAY K 8644 on inositol phospholipid breakdown in brain slices from ethanol-dependent rats, *Br. J. Pharmacol.,* 89, 623P, 1986.
61. **Little, H., Dolin, S., and Halsey, M.,** Calcium channel antagonists decrease the ethanol withdrawal syndrome in rats, *Life Sci.,* 39, 2059, 1986.
62. **Ramkumar, V. and El-Fakahany, E. E.,** Current status of the dihydropyridine calcium channel antagonist binding sites in the brain, *Trends Pharmacol. Sci.,* 7, 171, 1986.
63. **Cott, J., Carlsson, A., Engel, J., and Lindquist, M.,** Suppression of ethanol-induced locomotor stimulation by GABA-like drugs, *N.S. Arch. Pharmacol.,* 295, 203, 1976.
64. **Nestoros, J. N.,** Ethanol specifically potentiates GABA-mediated neurotransmission in feline cerebral cortex, *Science,* 209, 708, 1980.
65. **Davis, W. C. and Ticku, M. J.,** Ethanol enhances [$^3$H]Diazepam binding at the benzodiazepine-γ amino butyric acid receptor ionophore complex, *Mol. Pharmacol.,* 20, 287, 1981.
66. **Ramanjaneyulu, R. and Ticku, M. J.,** Binding characteristics and interactions of depressant drugs with [$^{35}$S]-t-butylbicyclophosphorothionate, a ligand that binds to the picrotoxin site, *J. Neurochem.,* 42, 221, 1984.
67. **Greenberg, D. A., Cooper, E. C., Gordon, A., and Diamond, I.,** Ethanol and the γ-aminobutyric acid-benzodiazepine receptor complex, *J. Neurochem.,* 42, 1062, 1984.
68. **Suzdak, P. D., Schwartz, R. D., Skolnick, P., and Paul, S. M.,** Ethanol stimulates γ-aminobutyric acid receptor-mediated chloride transport in rat brain synaptoneurosomes, *Proc. Natl. Acad. Sci. U.S.A.,* 83, 4071, 1986.
69. **Samuel, D., Lynch, M. A., and Littleton, J. M.,** Picrotoxin inhibits the effect of ethanol on the spontaneous efflux of [$^3$H]dopamine from superfused slices of rat corpus striatum, *Neuropharmacology,* 22, 1413, 1983.
70. **Yamamoto, H. A. and Harris, R. A.,** Calcium-dependent $^{86}$Rb efflux and ethanol intoxication: studies of human red blood cells and rodent brain synaptosomes, *Eur. J. Pharmacol.,* 88, 357, 1983.

Chapter 12

# TOLERANCE TO ALCOHOL AND ITS RELATIONSHIP TO DEPENDENCE

**Richard Laverty**

## TABLE OF CONTENTS

# I. INTRODUCTION AND DEFINITIONS

That continued use of alcohol can lead to a degree of tolerance, i.e., a greater amount is needed to produce the same effect, has been recognized in the scientific literature since before 1918.[1] The causes and consequences of this tolerance have since been the subject of much research, not all of which can be covered here. Much of the earlier literature has been reviewed,[2] so this chapter will attempt to highlight some recent developments and more significant advances.

Tolerance to ethanol, manifested either as a reduced response to a given dose or as an increased dose requirement to produce a given effect, is relatively easy to produce and to measure. However, this does not mean it is a simple or unitary event. Tolerance has been described even within a single dose; the "Mellanby effect" is the historic forerunner for observations of the acute development of tolerance to the effects of alcohol,[3,4] in that the behavioral effects of alcohol are greater for a given blood alcohol concentration before the peak blood level has been reached than at that concentration when the blood concentration is falling.

Tolerance also occurs following one or more doses of alcohol. Like acute tolerance, chronic tolerance may comprise a number of separate components.[5] Dispositional or metabolic tolerance is that due to changes in the pharmacokinetics of the drug. Changes in the absorbtion or metabolism of ethanol as a consequence of prior exposure may result in less alcohol reaching its site of action or it being removed more rapidly.

Functional or tissue tolerance is characterized by a reduction in the pharmacodynamic response of the cell or organ which acts as the pharmacological effector unit. In this case, for the same cellular concentration of alcohol there is a reduction in the effect produced.

One aspect of tolerance which is deservedly receiving increasing emphasis is the role of environmental factors[6,7] in the development of tolerance. An animal repeatedly tested with alcohol in a given environment becomes tolerant more rapidly than one exposed to the test environment only once. Thus, the animal learns to associate a situation with the exposure to alcohol and becomes tolerant more rapidly.

The distinctions and interrelationships between various manifestations of tolerance still remain to be fully clarified. The mechanisms of acute tolerance may apply to chronic tolerance, especially to functional tolerance. As will be discussed later, environmental factors may be shown to affect both dispositional and functional tolerance. However, in this brief review, it is considered that environmental factors act predominantly on the whole animal through behavioral or learned mechanisms. Such a clear-cut distinction between environmentally determined and functional (cellular) aspects of tolerance may not be acceptable to all those interested in this topic. It is possible to distinguish between dispositional and functional tolerance under conditions in which environmentally induced tolerance should be minimal.[8] However, in the research literature these interactions have not always been clearly allowed for, and much of the confusion and apparent irreproducibility between reports about the nature of tolerance are due to a failure to appreciate these complexities and the interdependence of different aspects of tolerance.

Dependence is more difficult to define and to measure than tolerance. It is usually defined in terms of effects seen when consumption of the drug is stopped. Physical dependence is estimated by the physical changes in body parameters such as pulse rate, vasomotor tone, body posture, or sensitivity to stimuli. Psychological dependence attempts to assess the psychological consequences of drug withdrawal and is measured, if possible, by mood scores, questionnaires, intensity of drug-seeking behaviors, etc. Since it is impossible to eliminate interaction between psychological states and physical manifestations in body parameters, it is probable that these distinctions are more of semantic rather than of practical use. The factors influencing the measurement and production of physical dependence, based largely on animal studies, have been reviewed.[3,5,6,9]

# II. TOLERANCE TO ALCOHOL

## A. Acute Tolerance

As already mentioned, there is well-established literature describing the development of tolerance, i.e., a reduction in effect for a given blood or brain alcohol concentration, within the confines of a single administration of alcohol.[10-12] While some of this effect may be the consequence of a delay between the concentration curve of ethanol in blood with that in brain during the absorptive phase,[3] this can be eliminated experimentally so that a definite tolerance due to changes in effect can be established over a time period measured in hours.

Acute tolerance to the effects of alcohol in man has been shown by psychomotor and cognitive tests and by written or instrumental assessments of euphoria and intoxication.[13,14] In almost all cases, the tolerance has been shown by comparing the effects of ethanol at a given blood level when the blood alcohol level is rising and then falling following a single administration of alcohol. However, in a recent experiment[15] on six nonalcoholic men, the blood alcohol level was held constant by giving maintenance doses of ethanol. Under these circumstances no acute tolerance was measured over a 6-h period. Whether this reflects the problems of experimental design and intersubject variability, or is a valid observation that a change in blood ethanol level is necessary to produce tolerance, will require further experimentation. It may be that a fall in concentration of alcohol at the site of action of alcohol is necessary before tolerance is achieved.

The significance of studies on acute tolerance lies in their application to mechanisms of tolerance, in general, and to the occasional social user of alcohol. It has been suggested that rapid development of acute tolerance may be paralleled by relatively rapid development of tolerance following chronic use of alcohol, i.e., that both forms of tolerance have a common basis.[16] This has been shown recently to apply to a group of male social drinkers whose development of acute tolerance during the first session of ethanol consumption correlated with the tolerance developing over four such sessions.[17] The social significance of acute tolerance is that alcohol-induced impairment of driving may be reduced if the driver were to wait until after the peak of the blood alcohol concentration curve.[18] However, evidence for the effectiveness of such behavior on accident rates or encouragement by changes in drink-driving legislation has yet to be observed.

Recent developments of a specific drug which can reverse the behavioral effects of alcohol[19] may provide a powerful tool for the study of acute and chronic tolerance, by enabling the effects of alcohol at one site to be separated from other, perhaps consequential effects occurring elsewhere which may influence the development of tolerance.

## B. Dispositional Tolerance

There has been extensive literature on alcohol disposition, particularly metabolism,[20] and several chapters of this volume are devoted to this topic. However, the literature on alcohol disposition in tolerant animals still remains confused.

Absorbtion of alcohol taken orally is one major experimental variable.[11,21] First-pass metabolism of alcohol by the liver or intestines may reduce the bioavailability of oral alcohol. I.p. administration of increasing doses of alcohol does not produce a dose-dependent increase in blood alcohol levels.[22] Thus, absorbtion and first-pass metabolism may be dose dependent.

First-pass metabolism is dependent on a variety of factors, and quite divergent results can be obtained if experimental conditions are varied.[23,24] Fructose given with ethanol appeared to reduce absorption, while fructose given 1 h after ethanol increased the metabolic clearance rate.[25] The nature of the alcoholic drink and the rate at which it is consumed[21] can also have marked effects on the absorption rate and, hence, the peak blood alcohol concentration. Other factors such as age,[26] sex, and menstrual cycle[21,27] may affect the peak blood alcohol level by altering the apparent volume of distribution.

The question of tolerance developing in chronic alcohol consumers due to an increased metabolic rate has been discussed for some time[1,20] without complete resolution. While hard scientific evidence is relatively lacking, anecdotal evidence for increased metabolism in heavy drinkers is strong. One young male at the end of a week's binge consumed over 500 g ethanol in a 12-h period and was an hour later in a "moderately advanced state of intoxication".[119] One study[28] indicated that ethanol metabolism in alcoholics is 50% higher at high blood levels (over 40 mmol/l) than at low (below 10 mmol/1). It was suggested that measurement of metabolism at low blood levels may misrepresent the actual metabolic capacity of heavy drinkers. Underestimation of the metabolic rate in drinkers may lie in the experimental design used by many studies.[29] Most subjects are given doses of ethanol to produce ethically acceptable blood levels (below 80 to 100 mg%, i.e., 20 mmol/l). However, absorption and distribution of such a dose may not be complete before 2 h or more, depending on the diluent used, while metabolism ceases to be kinetically zero order below 10 mmol/l, which is reached at about 5 h after starting consumption. Hence, there may be only about a 2- to 3-h period during which linear kinetics may be recorded. Measurement of metabolism without allowing complete absorption or after nonlinear clearance has commenced will result in an underestimate of the true metabolic capacity. (See also Volume I, Chapter 2.)

Whether continued alcohol use increases its metabolic rate remains to be determined. Careful consideration of problems of absorption and distribution as well as of metabolism is needed in designing experiments to determine metabolic rates in both alcoholics and nonalcohol users.

## C. Environmental and Behavioral Factors

It is well established that tolerance to alcohol can be influenced by a range of environmental or behavioral factors.[30-33] What remains in doubt is whether the environmentally dependent component of alcohol tolerance arises by classic conditioning,[34,35] in which cues in the environmental surrounds of the experiment trigger compensatory behavioral responses, by a learning process[36] of an operant or instrumental nature, or by habituation[37] in which the subject becomes accustomed to the stimuli without the development of compensatory behavior or responses.

While this debate continues,[38] the need for carefully designed and controlled experiments remains. It is most important that the environmentally dependent component[6,7] of tolerance be allowed for in all experiments. It must also be appreciated that the various contributing factors may not be independent, but may be interrelated. For example, it has been shown recently[39] that environment-dependent tolerance to ethanol in mice is due, at least in part, to dispositional changes induced by environmental cues. Mice made tolerant by repeated alcohol injections in a given environment had lower ethanol levels in brain and blood, suggesting some reduction in absorption had occurred. However, these mice, when given alcohol directly into the brain, still showed a reduced response, indicating that functional tolerance occurred as well as dispositional changes. Similarly, mice given alcohol in a liquid diet became more tolerant than injected mice, but this tolerance was not environment dependent. Thus, in this one series of experiments, dispositional, environmentally dependent, and functional tolerance were all identified. Discussion of the detailed mechanisms of tolerance becomes redundant if the major factors in tolerance, drug disposition and functional and behavioral changes, have not been measured and allowed for simultaneously.

Environmentally dependent tolerance to the hypothermic effects of alcohol has been much studied. In two recent papers,[40,41] interpretations differ on whether removal of the hypothermic experience causes the abolition of tolerance. When rats were treated with alcohol,[40] tolerance developed more slowly if the ambient temperature prevented a hypothermic response. However, after 20 d of treatment, tolerance could be shown to develop. When hypothermia was prevented by microwave treatment,[41] no tolerance was observed. Unfortunately, this treatment was maintained for only 7 d, so that it is possible that tolerance might have developed after longer treatment. Neither group tested for changes in alcohol disposition.

One difficulty in assessing the importance of environmental factors in tolerance is that tolerance development is not the same for all effects of alcohol.[1] In the work cited[41] tolerance to the ataxic effects could still be demonstrated even though tolerance to the hypothermic effect was suppressed. In other studies[42] tolerance to alcohol in mice was shown to occur more rapidly to the hypothermic effect than to the hypnotic effect. Tolerance to the effects of alcohol on an operant task could be modified behaviorally,[43] but not the effects on locomotion; some behavioral modification of tolerance to a tail-flick response to pain was observed.

One suggestion is that there are different mechanisms for tolerance development depending on whether the measured effect of ethanol is activating or depressant.[44] In mice, low doses of alcohol increase motor activity, but repeated alcohol administration does not produce tolerance to this effect, even if treatment is extended to 5 months.[45] A corresponding effect was observed in rats[46] where tolerance to hypothermic effects of alcohol could be produced, but not to the tachycardia effects. However, in man tolerance to the tachycardia produced by alcohol has been demonstrated[47] and is under environmental control. It would appear, therefore, that there is a marked species effect on tolerance production and it may be unwise to extrapolate from one species to another.

Tolerance may be influenced not only by geographical or spatial cues, but also by pharmacological effects. A low dose of alcohol can act as a conditional stimulus[48] to influence the response of rats, so that conditioned tolerance to the hypothermic effects of a higher dose of alcohol is produced.

In man tolerance development has been shown experimentally to be under strong behavioral control. Classical conditioning using alcohol-associated cues,[49,50] operant conditioning using contingent rewards,[49] and learning by means of mental rehearsal[51,52] have all been shown to increase the degree of tolerance developed over a short series of exposures to alcohol.

### D. Membrane Aspects of Tolerance

Functional tolerance, also known as cellular tolerance, arises due to some process which occurs in cells, presumably neurons, as a consequence of prolonged exposure to alcohol. Nonspecific depressant drugs, like general anesthetics[53,54] and alcohol, have for many years been thought to act by a physicochemical effect on membranes due to their ability to dissolve in the lipid phase of the cell membrane.[55] Over the past decade, however, there has been an increasing emphasis on the interaction of alcohol with membrane proteins,[56] such as enzymes, ion channels, or receptors.

It has been shown that ethanol *in vitro* increased the fluidity of both natural and artificial lipid membranes[57,58] and that membranes from animals made tolerant to ethanol are resistant to the fluidizing effects of ethanol.[59,60] Thus, there must be changes in the structure of the membrane, probably in the lipid components, due to exposure to alcohol.

Measurements of membrane fatty acids, cholesterol, phospholipids, and other lipid components from cells of alcohol-tolerant animals have shown variable and inconsistent changes.[60] For example, tolerant mice have membranes which contain more cholesterol[61] or less cholesterol;[62] in rats there was an increase[63] or no change[64] in the proportion of cholesterol. Alcohol tolerance tended to increase the proportion of saturated fatty acids compared with unsaturated[62,64,65] in some experiments, but not in others.[61,63] The causes of these variations in results are not clear. Many different tissues were used as a source of membranes, and their methods of purification differed considerably. Certainly the changes are small, though interestingly, in most cases, the membranes from tolerant animals are less fluid or are resistant to the fluidizing effect of alcohol.[62-64] Diet or nutritional status is one obvious possibility, as changing the fatty acid composition of membranes by dietary changes[66] can affect the sensitivity of the animal to alcohol.

The increase in emphasis on a specific effect of alcohol on membrane proteins,[67] particularly on ion channels,[19] may stimulate further new research. It can be shown that alcohol affects proteins as well as lipids in membranes[68,69] when added *in vitro*. Longer-term studies using alcohol-tolerant rats are obviously required.

One major problem in the lipid and membrane studies carried out to date is the smallness of the effects produced by alcohol. For example, the fluidity change produced by a 1° rise in temperature is the same as that produced by lethal concentrations of ethanol.[69] The membrane lipid changes produced in alcohol-tolerant animals are, thus, probably too small to account for the observed behavioral effects. This emphasizes the need for a major revision of our thinking about the site of action of alcohol both acutely and following prolonged use (for further details of ethanol-membrane interactions, see this volume, Chapter 11).

### E. Neurochemical Aspects of Tolerance

In a recent review, it was stated that "there is still no general agreement on the mechanism of alcohol intoxication, let alone tolerance and dependence."[70] It would appear, then, that the search for a neurochemical basis for tolerance to the effects of alcohol on the brain has not been particularly rewarding. Changes in brain neurochemistry could affect functional tolerance at a neuronal level, but also environmentally dependent tolerance by altering behavioral or learned responses. The biochemistry and neuropharmacology of the catecholamine and indoleamine neurotransmitters have been well studied in recent years, so the emphasis has naturally concentrated on them.[3,71-75] However, as these are only 3 of the 50 or so possible brain neurotransmitters, such a limited emphasis is very likely to lead to very limited results. In recent years, a greater diversity of methodologies has caused the emphasis to shift; recent developments[19] will stimulate interest in other transmitters and more specific effects on neurotransmitter systems.

Alcohol given acutely affects neurotransmitter release, but the effects may vary with the transmitter or the region studied.[3,71,75] For example, acetylcholine release is increased by alcohol in peripheral nerve terminals, but decreased in the brain. While some clear-cut results on specific neurotransmitters have been obtained, it is probable that they represent the consequence of other effects of alcohol rather than reflecting a causal mechanism.

Similarly, with studies in tolerant animals there has been considerable variation in the effects observed on various neurotransmitter receptors, release mechanisms, and metabolism.[71,75] For instance, depletion of brain catecholamines[71,76] or indoleamines[72,73,77] failed to produce consistent effects on the development of tolerance. Here, again, apart from species and strain differences, the variation may be due to the interaction of behavioral or learning changes induced by amine depletions with the experimental conditions associated with the production and measurement of the tolerance.

Research into the interaction of alcohol with the γ-amino-butyric acid (GABA) neurotransmitter system has been reviewed.[71,78] However, the GABA receptor, with its associated receptors for benzodiazepines, barbiturates, and other endogenous and exogenous pharmacological and toxicological agents, is currently the center of a storm of research, so that it would be unwise to consider it here until the dust and debris settle drastically. The recent study already cited[19] showing that a benzodiazepine-type drug can specifically antagonize the effects of alcohol on the GABA-mediated chloride uptake mechanism illustrates the type of sudden breakthrough that may occur when the appropriate investigative tools become available.

Also of current interest are the brain neuropeptides and their possible role in transmission and drug effects. Only 1 or 2 of the 40 or more putative peptide neurotransmitters have been studied in connection with alcohol tolerance. Thus, here is a very large potential area of research.

Vasopressin (AVP), a pituitary hormone and brain neuropeptide known to be affected by alcohol and to affect animal learning behavior, has been studied for its involvement in alcohol tolerance.[79,80] AVP and related peptides are involved in the maintenance of tolerance, though whether specifically or through mediation by indoleamine[81] or catecholamine[82] systems remains to be clarified. AVP inhibits the acquisition of an environmentally dependent tolerance to alcohol in mice;[83] it also affects the disposition of the injected alcohol. As commented, "it will be of interest to ascertain the effect of AVP on the acquisition of environment-independent ethanol tolerance," especially if the dispositional effects could be controlled.

Of the other brain neuropeptides, only relatively little is known. Neurotensin, bombesin, and β-endorphin potentiate the acute effects of alcohol,[84] while the corticotropin-releasing factor and thyroid- stimulating hormone are antagonistic; these effects may be strain selective.

The acute effects of alcohol on the cellular mechanisms mediating changes in neurotransmitter release and receptor mechanisms have been studied mainly through the effects on $Ca^{2+}$ transport and $Ca^{2+}$-mediated enzyme systems. Alcohol inhibits $Ca^{2+}$ uptake into nerve endings,[85] though the effect may be region or temperature dependent.[86] This may be due to the membrane disordering produced by alcohol. However, the disordering produced by various agents correlated better with changes in sodium transport[87] than with calcium transport. Thus, the effect of alcohol on calcium transport may not be just a nonspecific effect of alcohol.

Alcohol-tolerant or -dependent animals also show changes in $Ca^{2+}$ transport and sensitivity. In slices of rat striatal tissue, alcohol acutely reduces dopamine release presumably due to reduced $Ca^{2+}$ entry; in tolerant animals, release was increased[88] suggesting some compensatory mechanism. This is due to an increased response of the amine release mechanism[89,90] to a given amount of $Ca^{2+}$. A similarly increased release of noradrenaline from cortical or hypothalamic tissue from tolerant animals has also been reported,[91,92] though there were differences between brain regions.

Calcium-dependent enzymes in synaptic membranes are also altered in tolerant animals. Phospholipase $A_2$ and other enzymes affecting phospholipid metabolism in nerve cells are $Ca^{2+}$ dependent and may also, through phosphatidylinositol and arachidonic acid metabolism, affect intracellular $Ca^{2+}$, neurotransmitter release, and receptor functions. In tolerant animals, these enzymes are no longer inhibited by alcohol.[93] Similarly, $Ca^{2+}$-dependent phosphorylation of presynaptic nerve membranes was increased in alcohol-tolerant animals.[94] There would appear to be biochemical correlates of the increased sensitivity of nerve terminals to $Ca^{2+}$ in tolerant animals.

There have also been changes in the $Ca^{2+}$ receptor sites reported in alcohol-tolerant tissues. Cloned neuronal cells show reduced $Ca^{2+}$ uptake due to alcohol, but on continued exposure, this effect was reversed. Accompanying this tolerance to the effects on uptake was an increase in the number of sites binding the $Ca^{2+}$ channel-blocking drug nitrendipine.[95] Thus, it would appear that tolerance is accompanied by an increase in $Ca^{2+}$ channel receptor sites. These observations have possible clinical relevance in the use of calcium channel antagonists to decrease the severity of ethanol withdrawal.[96] Conversely, a calcium channel activator has been shown to potentiate the depressant effects of acute alcohol.[97]

Tolerance to alcohol appears then to be accompanied by compensatory effects to overcome the reductions in calcium inflow and neurotransmitter outflow which are produced by acute alcohol treatment. That this chain of events is not found uniformly in all experiments, tissues, and with all transmitters probably reflects the complexities of the central nervous system as well as those of tolerance itself.

## III. CROSS-TOLERANCE BETWEEN ALCOHOL AND OTHER DRUGS

In 1916 it was possible to write "It is a well recognized clinical fact that individuals addicted to the use of alcohol are more resistant to the narcotic effects of ether."[98] In 1981 the statement was made: "Common lore claims that alcoholics, when sober, are unusually resistant to hypnotic and sedative drugs."[99] In a 1982 review of cross-tolerance[100] it was observed that "that cross-tolerance occurs among drugs which exert similar effects possibly via a common mechanism," but that "there are conflicting data which cannot be ignored."

The reasons why progress into cross-tolerance has been slow are apparent when one considers the problems with the concept of tolerance discussed in the previous sections. It is necessary to allow for the presence of dispositional, functional, and behaviorally modified components of tolerance for each drug and for the interactions between drugs. Unfortunately, in much of the experimental work carried out, these variables have often been left uncontrolled, so that variability in results and difficulties in interpretation are bound to occur.

Functional cross-tolerance could result as a consequence of changes in cell membranes. The differences in membrane fluidity and partition coefficient in membranes from ethanol-fed rats were much greater with alcohol than with halothane and phenobarbital,[101] but the latter depressant drugs did produce effects that would be consistent with a membrane site for cross-tolerance. However, cross-tolerance with halothane or pentobarbitone is not always observed in alcohol-tolerant animals,[102] and at a molecular level, the mechanisms by which pentobarbitone causes membrane disordering appear to be different from those of alcohol.[69] This would be consistent with the recent observations[19] of a specific benzodiazepine-type antagonist for the effects of alcohol, but not of pentobarbitone.

The observations that alcohol-tolerant animals may become cross-tolerant to barbital, but not to pentobarbital,[103] are difficult to explain. Interestingly, this same research group at about the same time reported that alcohol-tolerant animals were cross-tolerant to pentobarbital in a motor activity test;[104] these animals also showed some cross-tolerance to a benzodiazepine.

The learning or behavioral components of alcohol tolerance have already been discussed and their importance stressed. Similarly, cross-tolerance has important learned components.[105,106] Pentobarbitone-treated rats appear to be cross-tolerant to ethanol only if they learn to control their ataxia while under the effects of pentobarbitone,[105] so that they can behave more normally when treated with alcohol. Similarly, rats learned to be tolerant to the analgesic effects of alcohol and then showed cross-tolerance to the analgesic effects of morphine and clonidine.[106] In this latter experiment it would seem unlikely that there would be any common receptor or membrane effect among these pharmacologically dissimilar analgesic agents.

Even if the learning component is minor or absent it is possible to achieve cross-tolerance under some circumstances, but not others. For example, when rats and mice were exposed to chronic alcohol treatment by a liquid diet, mice showed tolerance to ethanol and cross-tolerance to pentobarbitone, but not barbitone,[107] whereas rats showed no cross-tolerance to either pentobarbitone or barbitone, yet were tolerant to ethanol. Hence, there are obviously many factors which may influence both tolerance and cross-tolerance that are required to be identified.

## IV. THE RELATIONSHIP BETWEEN TOLERANCE AND DEPENDENCE

There have been many review papers discussing the relationships between tolerance and dependence and their connection, if any, with behaviors associated with alcohol consumption.[5,34,108-112] It would, therefore, seem unrewarding to detail in this limited review the many ideas and results that have been advanced, but an attempt will be made to summarize the main themes.

Tolerance, though a many-faceted concept, lends itself more easily to measurement. Physical dependence, being defined in terms of the severity of behaviors observed on withdrawal of the drug, is more difficult to quantify, but can be measured; however, dependence, especially psychological dependence, overlaps into the more vague areas of "addiction" and "craving", where moral overtones tend to intrude and definitions are often cyclical in argument.

Tolerance may be considered to be a process of adaptation to the presence of alcohol. It may be a change in absorption or metabolism (dispositional), in the behavioral effect of the drug (environmentally dependent), or at the neuronal or cellular level (functional). Physical dependence also may be viewed as a consequence of this adaptation; when the drug is removed, the perturbed system is no longer in equilibrium. This loss of equilibrium becomes manifest as the withdrawal syndrome of behavioral changes. One objection to this[109] is that withdrawal might be thought to occur only as the blood alcohol levels decrease, whereas the signs of physical dependence may not become maximal until some hours or even days after the blood alcohol levels have dropped to zero.[113] However, if the perturbation induced by the alcohol involves the synthesis or replacement of key proteins, e.g., receptors or ion channels, then removal of the drug could be complete in hours, but removal of the protein perturbation may take days. Thus, the time course of physical dependence does not necessarily have to relate to the time course of drug removal.

Dependence being measured in terms of a behavioral change is subject to the complexities of the processes that control behavior. Dependence may result from behavioral changes learned during exposure to the alcohol.[34] Experimentally, it has been possible to dissociate tolerance and dependence in some animals by neurochemical lesions.[74] However, these lesions, by interfering with brain neurotransmitter pathways, also affect the animals' behavioral responses,[72,76] so that it may be the manifestations of dependence that are altered, rather than the underlying processes that produce dependence. Equally, it is possible to influence the apparent severity of the dependence by suppressing the withdrawal reaction with agents such as clonidine[75,114,115] that modify the noradrenergically mediated manifestations of dependence.

Conversely, measurement of dependence without consideration of the concurrent development of tolerance can lead to misinterpretation of results. In strains of mice selectively bred for severity of withdrawal symptoms, the genetic difference appeared to be in the dispositional tolerance developed,[116] and other measures of tolerance and withdrawal severity were identical between strains if this factor were allowed for.

The relationships between tolerance and physical dependence and the behavioral problems referred to as alcohol abuse or addiction are even more complex and difficult to establish. The simplistic idea that dependent drinkers drink to prevent withdrawal symptoms has not been supported in practice, as human addicts have been observed to vary their intake and even go into spontaneous withdrawal of their own volition.[5,108] "Binge" drinkers are driven by powerful drives to drink, yet cease drinking quite independently of any development of tolerance or dependence. The term "craving" includes most of the ideas of psychological dependence, but cannot be exactly defined or measured and so, while commonly accepted and used, should be strenuously avoided as a scientific concept.

Behavioral or environmentally dependent tolerance plays a significant role in alcohol dependence. It is a common experience that drinking is encouraged and increased in familiar surroundings, at a particular time of day, or by certain tastes, flavors, company, or other associated factors.[34] Learning to control one's behavior to compensate for the effects of alcohol is common practice and produces apparent tolerance; it is not clear why drinkers increase their consumption to reinstate the loss of control.

It is commonly regarded that alcohol is consumed to relieve anxiety or to abolish unpleasant memories. Experimental evidence does not support this. Studies on autonomic function[75] are consistent with arousal both during consumption and also during withdrawal. In animals there would appear to be a complex, dose-dependent interrelationship[117,118] between stress and alcohol effects. A more appealing theory is that alcohol is consumed to produce a change in state:[108] if depressed one seeks arousal, if overexcited one seeks calming. This concept would have general applicability, since with other abused drugs which induce mood changes, the more rapid the change the more likely they are to be abused. However, as a basis for further experimental work this theory has little to offer in clarifying the causes of excessive alcohol consumption or the problems of tolerance and dependence.

Whatever the mechanism it is clear that factors in the genetic, behavioral, or environmental makeup of an individual can cause that person to consume an excessive amount of alcohol. It is not tolerance or dependence which is the problem, as patients can be tolerant to, or dependent on, a range of therapeutic agents, e.g., insulin, hexamethonium, or clonidine, without any major problems or hazard to life. The problem is the toxicity of alcohol in high doses on various organ systems of the body, including the brain. The combination of the nutritional deficits resulting from large doses of an unhealthy calorie source with more direct toxic effects produces a potentially damaging situation. If ethanol were a more potent drug it would perhaps be used more carefully and, therefore, cause less damage; the problems would be those of acute toxicity rather than those that appear to accompany the development of tolerance and dependence.

## V. CONCLUSIONS

It is misleading advertising even to offer any conclusions from this brief overview of a large and complex topic. Many of the earlier reviews already cited have provided summaries of past work and indications for future research. I will merely indicate here some of my own thoughts on the problems of tolerance and dependence and their relationship with alcohol abuse.

1.  Tolerance can and must be measured, not only as an entity, but also separately as its dispositional, environmentally dependent, and functional components. The design of experiments studying tolerance needs to consider the separate components of tolerance and their possible interrelationships.
2.  Dependence may be measured, as physical dependence, by means of behavioral or physiological responses to withdrawal of the drug. The relationships of physical dependence to tolerance remain to be clarified.
3.  Psychological tolerance, while a useful concept, seems to have eluded definition or measurement and so lacks scientific worth.
4.  Drinking behavior in man seems to be independent of either alcohol tolerance or physical dependence. While researchers may wish to persuade grant-giving bodies that research into tolerance and dependence could provide a cure for alcoholism, they should avoid this tender trap and design their experiments to establish the exact nature and interrelationships of tolerance and dependence as entities in their own right.

## REFERENCES

1. Advisory Committee to the Central Control Board (Liquor Traffic), Lord D'Abernon, Chairman, *Alcohol: Its Action on the Human Organism,* His Majesty's Stationery Office, London, 1918, 92.
2. **Rigter, H. and Crabbe, J. C.,** *Alcohol Tolerance and Dependence,* Elsevier, Amsterdam, 1980.
3. **Goldstein, D. B.,** *Pharmacology of Alcohol,* Oxford University Press, New York, 1983, chap. 7.

4. **Cicero, T. J.,** Alcohol self-administration, tolerance and withdrawal in humans and animals: theoretical and methodological issues, in *Alcohol Tolerance and Dependence,* Rigter, H. and Crabbe, J. C., Eds., Elsevier, Amsterdam, 1980, chap. 1.
5. **Cappell, H. and LeBlanc, A. E.,** The relationship of tolerance and physical dependence to alcohol abuse and alcohol problems, in *The Biology of Alcoholism,* Vol. 7, Kissin, B. and Begleiter, H., Eds., Plenum Press, New York, 1983, chap. 10.
6. **Tabakoff, B., Melchior, C. L., and Hoffman, P. L.,** Commentary on ethanol tolerance, *Alcoholism Clin. Exp. Res.,* 6, 252, 1982.
7. **Tabakoff, B. and Rothstein, J. D.,** Biology of tolerance and dependence, in *Medical and Social Aspects of Alcohol Abuse,* Tabakoff, B., Sutker, P. B., and Randall, C. L., Eds., Plenum Press, New York, 1983, chap. 7.
8. **Wood, J. M. and Laverty, R.,** Metabolic and pharmacodynamic tolerance to ethanol in rats, *Pharmacol. Biochem. Behav.,* 10, 871, 1979.
9. **Friedman, H. J.,** Assessment of physical dependence on and withdrawal from ethanol in animals, in *Alcohol Tolerance and Dependence,* Rigter, H. and Crabbe, J. C., Eds., Elsevier, Amsterdam, 1980, chap. 4.
10. **Mirsky, I. A., Piker, P., Rosenbaum, M., and Lederer, H.,** ''Adaptation'' of the central nervous system to varying concentrations of alcohol in the blood, *Q. J. Stud. Alcohol,* 2, 35, 1941.
11. **Goldberg, L.,** Quantitative studies on alcohol tolerance in man, *Acta Physiol. Scand.,* 5(Suppl. 16), 1945.
12. **Littleton, J. M.,** The assessment of rapid tolerance to ethanol, in *Alcohol Tolerance and Dependence,* Rigter, H. and Crabbe, J. C., Eds., Elsevier, Amsterdam, 1980, chap. 2.
13. **Radlow, R. and Hurst, P. M.,** Temporal relations between blood alcohol concentration and alcohol effect: an experiment with human subjects, *Psychopharmacology,* 85, 260, 1985.
14. **Lukas, S. E., Mendelson, J. H., and Benedikt, R. A.,** Instrumental analysis of ethanol-induced intoxication in human males, *Psychopharmacology,* 89, 8, 1986.
15. **Kaplan, H. L., Sellers, E. M., Hamilton, C., Naranjo, C. A., and Dorian, P.,** Is there acute tolerance to alcohol at steady state?, *J. Stud. Alcohol,* 46, 253, 1985.
16. **Khanna, J. M., Le, A. D., LeBlanc, A. E., and Shah, G.,** Initial sensitivity versus acquired tolerance to ethanol in rats selectively bred for ethanol sensitivity, *Psychopharmacology,* 86, 302, 1985.
17. **Beirness, D. and Vogel-Sprott, M.,** The development of alcohol tolerance: acute recovery as a predictor, *Psychopharmacology,* 84, 398, 1984.
18. **Mitchell, M. C.,** Alcohol-induced impairment of central nervous system function: behavioural skills involved in driving, *J. Stud. Alcohol,* Suppl. 10, 109, 1985.
19. **Suzdak, P. D., Glowa, J. R., Crawley, J. N., Schwartz, R. D., Skolnick, P., and Paul, S. M.,** A selective imidazobenzodiazepine antagonist of ethanol in the rat, *Science,* 234, 1243, 1986.
20. **Carpenter, T. M.,** The metabolism of alcohol: a review, *Q. J. Stud. Alcohol,* 1, 201, 1940.
21. **Dubowski, K.,** Absorption, distribution and elimination of alcohol: highway safety aspects, *J. Stud. Alcohol,* Suppl. 10, 98, 1985.
22. **Bloom, F., Lad, P., Pittman, Q., and Rogers, J.,** Blood alcohol levels in rats: non-uniform yields from intraperitoneal doses based on body weight, *Br. J. Pharmacol.,* 75, 251, 1982.
23. **Julkunen, R. J. K., Padova, C. D., and Lieber, C. S.,** First pass metabolism of ethanol — a gastrointestinal barrier against the systemic toxicity of ethanol, *Life Sci.,* 37, 567, 1985.
24. **Wagner, J. G.,** Lack of first-pass metabolism of ethanol at blood concentrations in the social drinking range, *Life Sci.,* 39, 407, 1986.
25. **Crownover, B. P., La Dine, J., Bradford, B., Glassman, E., Forman, D., Schneider, H., and Thurman, R. G.,** Activation of ethanol metabolism in humans by fructose: importance of experimental design, *J. Pharmacol. Exp. Ther.,* 236, 574, 1986.
26. **Vogel-Sprott, M. and Barrett, P.,** Age, drinking habits and the effects of alcohol, *J. Stud. Alcohol,* 45, 517, 1984.
27. **Boss, A. M., Kingstone, D., Marshall, A. W., Morgan, M. Y., and Sherlock, S.,** Ethanol elimination: influence of sex and menstrual cycle, *J. Physiol.,* 336, 76P, 1983.
28. **Keiding, S., Christensen, N. J., Damgaard, S. E., Dejgard, A., Iversen, H. L., Jacobsen, A., Johansen, S., Lundquist, F., Rubenstein, E., and Winkler, K.,** Ethanol metabolism in heavy drinkers after massive and moderate intake, *Biochem. Pharmacol.,* 32, 3097, 1983.
29. **Couchman, K. G.,** Do alcoholics metabolize alcohol faster than non-alcoholics?, *J. N. Z. Med. Soc. Alcohol Alcoholism,* 4, 138, 1981.
30. **Chen, C. S.,** Acquisition of behavioural tolerance to ethanol as a function of reinforced practice in rats, *Psychopharmacology,* 63, 285, 1979.
31. **Le, A. D., Poulos, C. X., and Cappell, H.,** Conditioned tolerance to the hypothermic effect of ethyl alcohol, *Science,* 206, 1109, 1970.
32. **Wenger, J. R., Tiffany, T. M., Bombadier, C., Nicholls, K., and Woods, S. C.,** Ethanol tolerance in the rat is learned, *Science,* 213, 575, 1981.

33. **Crowell, C. R., Hinson, R. E., and Siegel, S.,** The role of conditional drug responses in tolerance to the hypothermic effects of ethanol, *Psychopharmacology,* 73, 51, 1981.
34. **Hinson, R. E. and Siegel, S.,** The contribution of Pavlovian conditioning to ethanol tolerance and dependence, in *Alcohol Tolerance and Dependence,* Rigter, H. and Crabbe, J. C., Eds., Elsevier, Amsterdam, 1980, chap. 7.
35. **Siegel, S. and MacRae, J.,** Environmental specificity of tolerance, *Trends Neurol. Sci.,* 7, 140, 1984.
36. **Wenger, J. R., Berlin, V., and Woods, S. C.,** Learned tolerance to the behaviourally disruptive effects of ethanol, *Behav. Neural Biol.,* 28, 418, 1980.
37. **Goudie, A. J. and Griffiths, J. W.,** Behavioural factors in drug tolerance, *Trends Pharmacol. Sci.,* 7, 192, 1986.
38. **Tabakoff, B., Melchior, C. L., and Hoffman, P.,** Factors in ethanol tolerance, and reply by Wenger, J. R. and Woods, S. C., *Science,* 224, 523, 1984.
39. **Melchior, C. L. and Tabakoff, B.,** Features of environment-dependent tolerance to ethanol, *Psychopharmacology,* 87, 94, 1985.
40. **Le, A. D., Kalant, H., and Khanna, J. M.,** Influence of ambient temperature on the development and maintenance of tolerance to ethanol-induced hypothermia, *Pharmacol. Biochem. Behav.,* 25, 667, 1986.
41. **Hjeresen, D. L., Reed, D. R., and Woods, S. C.,** Tolerance to hypothermia induced by ethanol depends on specific drug effects, *Psychopharmacology,* 89, 45, 1986.
42. **Melchior, C. L. and Tabakoff, B.,** Modification of environmentally cued tolerance to ethanol in mice, *J. Pharmacol. Exp. Ther.,* 219, 175, 1981.
43. **Wigell, A. H. and Overstreet, D. H.,** Acquisition of behaviourally augmented tolerance to ethanol and its relationship to muscarinic receptors, *Psychopharmacology,* 83, 88, 1984.
44. **Tabakoff, B. and Kiianmaa, K.,** Does tolerance develop to the activating, as well as the depressant, effects of ethanol?, *Pharmacol. Biochem. Behav.,* 17, 1073, 1982.
45. **Masur, J., de Souza, M. L. O., and Zwicker, A. P.,** The excitatory effects of ethanol: absence in rats, no tolerance and increased sensitivity in mice, *Pharmacol. Biochem. Behav.,* 24, 1225, 1986.
46. **Peris, J. and Cunningham, C. L.,** Dissociation of tolerance to the hypothermic and tachycardic effects of ethanol, *Pharmacol. Biochem. Behav.,* 22, 973, 1985.
47. **Dafters, R. and Anderson, G.,** Conditioned tolerance to the tachycardia effect of ethanol in humans, *Psychopharmacology,* 78, 365, 1982.
48. **Greeley, J., Le, A. D., Poulos, C. X., and Cappell, H.,** Alcohol is an effective cue in the conditional control of tolerance to alcohol, *Psychopharmacology,* 83, 159, 1984.
49. **Beirness, D. and Vogel-Sprott, M.,** Alcohol tolerance in social drinkers: operant and classical conditioning effects, *Psychopharmacology,* 84, 393, 1984.
50. **Shapiro, A. P. and Nathan, P. E.,** Human tolerance to alcohol: the role of Pavlovian conditioning processes, *Psychopharmacology,* 88, 90, 1986.
51. **Vogel-Sprott, M., Rawana, E., and Webster, R.,** Mental rehearsal of a task under ethanol facilitates tolerance, *Pharmacol. Biochem. Behav.,* 21, 329, 1984.
52. **Annear, W. C. and Vogel-Sprott, M.,** Mental rehearsal and classical conditioning contribute to ethanol tolerance in humans, *Psychopharmacology,* 87, 90, 1985.
53. **Kaufman, R. D.,** Biophysical mechanisms of anaesthetic action, *Anaesthesiology,* 46, 49, 1977.
54. **Miller, K. W.,** The nature of the site of general anaesthesia, *Int. Rev. Neurobiol.,* 27, 1, 1985.
55. **Littleton, J. M.,** The effects of alcohol on the cell membrane: a possible basis for tolerance and dependence, in *Addiction and Brain Damage,* Richter, D., Ed., Croom Helm, London, 1980, chap. 3.
56. **Melgaard, B.,** The neurotoxicity of ethanol, *Acta Neurol. Scand.,* 67, 131, 1983.
57. **Goldstein, D. B.,** *Pharmacology of Alcohol,* Oxford University Press, New York, 1983, chap. 4.
58. **Goldstein, D. B.,** The effects of drugs on membrane fluidity, *Annu. Rev. Pharmacol. Toxicol.,* 24, 43, 1984.
59. **Rubin, E. and Rottenberg, H.,** Ethanol-induced injury and adaptation in biological membranes, *Fed. Proc.,* 41, 2465, 1982.
60. **Sun, G. Y. and Sun, A. Y.,** Ethanol and membrane lipids, *Alcoholism Clin. Exp. Res.,* 9, 164, 1985.
61. **Parsons, L. M., Gallaher, E. J., and Goldstein, D. B.,** Rapidly developing functional tolerance to ethanol is accompanied by increased erythrocyte cholesterol in mice, *J. Pharmacol. Exp. Ther.,* 223, 472, 1982.
62. **Harris, R. A., Baxter, D. M., Mitchell, M. A., and Hitzemann, R. J.,** Physical properties and lipid composition of brain membranes from ethanol tolerant-dependent mice, *Mol. Pharmacol.,* 25, 401, 1984.
63. **Crews, F. T., Majchrowicz, E., and Meeks, R.,** Changes in cortical synaptosomal plasma membrane fluidity and composition in ethanol-dependent rats, *Psychopharmacology,* 81, 208, 1983.
64. **La Droitte, P., Lamboeuf, Y., and de Saint-Blanquat, G.,** Lipid composition of the synaptosome and erythrocyte membranes during chronic ethanol treatment and withdrawal in the rat, *Biochem. Pharmacol.,* 33, 615, 1984.

65. **Aloia, R. C., Paxton, J., Daviau, J. S., van Gelb, O., Mlekusch, W., Truppe, W., Meyer, J. A., and Brauer, F. S.,** Effect of chronic alcohol consumption on rat brain microsome lipid composition, membrane fluidity and $Na^+$-$K^+$-ATPase activity, *Life Sci.,* 36, 1003, 1985.

66. **John, G. R., Littleton, J. M., and Jones, P. A.,** Membrane lipids and ethanol tolerance in the mouse. The influence of dietary fatty acid composition, *Life Sci.,* 27, 545, 1980.

67. **Franks, N. P. and Lieb, W. R.,** The pharmacology of simple molecules, *Arch. Toxicol.,* Suppl. 9, 27, 1986.

68. **Logan, B. J., Laverty, R., and Peake, B. M.,** ESR measurements on the effects of ethanol on the lipid and protein conformation in biological membranes, *Pharmacol. Biochem. Behav.,* 18 (Suppl. 1), 31, 1983.

69. **Logan, B. J. and Laverty, R.,** Comparative effects of ethanol and other depressant drugs on membrane order in rat synaptosomes using ESR spectroscopy, *Alcohol Drug Res.,* 7, 11, 1986.

70. **Littleton, J.,** Basic science and alcoholism, *Br. J. Addict.,* 81, 450, 1986.

71. **Tabakoff, B. and Hoffman, P. L.,** Alcohol and neurotransmitters, in *Alcohol Tolerance and Dependence,* Rigter, H. and Crabbe, J. C., Eds., Elsevier, Amsterdam, 1980, chap. 8.

72. **Wood, J. M.,** Effect of depletion of brain 5-hydroxytryptamine by 5,7-dihydroxytryptamine on ethanol tolerance and dependence in the rat, *Psychopharmacology,* 67, 67, 1980.

73. **Khanna, J. M., Kalant, H., Le, A. D., and LeBlanc, A. E.,** Role of serotonergic and adrenergic systems in alcohol tolerance, *Prog. Neuropsychopharmacol.,* 5, 459, 1981.

74. **Tabakoff, B. and Hoffman, P. L.,** Neurochemical aspects of tolerance to and physical dependence on alcohol, in *The Biology of Alcoholism,* Vol. 7, Kissin, B, and Begleiter, H., Eds., Plenum Press, New York, 1983, chap. 7.

75. **Nutt, D. and Glue, P.,** Monoamines and alcohol, *Br. J. Addict.,* 81, 327, 1986.

76. **Wood, J. M. and Laverty, R.,** Effect of depletion of brain catecholamines on ethanol tolerance and dependence, *Eur. J. Pharmacol.,* 58, 285, 1979.

77. **Le, A. D., Khanna, J. M., Kalant, H., and LeBlanc, A. E.,** Effect of modification of brain serotonin, norepinephrine and dopamine on ethanol tolerance, *Psychopharmacology,* 75, 231, 1981.

78. **Kulonen, E.,** Ethanol and GABA, *Med. Biol.,* 61, 147, 1983.

79. **Crabbe, J. C. and Rigter, H.,** Hormones, peptides and ethanol responses, in *Alcohol Tolerance and Dependence,* Rigter, H. and Crabbe, J. C., Eds., Elsevier, Amsterdam, 1980, chap. 12.

80. **Hoffman, P. L. and Tabakoff, B.,** Neurohypophyseal peptides maintain tolerance to the incoordinating effects of ethanol, *Pharmacol. Biochem. Behav.,* 21, 539, 1984.

81. **Le, A. D., Kalant, H., and Khanna, J. M.,** Interaction between des-glycinamine-arg-vasopressin and serotonin on ethanol tolerance, *Eur. J. Pharmacol.,* 80, 337, 1982.

82. **Hoffman, P. L., Melchior, C. L., and Tabakoff, B.,** Vasopressin maintenance of ethanol tolerance requires intact brain noradrenergic systems, *Life Sci.,* 32, 1065, 1983.

83. **Mannix, S. A., Hoffman, P. L., and Melchior, C. L.,** Intraventricular arginine-vasopressin blocks the acquisition of ethanol tolerance in mice, *Eur. J. Pharmacol.,* 128, 137, 1986.

84. **Erwin, V. G., Korte, A., and Marty, M.,** Neurotensin selectively alters ethanol-induced anaesthesia in LS/Ibg and SS/Ibg lines of mice, *Brain Res.,* 400, 80, 1987.

85. **Stokes, J. A. and Harris, R. A.,** Alcohols and synaptosomal calcium transport, *Mol. Pharmacol.,* 22, 99, 1982.

86. **Michaelis, M. L., Michaelis, E. K., and Tehan, T.,** Alcohol effects on synaptic membrane calcium ion fluxes, *Pharmacol. Biochem. Behav.,* 18 (Suppl. 1), 19, 1983.

87. **Harris, R. A. and Bruno, P.,** Membrane disordering by anaesthetic drugs: relationship to synaptosomal sodium and calcium fluxes, *J. Neurochem.,* 44, 1274, 1985.

88. **Lynch, M. A. and Littleton, J. M.,** Possible association of alcohol tolerance with increased $Ca^{2+}$ sensitivity, *Nature,* 303, 175, 1983.

89. **Lynch, M. A., Archer, E. R., and Littleton, J. M.,** Increased sensitivity of transmitter release to calcium in ethanol tolerance, *Biochem. Pharmacol.,* 35, 1207, 1986.

90. **Leslie, S. W., Woodward, J. J., Wilcox, R. E., and Farrar, R. P.,** Chronic ethanol treatment uncouples striatal calcium entry and endogenous dopamine release, *Brain Res.,* 368, 174, 1986.

91. **Lynch, M. A. and Littleton, J. M.,** Enhanced $^3$H-noradrenaline release in synaptosomes from ethanol-tolerant animals: the role of nerve terminal calcium ion concentrations, *Alcohol Alcoholism,* 20, 5, 1985.

92. **Daniell, L. C. and Leslie, S. W.,** Inhibition of fast phase calcium uptake and endogenous norepinephrine release in rat brain region synaptosomes by ethanol, *Brain Res.,* 377, 18, 1986.

93. **John, G. R., Littleton, J. M., and Nhamburo, P. T.,** Increased activity of $Ca^{2+}$-dependent enzymes of membrane liquid metabolism in synaptosomal preparations from ethanol-dependent rats, *J. Neurochem.,* 44, 1235, 1985.

94. **Shanley, B., Gurd, J., and Kalant, H.,** Ethanol tolerance and enhanced calcium/calmodulin-dependent phosphorylation of synaptic membrane proteins, *Neurosci. Lett.,* 58, 55, 1985.

95. **Messing, R. O., Carpenter, C. L., Diamond, I., and Greenberg, D. A.,** Ethanol regulates calcium channels in clonal neural cells, *Proc. Natl. Acad. Sci. U.S.A.,* 83, 6213, 1986.

96. **Dolin, S. J., Halsey, M. J., and Little, H. J.,** Calcium channel antagonists decrease the ethanol withdrawal syndrome, *Br. J. Pharmacol.,* 87, 40P, 1986.
97. **Dolin, S. J. and Little, H. J.,** The effects of Bay K 8644 on the general anaesthetic potencies of ethanol and argon, *Br. J. Pharmacol.,* 89, 622P, 1986.
98. **Myers, H. B.,** Cross tolerance. Altered susceptibility to codeine, heroin, cannabis-indica and chloral hydrate in dogs having an acquired tolerance for morphine, *J. Pharmacol. Exp. Ther.,* 8, 417, 1916.
99. **Lau, C. E., Tang, M., and Falk, J. L.,** Cross-tolerance to phenobarbital following chronic ethanol polydipsia, *Pharmacol. Biochem. Behav.,* 15, 471, 1981.
100. **Khanna, J. M. and Mayer, J. M.,** An analysis of cross-tolerance among ethanol, other general depressants and opioids, *Subst. Alcohol Actions Misuse,* 3, 243, 1982.
101. **Rottenberg, H., Waring, A., and Rubin, E.,** Tolerance and cross-tolerance in chronic alcoholics: reduced membrane binding of ethanol and other drugs, *Science,* 213, 583, 1981.
102. **Wood, J. M. and Laverty, R.,** Cross-tolerance between ethanol and other anaesthetic agents, *Proc. Univ. Otago Med. Sch.,* 56, 108, 1978.
103. **Gougos, A., Khanna, J. M., Le, A. D., and Kalant, H.,** Tolerance to ethanol and cross-tolerance to pentobarbital and barbital, *Pharmacol. Biochem. Behav.,* 24, 801, 1986.
104. **Le, A. D., Khanna, J. M., Kalant, H., and Grossi, F.,** Tolerance to and cross-tolerance among ethanol, pentobarbital and chlordiazepoxide, *Pharmacol. Biochem. Behav.,* 24, 93, 1986.
105. **Wenger, J. R., McEvoy, P. M., and Woods, S. C.,** Sodium pentobarbital-induced cross-tolerance to ethanol is learned in the rat, *Pharmacol. Biochem. Behav.,* 25, 35, 1986.
106. **Jorgensen, H. A., Fasmer, O. B., and Hole, K.,** Learned and pharmacologically-induced tolerance to ethanol and cross-tolerance to morphine and clonidine, *Pharmacol. Biochem. Behav.,* 24, 1083, 1986.
107. **Logan, B. J., Khanna, J. M., and Laverty, R.,** Species difference in cross-tolerance between ethanol and pentobarbitone, *Proc. Univ. Otago Med. Sch.,* 62, 41, 1984.
108. **Mello, N. K. and Mendelson, J. H.,** Alcohol and human behaviour, in *Drugs of Abuse, Handbook of Psychopharmacology,* Vol. 12, Iversen, L. L., Iversen, S. D., and Snyder, S. H., Eds., Plenum Press, New York, 1978, chap. 5.
109. **Goldstein, D. B.,** Physical dependence on ethanol: its relation to tolerance, *Drug Alcohol Depend.,* 4, 33, 1979.
110. **Littleton, J. M.,** Tolerance and physical dependence on alcohol at the level of synaptic membranes: a review, *J. R. Soc. Med.,* 76, 593, 1983.
111. **Tabakoff, B.,** A neurobiological hypothesis for prevention of alcoholism, in *Currents in Alcohol Research and the Prevention of Alcohol Problems,* von Wartburg, J.-P., Magnenat, P., Muller, R., and Wyss, S., Eds., Hans Huber, Berne, 1985, 33.
112. **Skinner, H. A.,** Alcohol dependence: how does it come about? Comment on Griffith Edward's "The Alcohol Dependence Syndrome: concept as a stimulus to enquiry", *Br. J. Addict.,* 81, 193, 1986.
113. **Wood, J. M. and Laverty, R.,** Physical dependence following prolonged ethanol or t-butanol administration to rats, *Pharmacol. Biochem. Behav.,* 10, 113, 1979.
114. **Bjorkqvist, S. E.,** Clonidine in alcohol withdrawal, *Acta Psychiatr. Scand.,* 52, 256, 1975.
115. **Parale, M. P. and Kulkarni, S. K.,** Studies with $a_2$-adrenoceptor agonists and alcohol abstinence syndrome in rats, *Psychopharmacology,* 88, 237, 1986.
116. **Crabbe, J. and Kosobud, A.,** Sensitivity and tolerance to ethanol in mice bred to be genetically prone or resistant to ethanol withdrawal seizures, *J. Pharmacol. Exp. Ther.,* 239, 327, 1986.
117. **Littleton, J. M.,** Alcohol, alcoholism and affect. The biological connection, in *Pharmacological Treatments for Alcoholism,* Edwards, G. and Littleton, J., Eds., Croom Helm, London, 1984, chap. 22.
118. **Vogel, W. H., De Turk, K., and Miller, J. M.,** Differential effects of ethanol on plasma catecholamine levels in rats, *Biochem. Pharmacol.,* 35, 3983, 1986.
119. **Laverty, R.,** personal communication.

Chapter 13

EFFECTS OF ETHANOL ON AUTONOMIC FUNCTION

**Ralph H. Johnson, David G. Lambie, and Graeme Eisenhofer**

TABLE OF CONTENTS

# I. INTRODUCTION

The most obvious effects of ethanol on the brain relate to higher nervous system dysfunction. It is, however, becoming increasingly recognized that effects of ethanol on the autonomic nervous system are also of significance, involving both acute and chronic effects on the sympathetic and the parasympathetic branches.

# II. ACUTE EFFECTS

## A. Sympathetic Nervous System

Ethanol increases urinary excretion of catecholamines[1-6] and plasma concentrations of catecholamines[7-9] (Figure 1), suggesting an acute stimulatory effect of ethanol on the sympathetic nervous system. Other studies, however, have shown that ethanol inhibits norepinephrine release from peripheral nerves,[10,11] although spontaneous release from the rat vas deferens was found to be increased.[12] Ethanol ingestion has been shown to depress the circulatory clearance of norepinephrine, suggesting that raised plasma norepinephrine concentrations after ethanol may result from an inhibitory action on uptake mechanisms and/or metabolism of norepinephrine.[7] Findings that ethanol increases plasma norepinephrine while reducing plasma concentrations of the intraneuronal norepinephrine metabolite dihydroxyphenylglycol[8] are consistent with an inhibitory action of ethanol on neuronal uptake and/or oxidative metabolism of norepinephrine that might raise plasma norepinephrine concentrations. Further studies are required to assess the acute effects of ethanol on the release and clearance of catecholamines across particular vascular beds and in particular organs and tissues, and also to establish whether chronic exposure to ethanol modifies catecholamine release and clearance in response to acute ethanol ingestion.[13] In one study in chronic alcoholics, enhanced urinary catecholamine excretion was found to continue throughout the course of long-term (up to 3 weeks) ethanol intake, without any adaptive response to indicate increased adrenergic tolerance to ethanol.[14] A shift from oxidative to reduced modes of catecholamine metabolism was found, but this has also been reported to occur after a single dose of ethanol.[15]

Changes in plasma catecholamines and sympathetic nervous function after acute ethanol ingestion may be responsible for some of the cardiovascular effects of ethanol, such as increase in heart rate, changes in blood pressure, and vasodilatation. Attempts to establish links between sympathetic nervous activity and cardiovascular changes after ethanol ingestion have, however, proved inconclusive. A pressor effect of acute ethanol ingestion has been emphasized by some investigators, but this effect was shown to be of short duration, limited to systolic blood pressure and unrelated to changes in plasma catecholamines.[9] A more protracted effect of ethanol ingestion is a lowering of blood pressure, especially in the standing position or during lower body negative pressure.[16] The increase in heart rate and catecholamines after ethanol may reflect an autonomic response to maintain blood pressure in the face of peripheral vasodilatation. Cardiovascular effects of ethanol may also involve an autonomically mediated response to the action of ethanol in reducing cardiac contractility.[17,18]

It appears likely that changes in blood pressure after ethanol may be related to the cutaneous vasodilatory action of ethanol;[19,20] however, vasoconstriction has been observed in some vascular beds, most commonly when ethanol is infused intravenously.[19,20] The vascular effects of ethanol may depend not only on the particular vascular bed and mode of administration, but also on the dose of ethanol and on the relative concentrations of ethanol and of acetaldehyde as its metabolic product. The vasodilator effect of ethanol was previously considered to be mediated by actions on central nervous structures, with peripheral effects considered insignificant.[21] Evidence now indicates that ethanol has direct inhibitory effects

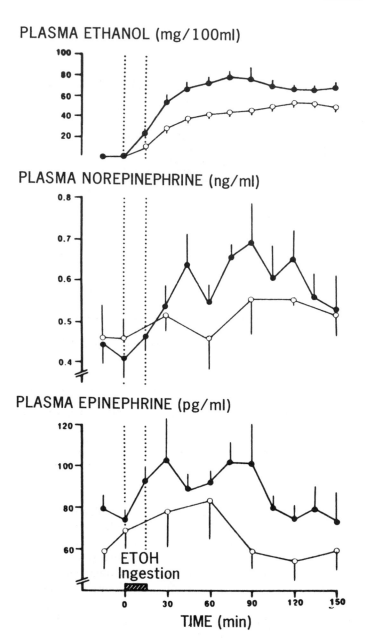

FIGURE 1.   Plasma ethanol, norepinephrine, and epinephrine (mean ± SE) concentrations in nine normal males before and after consumption of 0.5 (○) and 1 (●) ml/kg ethanol (ETOH). (From Eisenhofer, G., Lambie, D. G., and Johnson, R. H., *Clin. Pharmacol. Ther.*, 34, 143, 1983. With permission.)

on adrenergic postsynaptic mechanisms[22] and depresses blood pressure responses to α-adrenergic stimulation[23,24] (Figure 2). It is likely that this latter effect is not specific to α-adrenergic mechanisms, but is related to a generalized action of ethanol in inhibiting smooth muscle contractile responses to a variety of vasoactive agents.[25,26] This may be the cause of impaired blood pressure control[8,16] and of reduced pressor responses to physical[27] and mental[16,28] stress after acute ethanol ingestion.

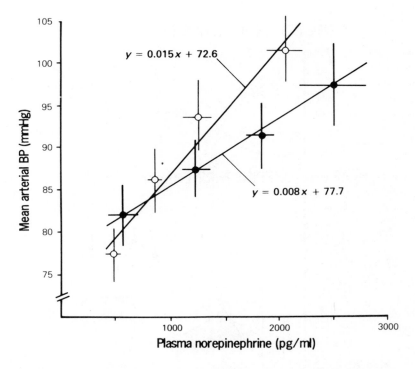

FIGURE 2.  Relationships between mean arterial blood pressure (y) and plasma nor-epinephrine (x) in eight normal males with (●) and without (○) prior consumption of ethanol during graded i.v. infusions of norepinephrine (24, 48, 90 mg kg$^{-1}$min$^{-1}$). The four points for each relationship represent the mean ($\pm$ SE) of all blood pressure and norepinephrine determinations for the eight subjects studied, under basal conditions and during each of the three norepinephrine infusion periods. Ethanol consumption significantly ($p < 0.01$) reduced the slope of the relationship. (From Eisenhofer, G., Lambie, D. G., and Johnson, R. H., *Br. J. Clin. Pharmacol.*, 18, 581, 1984. With permission.)

*1. Micturition Syncope*

Peripheral vasodilatation may combine with low blood volume due to the diuretic effect of ethanol to dispose the alcohol drinker to orthostatic hypotension and syncope. Also, afferent pathways from baroreceptors may be blocked by ethanol.[29]

Syncope following ethanol ingestion is particularly associated with micturition. Micturition syncope is a disorder found predominantly in men, often young adults, and syncope occurs at the end of micturition or soon after. Loss of consciousness is abrupt and recovery rapid and complete.[30] In most subjects it occurs only occasionally, although in a few it is recurrent.[31,32] In micturition syncope a full bladder is emptied quickly and vasoconstriction which occurs as a result of distension of viscera, including the bladder, gives way to dilatation. In a few subjects this dilatation and erect posture are enough to lower the blood pressure sufficiently to cause a faint.

*2. Accidental Hypothermia*

Another situation in which peripheral dilatation due to ethanol might be hazardous occurs when ethanol leads to an impaired level of consciousness with the subject exposed to cold. Accidental hypothermia is common in alcoholic populations;[33] however, the importance of vasodilatation in alcoholic hypothermia appears doubtful. The vasodilatory response to ethanol is markedly reduced at low environmental temperature.[34]

The susceptibility to hypothermia in alcoholics might be related to a central depressant effect of ethanol on temperature regulating centers in the hypothalamus, causing peripheral vasodilatation and inhibition of shivering. How ethanol inhibits shivering requires elucidation.[35]

Neuropeptides, including vasopressin, appear to be involved in the development or maintenance of tolerance to the hypothermic effect of ethanol.[36,37] Ethanol inhibits plasma vasopressin release by decreasing the response of the osmosodium receptors to changes in plasma tonicity.[38] The depression of vasopressin by ethanol may therefore contribute to the development of hypothermia.

Structural damage in the hypothalamus, discussed later, plays an important role in causing hypothermia in some alcoholic individuals.

*3. Facial Flushing*

Facial flushing is described in detail in Volume II, Chapter 16. A summary only is given here.

Facial flushing after drinking ethanol may be particularly observed in some individuals of Oriental origin.[39-41] Facial flush appears to be related to high blood acetaldehyde concentrations, associated with a deficiency in an isoenzyme of aldehyde dehydrogenase which has a low $K_m$ for acetaldehyde.[42] There is evidence of increased response of plasma catecholamines to ethanol in such individuals. Subjects with facial flushing also show marked increases in heart rate and decreases in diastolic blood pressure after ethanol.[42] Individuals who may be at risk of future development of alcoholism have been shown to have a significantly increased incidence of ethanol-induced facial flushing, suggesting a possible basis for a genetically inherited risk for future development of alcoholism.[43] The cardiovascular responses to ingestion of ethanol during treatment with disulfiram also include facial flushing and tachycardia, associated with elevation of blood acetaldehyde and plasma catecholamines. Disulfiram administered alone may raise plasma noradrenaline and blood pressure.[44]

Facial flush may be seen following even small amounts of ethanol in some noninsulin-dependent diabetic patients on chlorpropamide therapy.[45] This has been suggested to be an autosomal-dominant, inherited trait,[46] and diabetics with chlorpropamide-ethanol flush have been reported to show relatively low frequencies of diabetic retinopathy[47] and large-vessel disease.[48] The mechanism probably involves prostaglandins[48,49] rather than an effect on catecholamine release from sympathetic fibers.

## B. Parasympathetic Nervous System

Tachycardia may be associated with acute ethanol ingestion,[50] probably resulting from increased sympathetic activity. A role of decreased vagal activity appears unlikely, as ethanol does not affect the heart rate response to atropine.[51]

# III. CHRONIC EFFECTS

## A. Sympathetic Nervous System

Alcoholic patients may exhibit sympathetic dysfunction, and this may affect both blood pressure regulation and thermoregulation.

*1. Blood Pressure Regulation*

An association between heavy ethanol consumption and raised blood pressure may account for 30% of hypertension in affluent countries.[52-54] One study suggested that regular use of three or more drinks per day containing ethanol is a risk factor for hypertension.[54] Hypertension is common before and during withdrawal, but it is relatively rare in alcoholics who have ceased drinking.[55,56] The cause of the association remains uncertain and there are probably a variety of factors. Sympathetic overactivity may play a part. A study of randomly selected heavy drinkers showed that the resting plasma norepinephrine concentrations were slightly greater than in individuals with low ethanol intake, although there was no correlation between plasma norepinephrine and blood pressure.[57] It has been suggested that an acute effect of ethanol in raising circulating epinephrine concentrations may dispose to hypertension,[9,58] although a difference in sympathetic activity between drinkers and nondrinkers remains to be established. Plasma concentrations of free and conjugated catecholamines were not found to be different between nondrinkers and drinkers with higher blood pressures.[59]

Another possibility is that hypertension after chronic ethanol consumption may reflect the result of adaptive changes in the mechanisms controlling blood pressure following prolonged exposure to the acute vasodepressant effects of ethanol. This is supported by findings of tolerance to the vasodepressant effects of acute ethanol after prolonged exposure in rats.[25] However, findings of reduced pressor responsiveness to norepinephrine despite raised blood pressure, after 4 d of moderate (80 g) ethanol consumption, indicate that vasoconstrictor mechanisms are not potentiated by continued ethanol consumption over this time period.[24]

Increased sympathetic nervous activity may contribute to hypertension and other symptoms (including tremor, sweating, and tachycardia) during ethanol withdrawal,[60] when catecholamines and catecholamine metabolites are substantially raised in cerebrospinal fluid,[61-64] blood,[65-69] and urine[4,61] (Figure 3). There is also evidence that during alcohol withdrawal there is greater sodium retention in hypertensive subjects.[70] It is therefore possible that sodium retention contributes to the hypertension and that the raised blood pressure is not solely dependent upon sympathetic nervous system activity as has been suggested.[71]

α-Adrenoreceptor densities have been shown to be reduced both centrally[72] and peripherally[73] after acute ethanol administration and during withdrawal from ethanol in the rat, possibly as a response to raised catecholamine concentrations. Evidence for reduced α-adrenoreceptor-mediated responses has been found in normal subjects after 4 d of ethanol consumption[24] (Figure 4) and in alcoholic subjects.[74] Increased β-adrenoreceptor activity was found in a rat heart preparation associated with ethanol withdrawal,[75] although chronic ethanol ingestion may impair β-adrenoreceptor-mediated circulatory responses as indicated by impaired reactivity to β-adrenoreceptor agonists in alcoholic patients[76] and rats.[77] More work is required to determine the involvement of alterations of central and peripheral α-adrenoreceptor and β-adrenoreceptor activity in the clinical symptoms of ethanol withdrawal. It is possible that the increase in central sympathetic activity could be the result of central β-adrenoreceptor supersensitivity.

Orthostatic hypotension occurs occasionally in withdrawing alcoholic patients,[66] although it is generally considered rare.[78,79] One possible explanation is that orthostatic hypotension during withdrawal reflects reduced α-adrenergic receptor activity as a response to increased sympathetic nerve activity. Dehydration may also be a factor. It has also been suggested that orthostatic hypotension may be related to thiamine administration causing peripheral vasodilatation,[80] but this remains to be confirmed.

In some chronic alcoholics who have orthostatic hypotension following withdrawal from ethanol, there may be impaired catecholamine secretion.[64] Orthostatic hypotension in these patients is, therefore, probably due to sympathetic nervous system dysfunction.[81] Sympathetic failure with ethanol was suggested to be due to a block of the baroreceptors on the afferent side of the reflex arc.[29] However, subsequent studies have emphasized efferent sympathetic abnormalities, as shown by absence of blood pressure responses to noise and mental arithmetic and by absence of the sympathetic venoarteriolar axon reflex.[82] There has also been pathological evidence of abnormal ganglion cells and myelin degeneration in sympathetic chains.[83,84] A pathological study of patients with alcoholic neuropathy, nevertheless, failed to reveal degeneration in the splanchnic nerves.[78]

Plasma norepinephrine concentrations are raised in patients with liver cirrhosis.[85,86] This

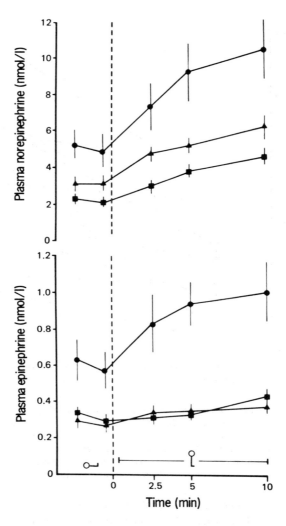

FIGURE 3. Plasma catecholamine concentrations (mean + SE) in the supine position ( ) and in response to standing ( ) in ten withdrawing alcoholics (●), ten abstinent alcoholics (▲), and ten control subjects (■). Withdrawing alcoholics had supine and standing norepinephrine and epinephrine concentrations which were significantly ($p < 0.05$) higher than abstinent alcoholics or control subjects. Abstinent alcoholics had supine norepinephrine concentrations which were significantly ($p < 0.05$) higher than control subjects. Plasma epinephrine concentrations 10 min after standing were significantly ($p < 0.01$) higher than supine concentrations in all groups. The withdrawing alcoholics had a fall of blood pressure from 146/84 to 126/82 mmHg during the procedure, whereas there was little or no change in the other groups. (From Eisenhofer, G., Whiteside, E. A., and Johnson, R. H., *Clin. Sci.*, 68, 71, 1985. With permission.)

is not due to reduced splanchnic elimination,[87,88] but may be due to increased renal release, causing sodium and water retention as a response to maintain blood pressure. Negative correlations were shown in cirrhotic patients between plasma norepinephrine and urinary sodium excretion,[89] glomerular filtration rate,[90] and renal blood flow.[91]

*2. Thermoregulation*

Peripheral damage to sympathetic nerves may affect pathways subserving sweating in chronic alcoholics. Sweating loss is usually distal in a glove-and-stocking distribution.[78] The consequence of this might be expected to be reduced fluid loss and thermal instability,

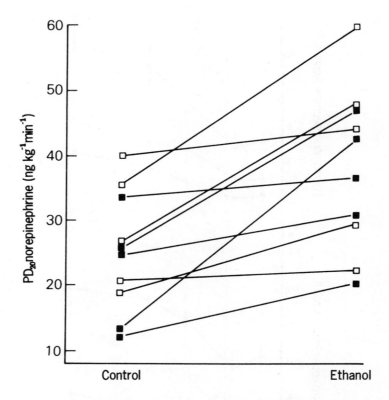

FIGURE 4.   $PD_{20}$ for norepinephrine (i.e., infusion dose of agonist required to raise mean arterial pressure by 20 mmHg) following 4 d of consuming 80 g of ethanol per day (ethanol) and 4 d of abstaining from ethanol (control). The lines join values from individual nonalcoholic subjects: ■ — males, □ — females. Vascular responsiveness to norepinephrine infusions was reduced. (From Howes, L. G. and Reid, J. L., *Br. J. Clin. Pharmacol.*, 20, 669, 1985. With permission.)

with hyperthermia being more easily precipitated. Surprisingly, it was reported that chronic alcoholics have a higher loss of fluid during heat stress compared to control subjects.[92] This could be due to the development of compensatory sweating, which has been observed in diabetics with incomplete anhidrosis.[93] In a study of the thermoregulatory responses to acute heat stress, however, alcoholic subjects were found to have higher temperatures after heat exposure. They had reduced weight loss, impaired distal sweating, and there was no evidence of compensatory sweating.[94]

A further problem of thermoregulation in chronic alcoholism may be hypothermia.[33] This may occur as a complication of Wernicke's encephalopathy. This disorder is due to thiamine deficiency, and small petechial hemorrhages occur in the hypothalamus and upper brain stem including the corpora mammalaria. Characteristically, there are disturbances of consciousness and memory, abnormalities of eye movements, ataxia, and often also polyneuropathy. Hypothermia[95] in this disorder may be the result of a hypothalmic lesion. One of Wernicke's original series in 1881 developed hypothermia[96] and it has been reported in a number of patients since then.[97-99] It is probable that hypothermia may frequently be missed unless expressly looked for and, since Wernicke's encephalopathy responds dramatically to thiamin, it is a condition which it is important to diagnose. Both absence of thermoregulatory responses and lack of discomfort to thermal stress may cause hypothermia in patients with Wernicke's encephalopathy.[100] The acute effects of ethanol in causing peripheral vasodilatation, described previously (Figure 5), might also contribute to hypothermia in such patients.

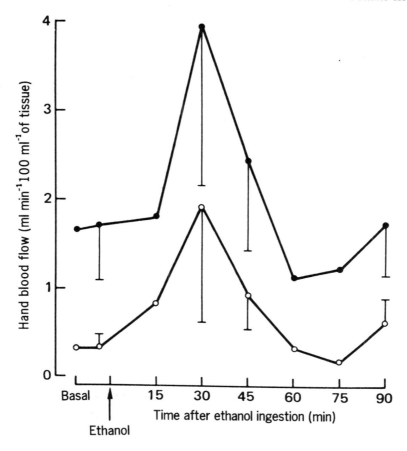

FIGURE 5. Effect of ethanol ingestion in four normal males, 0.5 g/kg body weight, at 21°C on hand blood flow: plethysmograph temperature 34°C, ●—●, left hand; 28°C, ○—○, right hand. Values are means ± SE. Although basal hand blood flows were smaller at 28°C, increases were observed in hand blood flow in both hands after ethanol ingestion. (From Fellows, I. W., MacDonald, I. A., and Bennett, T., *Clin. Sci.*, 66, 733, 1984. With permission.)

## B. Parasympathetic Nervous System

Vagal nerve degeneration was found in neuropathological studies of alcoholic patients with severe nervous damage, in whom dysphagia and dysphonia were prominent symptoms.[101] Disorders of gastric motility have also been reported.[102] We have investigated whether vagal neuropathy might affect cardiac function in such patients. Two groups of patients were studied according to the severity of their neurological damage and it was found that parasympathetically mediated heart rate responses, for example, those to i.v. atropine and Valsalva's maneuver, were depressed in those patients with marked nervous system damage[103] (Figure 6). Other workers have subsequently confirmed that heavy drinkers may show depressed reflex heart rate responses.[104,105] Damage to the vagus can be shown by various tests which include depressed heart rate responses to standing, sustained hand grip exercise, atropine, submersion of the face, Valsalva's maneuver, and increasing the blood pressure with pressor drugs such as phenylephrine. It has been suggested that acute death in diabetics may be related to vagal neuropathy.[106] It remains to be demonstrated in alcoholics whether autonomic dysfunction contributes to death rate, although cardiomyopathy in alcoholism may be related to autonomic dysfunction.[107,108] Vagal neuropathy may be reversible with abstinence and/or improved nutrition.[109]

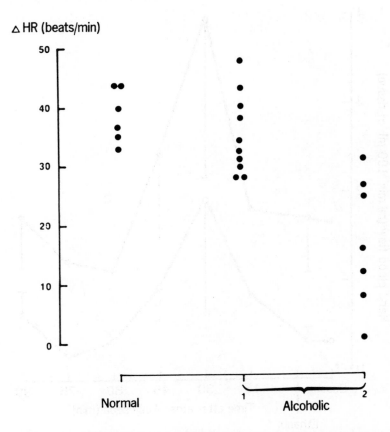

FIGURE 6.    Heart rate response to atropine (1.8 mg i.v.) in normal subjects and group 1 (with little or no evidence of general neurological damage) and group 2 (with major evidence of neurological damage) alcoholic patients. Normal range ≥30 beats/min. The heart rate response to atropine is reduced in alcoholics with a greater degree of peripheral and central nervous damage, suggesting that there is also chronic damage to the vagus nerve. (From Duncan, G., Johnson, R. H., Lambie, D. G., and Whiteside, E. A., *Lancet*, 2, 1053, 1980. With permission.)

Disordered breathing, probably central in origin, has been observed during sleep in subjects with autonomic neuropathy associated with Shy-Drager syndrome or familial dysautonomia.[110] Central sleep apnea and hypopnea were found to be common in chronic alcoholics[111] and showed significant positive associations with clinical evidence of central nervous system damage. Hypopnea also showed a significant association with cardiac vagal neuropathy. There is no evidence of altered chemosensitivity in alcoholic patients as an explanation of abnormal respiratory events during sleep.[112]

Parasympathetic nervous activity via the vagus has been implicated in the control of glucagon secretion, but there is no evidence of altered glucagon release in alcoholics with damage to the vagus nerves innervating the heart.[113]

Parasympathetic neuropathy in alcoholics may involve the pupil[114] (Figure 7) and the sacral innervation. Erectile impotence is common in alcoholics.[115] The rate of parasympathetic damage was studied in 13 alcoholics who complained of impotence.[116] Seven had normal nocturnal erections and their impotence was, therefore, probably psychogenic. Six had diminished or absent nocturnal erection and in this group there was greater evidence of neurological damage than in the group with psychogenic impotence. Two of the patients with organic impotence had parasympathetic nervous system dysfunction as adduced from

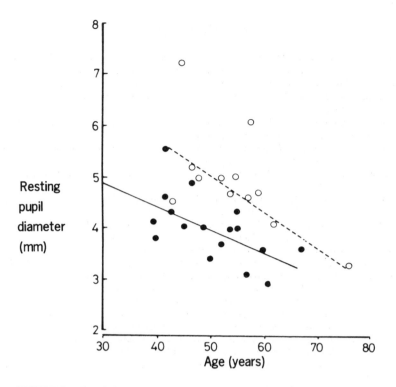

FIGURE 7.   Correlation between resting pupil size and age in chronic alcoholics: group 1, without vagal neuropathy ● and solid line (r = 0.64, p < 0.005); group 2, with vagal neuropathy ○ and dashed line (r = 0.61, p > 0.01). The resting pupillary diameters were larger in the chronic alcoholics with vagal neuropathy, suggesting that there is also a lesion in the parasympathetic supply to the pupil. (From Tan, E. T. H., Lambie, D. G., Johnson, R. H., and Whiteside, E. A., *J. Neurol. Neurosurg. Psychiatry,* 47, 61, 1984. With permission.)

studies of vagal activity affecting the heart. Some pituitary hormones (follicle-stimulating hormone, luteinizing hormone) were also elevated and may have contributed to the impotence. Parasympathetic damage should now be considered as a possible contributory factor in organic impotence associated with chronic alcoholism. Previous explanations depended solely upon an action of alcohol on both the testes and the hypothalamus to cause decreased production of testosterone. Alcoholic liver disease, possibly causing increased conversion of androgens to estrogens, also contributes to the feminine changes apparent in many chronic alcoholics.[117] (For further details on alcohol interactions with the reproductive system, see this volume, Chapter 6.)

## ACKNOWLEDGMENTS

We thank the Maurice and Phyllis Pykel Trust, the Medical Research Council of New Zealand, The Neurological Foundation, and the National Heart Foundation of New Zealand for support related to our personal work included in this review.

# REFERENCES

1. **Adams, M. A. and Hirst, M.,** Adrenal and urinary catecholamines during and after severe ethanol intoxication in rats: a profile of changes, *Pharmacol. Biochem. Behav.,* 21, 125, 1984.
2. **Anton, A. H.,** Ethanol and urinary catecholamines in man, *Clin. Pharmacol. Ther.,* 6, 452, 1965.
3. **Brohult, J., Levi, L., and Reichard, H.,** Urinary excretion of adrenal hormones in man, *Acta Med. Scand.,* 185, 5, 1970.
4. **Giacobin, E., Izikowitz, S., and Wegmann, A.,** The urinary excretion of noradrenaline and adrenaline during acute alcohol intoxication in alcoholic addicts, *Experientia,* 16, 467, 1960.
5. **Myrsten, A.-L., Post, B., and Frankenhaeuser, M.,** Catecholamine output during and after acute alcoholic intoxication, *Percept. Mot. Skills,* 33, 652, 1971.
6. **Perman, E. S.,** The effect of ethyl alcohol on the secretion from the adrenal medulla in man, *Acta Physiol. Scand.,* 44, 241, 1958.
7. **Eisenhofer, G., Lambie, D. G., and Johnson, R. H.,** Effects of ethanol on plasma catecholamines and norepinephrine clearance, *Clin. Pharmacol. Ther.,* 34, 143, 1983.
8. **Howes, L. G. and Reid, J. L.,** Changes in plasma free 3,4-dihydroxy-phenylethyleneglycol and noradrenaline levels after acute alcohol administration, *Clin. Sci.,* 69, 423, 1985.
9. **Ireland, M. A., Vandongen, R., Davidson, L., Beilin, L. J., and Rouse, I. L.,** Acute effects of moderate alcohol consumption on blood pressure and plasma catecholamines, *Clin. Sci.,* 66, 643, 1984.
10. **Gothert, M., Duhrsen, U., and Rieckesmann, J.-M.,** Ethanol, anaesthetics and other lipophilic drugs preferentially inhibit 5-hydroxy-tryptamine — and acetylcholine-induced noradrenaline release from sympathetic nerves, *Arch. Int. Pharmacodyn. Ther.,* 242, 196, 1979.
11. **Gothert, M. and Thielecke, G.,** Inhibition by ethanol of noradrenaline output from peripheral sympathetic nerves: possible interaction of ethanol with neuronal receptors, *Eur. J. Pharmacol.,* 37, 321, 1976.
12. **Degani, N. C., Sellers, E. M., and Kadzielawa, K.,** Ethanol-induced spontaneous norepinephrine release from the rat vas deferens, *J. Pharmacol. Exp. Ther.,* 210, 22, 1979.
13. **Guaza, C. and Borrell, S.,** Adrenomedullary responses to acute and chronic ethanol administration to rats, *Biochem. Pharmacol.,* 32, 3091, 1983.
14. **Ogata, M., Mendelson, J. H., Mello, N. K., and Majchrowicz, E.,** Adrenal function and alcoholism. II. Catecholamines, *Psychosom. Med.,* 32, 159, 1970.
15. **Smith, A. A. and Gitlow, S.,** Effect of disulfiram and ethanol on the catabolism of norepinephrine in man, in *Biochemical Factors in Alcoholism,* Maickel, R. P., Ed., Pergamon Press, Oxford, 1967, 53.
16. **Eisenhofer, G., Lambie, D. G., and Johnson, R. H.,** Effects of ethanol ingestion on alpha-adrenoceptor-mediated circulatory responses in man, *Br. J. Clin. Pharmacol.,* 18, 581, 1984.
17. **Child, J. S., Kovick, R. B., Levisman, J. A., and Pearce, M. L.,** Cardiac effects of acute ethanol ingestion unmasked by autonomic blockade, *Circulation,* 59, 120, 1979.
18. **Lang, R. M., Borow, K. M., Neuman, A., and Feldman, T.,** Adverse cardiac effects of acute alcohol ingestion in young adults, *Ann. Intern. Med.,* 102, 742, 1985.
19. **Fewings, J. D., Hanna, M. J. D., Walsh, J. A., and Whelan, R. F.,** The effects of ethyl alcohol on the blood vessels of the hand and forearm in man, *Br. J. Pharmacol. Chemother.,* 27, 93, 1966.
20. **Gillespie, J. A.,** Vasodilator properties of alcohol, *Br. Med. J.,* 2, 274, 1967.
21. **Ritchie, J. M.,** The aliphatic alcohols, in *The Pharmacological Basis of Therapeutics,* 6th ed., Goodman, L. S. and Gilman, A., Eds., Macmillan, New York, 1980, 376.
22. **Takeda, R. and Momose, Y.,** An inhibitory effect of ethanol on adrenergic neuromuscular transmission in the guinea-pig vas deferens, *Jpn. J. Pharmacol.,* 33, 757, 1983.
23. **Eisenhofer, G., Lambie, D. G., and Johnson, R. H.,** Effects of ethanol on cardiovascular and catecholamine responses to mental stress, *J. Psychosom. Med.,* 30, 93, 1986.
24. **Howes, L. G. and Reid, J. L.,** Decreased vascular responsiveness to noradrenaline following regular ethanol consumption, *Br. J. Clin. Pharmacol.,* 20, 669, 1985.
25. **Altura, B. M. and Altura, B. T.,** Microvascular and vascular smooth muscle and actions of ethanol, acetaldehyde, and acetate, *Fed. Proc.,* 41, 2447, 1982.
26. **Altura, B. M., Ogunkoya, A., Gebrewold, A., and Altura, B. T.,** Effects of ethanol on terminal arterioles and muscular venules: direct observations on the microcirculation, *J. Cardiovasc. Pharmacol.,* 1, 97, 1979.
27. **Giles, T. D., Cook, J. R., Sachitano, R. A., and Held, B. J.,** Influence of alcohol on the cardiovascular response to isometric exercise in normal subjects, *Angiology,* 33, 332, 1982.
28. **Zeichner, A., Feuerstein, M., Swartzman, L., and Reznick, E.,** Acute effects of alcohol on cardiovascular reactivity to stress in type A (coronary prone) businessmen, in *Stress and Alcohol Use,* Poherecky, L. A. and Brick, J., Eds., Elsevier, New York, 1983, 353.
29. **Barraclough, M. A. and Sharpey-Schafer, E. P.,** Hypotension from absent circulatory reflexes: effects of alcohol, barbiturates, psychotherapeutic drugs and other mechanisms, *Lancet,* 1, 1121, 1963.
30. **Lyle, C. B., Monroe, J. T., Flinn, D. E., and Lamb, L. E.,** Micturition syncope: report of 24 cases, *N. Engl. J. Med.,* 265, 982, 1961.

31. **Coggins, C. H., Lillington, G. A., and Gray, C. P.,** Micturition syncope, *Arch. Intern. Med.,* 113, 14, 1964.
32. **Lukash, W. M., Sawyer, G. T., and Davis, J. E.,** Micturition syncope produced by orthostasis and bladder distension, *N. Engl. J. Med.,* 270, 341, 1964.
33. **Weyman, A. E., Greenbaum, D. M., and Grace, W. J.,** Accidental hypothermia in an alcoholic population, *Am. J. Med.,* 56, 13, 1974.
34. **Fellows, I. W., MacDonald, I. A., and Bennett, T.,** The influence of environmental temperature upon the thermoregulatory responses to ethanol in man, *Clin. Sci.,* 66, 733, 1984.
35. **Martin, S. and Cooper, K. E.,** Alcohol and respiratory and body temperature changes during tepid water immersion, *J. Appl. Physiol.,* 44, 683, 1978.
36. **Hoffman, P. L., Ritzmann, R. F., Waller, R., and Tabakoff, B.,** Arginine vasopressin maintains ethanol tolerance, *Nature,* 276, 614, 1978.
37. **Pittman, Q. J., Rogers, J., and Bloom, F. E.,** Arginine vasopressin deficient Brattleboro rats fail to develop tolerance to the hypothermic effects of ethanol, *Regul. Pept.,* 4, 33, 1982.
38. **Eisenhofer, G. and Johnson, R. H.,** Effect of ethanol ingestion on plasma vasopressin and water balance in humans, *Am. J. Physiol.,* 242, R522, 1982.
39. **Seto, A., Tricomi, S., Goodwin, D. W., Kolodney, R., and Sullivan, T.,** Biochemical correlates of ethanol-induced flushing in orientals, *J. Stud. Alcohol,* 39, 1, 1978.
40. **Wolff, P. H.,** Ethnic differences in alcohol sensitivity, *Science,* 175, 449, 1972.
41. **Wolff, P. H.,** Vasomotor sensitivity to alcohol in diverse mongoloid populations, *Am. J. Hum. Genet.,* 25, 193, 1973.
42. **Adachi, J. and Mizoi, Y.,** Acetaldehyde-mediated alcohol sensitivity and elevation of plasma catecholamines in man, *Jpn. J. Pharmacol.,* 33, 531, 1983.
43. **Schuckit, M. A. and Duby, J.,** Alcohol-related flushing and the risk for alcoholism in sons of alcoholics, *J. Clin. Psychiatry,* 43, 415, 1982.
44. **Lake, C. R., Major, L. F., Ziegler, M. G., and Kopin, I. J.,** Increased sympathetic nervous system activity in alcoholic patients treated with disulfiram, *Am. J. Psychiatry,* 134, 1411, 1977.
45. **Wiles, P. G. and Pyke, D. A.,** The chlorpropamide alcohol flush, *Clin. Sci.,* 67, 375, 1984.
46. **Leslie, R. D. G. and Pyke, D. A.,** Chlorpropamide-alcohol flushing: a dominantly inherited trait associated with diabetes, *Br. Med. J.,* 2, 1519, 1978.
47. **Leslie, R. D. G., Barnett, A. H., and Pyke, D. A.,** Chlorpropamide alcohol flushing and diabetic retinopathy, *Lancet,* 1, 997, 1979.
48. **Barnett, A. H. and Pyke, D. A.,** Chlorpropamide-alcohol flushing and large-vessel disease in non-insulin-dependent diabetes, *Br. Med. J.,* 1, 261, 1980.
49. **Strakosch, C. R., Jefferys, D. B., and Keen, H.,** Blockade of chlorpropamide alcohol flush by aspirin, *Lancet,* 1, 394, 1980.
50. **Kupari, M.,** Acute cardiovascular effects of ethanol, a controlled non-invasive study, *Br. Heart J.,* 49, 174, 1983.
51. **Eisenhofer, G., Lambie, D. G., and Johnson, R. H.,** No effects of ethanol ingestion on beta-adrenoceptor-mediated circulatory responses to isoprenaline in man, *Br. J. Clin. Pharmacol.,* 20, 684, 1985.
52. **Mathews, J. D.,** Alcohol use, hypertension and coronary heart disease, *Clin. Sci. Mol. Med.,* 51, 661s, 1976.
53. **Ashley, M. J. and Rankin, J. G.,** Alcohol consumption and hypertension — the evidence from hazardous drinking and alcohol populations, *Aust. N. Z. J. Med.,* 9, 201, 1979.
54. **Klatsky, A. L., Friedman, G. D., Siegelaub, M. S., and Gerard, M. E.,** Alcohol consumption and blood pressure. Kaiser-Permanente multiphasic health examination data, *N. Engl. J. Med.,* 296, 1194, 1977.
55. **Potter, J. F. and Beevers, D. G.,** Pressor effect of alcohol in hypertension, *Lancet,* 1, 119, 1984.
56. **Saunders, J. B., Beevers, D. G., and Paton, A.,** Factors influencing blood pressure in chronic alcoholics, *Clin. Sci.,* 57, 295s, 1979.
57. **Ibsen, H., Christensen, N. J., Rasmussen, S., Hollnagel, H., Damkjaer Nielsen, M., and Giese, J.,** The influence of chronic high alcohol intake on blood pressure, plasma noradrenaline concentration and plasma renin concentration, *Clin. Sci.,* 61, 377s, 1981.
58. **Ireland, M., Vandongen, R., Davidson, L., Beilin, L. J., and Rouse, I.,** Pressor effect of moderate alcohol consumption in man: a proposed mechanism, *Clin. Exp. Pharmacol. Physiol.,* 10, 375, 1983.
59. **Arkwright, P. D., Beilin, L. J., Vandongen, R., Rouse, I. A., and Lalor, C.,** The pressor effect of moderate alcohol consumption in man: a search for mechanisms, *Circulation,* 66, 515, 1982.
60. **Kaysen, G. and Noth, R. H.,** The effects of alcohol on blood pressure and electrolytes, *Med. Clin. North Am.,* 68, 221, 1984.
61. **Borg, S., Czarnecka, A., Kvande, H., Mossberg, D., and Sedvall, G.,** Clinical conditions and concentrations of MOPEG in the cerebrospinal fluid and urine of male alcoholic patients during withdrawal, *Alcoholism,* 7, 411, 1983.

62. **Hawley, R. J., Major, L. F., Schulman, E. A., and Lake, C. R.,** CSF levels of norepinephrine during alcohol withdrawal, *Arch. Neurol.,* 38, 289, 1981.

63. **Kraemer, G. W., Lake, C. R., Ebert, M. H., and McKinney, W. T.,** Effects of alcohol on cerebrospinal fluid norepinephrine in rhesus monkeys, *Psychopharmacology,* 85, 444, 1985.

64. **Major, L. F., Hawley, R. J., and Linnoila, M.,** The role of the central noradrenergic nervous system in the mediation of the ethanol intoxication and ethanol withdrawal syndrome, *Psychopharmacol. Bull.,* 20, 487, 1984.

65. **Carlsson, C. and Haggendal, J.,** Arterial noradrenaline levels after ethanol withdrawal, *Lancet,* 2, 889, 1967.

66. **Eisenhofer, G., Whiteside, E. A., and Johnson, R. H.,** Plasma catecholamine responses to change of posture in alcoholics during withdrawal and after continued abstinence from alcohol, *Clin. Sci.,* 68, 71, 1985.

67. **Manhem, P., Nilsson, L. H., Moberg, A. L., Wadstein, J., and Hokfelt, B.,** Hypokalaemia in alcohol withdrawal caused by high circulating adrenaline levels, *Lancet,* 1, 679, 1984.

68. **Potter, J. F., Bannan, L. T., and Beevers, D. G.,** The effect of a non-selective lipophilic beta-blocker on blood pressure and noradrenaline, vasopressin, cortisol and renin release during alcohol withdrawal, *Clin. Exp. Hypertension,* A6, 1147, 1984.

69. **Clark, L. T. and Friedman, H. S.,** Hypertension associated with alcohol withdrawal: assessment of mechanisms and complications, *Alcoholism,* 9, 125, 1985.

70. **De Marchi, S. and Cecchin, E.,** Alcohol withdrawal and hypertension: evidence for a kidney abnormality, *Clin. Sci.,* 69, 239, 1985.

71. **Bannan, L. T., Potter, J. F., Beevers, D. G., Saunders, J. B., Walters, J. R. F., and Ingram, M. C.,** Effect of alcohol withdrawal on blood pressure, plasma renin activity, aldosterone, cortisol and dopamine-B-hydroxylase, *Clin. Sci.,* 66, 659, 1984.

72. **Ciofalo, F. R.,** Ethanol, neuroreceptors and postsynaptic membrane function, *Proc. West. Pharmacol. Soc.,* 23, 441, 1980.

73. **Lee, H., Hosein, E. A., and Rovinski, B.,** Effect of chronic alcohol feeding and withdrawal on rat liver plasma membrane structure and function: a study of binding of ($^3$H) prazosin to the membrane bound alpha-adrenergic receptor, *Biochem. Pharmacol.,* 32, 1321, 1983.

74. **Nygren, A. and Sunblad, L.,** Disturbed alpha-adrenergic modulation of insulin and growth hormone secretion in chronic alcoholics, *Diabetologia,* 18, 193, 1980.

75. **Banerjee, S. P., Sharma, V. K., and Khanna, J. M.,** Alterations in beta-adrenergic receptor binding during ethanol withdrawal, *Nature,* 276, 407, 1978.

76. **Hugues, F. C., Munera, Y., Julien, D., and Marche, J.,** Exploration de la fonction, bêta adrénergique chez l'homme l'épreuve a l'isoprénaline, *Coeur. Med. Interne,* 13, 527, 1974.

77. **Segel, L. D. and Mason, D. T.,** Beta-adrenergic receptors in chronic alcoholic rat hearts, *Cardiovasc. Res.,* 16, 34, 1982.

78. **Low, P. A., Walsh, J. C., Huana, C. Y., and McLeod, J. G.,** The sympathetic nervous system in alcoholic neuropathy: a clinical and pathological study, *Brain,* 98, 357, 1975.

79. **De Marchi, S. and Cecchin, E.,** Are orthostatic hypotension and impaired blood pressure control common features of the alcohol withdrawal syndrome?, *Clin. Sci.,* 70, 213, 1986.

80. **Gravallese, M. A., Jr. and Victor, M.,** Circulatory studies in Wernicke's encephalopathy, *Circulation,* 15, 836, 1957.

81. **Birchfield, R. I.,** Postural hypotension in Wernicke's disease, *Am. J. Med.,* 36, 404, 1964.

82. **Jensen, K., Andersen, K., Smith, T., Henriksen, O., and Melgaard, B.,** Sympathetic vasoconstrictor nerve function in alcoholic neuropathy, *Clin. Physiol.,* 4, 253, 1984.

83. **Appenzeller, O. and Ogin, G.,** Myelinated fibres in human paravertebral sympathetic chain: white rami communicantes in alcoholic and diabetic patients, *J. Neurol. Neurosurg. Psychiatry,* 37, 1155, 1974.

84. **Appenzeller, O. and Richardson, E. P., Jr.,** The sympathetic chain in patients with diabetic and alcoholic polyneuropathy, *Neurology,* 16, 1205, 1966.

85. **Bernardi, M., Trevisani, F., Santini, C., Ligabue, A., Capelli, M., and Gasparrini, G.,** Impairment of blood pressure control in patients with liver cirrhosis during tilting: study on adrenergic and renin-angiotensin systems, *Digestion,* 25, 124, 1982.

86. **Henriksen, J. H., Ring-Larsen, H., Kanstrup, I.-L., and Christensen, N. J.,** Splanchnic and renal elimination and release of catecholamines in cirrhosis. Evidence of enhanced sympathetic activity in patients with decompensated cirrhosis, *Gut,* 25, 1034, 1984.

87. **Keller, U., Gerber, P. P. G., Buhler, F. R., and Stauffacher, W.,** Role of the splanchnic bed in extracting circulating adrenaline and noradrenaline in normal subjects and in patients with cirrhosis of the liver, *Clin. Sci.,* 67, 45, 1984.

88. **Nicholls, K. M., Shapiro, M. D., Van Putten, V. J., Kluge, R., Chung, H.-M., Bichet, D. G., and Schrier, R. W.,** Elevated plasma norepinephrine concentrations in decompensated cirrhosis, *Circ. Res.,* 56, 457, 1985.

89. **Bichet, D. G., Van Putten, V. J., and Schrier, R. W.,** Potential role of increased sympathetic activity in impaired sodium and water excretion in cirrhosis, *N. Engl. J. Med.,* 307, 1552, 1982.

90. **Arryo, V., Planas, R., Gaya, J., Deulofeu, R., Rimola, A., Perez-Ayusa, R. M., Rivera, F., and Rodes, J.,** Sympathetic nervous activity, renin-angiotensin system and renal excretion of prostaglandin E₂ in cirrhosis. Relationship to functional renal failure and sodium and water excretion, *Eur. J. Clin. Invest.,* 13, 271, 1983.

91. **Ring-Larsen, H., Henriksen, J. H., and Christensen, N. J.,** Increased sympathetic activity in cirrhosis, *N. Engl. J. Med.,* 308, 1029, 1983.

92. **Zazgornik, J., Irsigler, K., Kline, E., and Kryspin-Exner, K.,** The accrine sweat-gland function of chronic alcoholics, *Nutr. Metab.,* 14, 307, 1972.

93. **Goodman, J. I.,** Diabetic anhidrosis, *Am. J. Med.,* 41, 831, 1966.

94. **Robinson, B. J., Johnson, R. H., Lambie, D. G., and Whiteside, E. A.,** Thermoregulatory responses in alcoholism, *Aust. Alcohol Drug Rev.,* 4, 157, 1985.

95. **Philip, G. and Smith, J. F.,** Diencephalic autonomic epilepsy, *Arch. Neurol. Psychiatry,* 22, 358, 1973.

96. **Wernicke, C.,** *Lehrbuch der Gehirnkrankheiten fur Aertze und Studierende,* Vol. 2, Fischer, Berlin, 1881—1883, 229.

97. **Hunter, J. M.,** Hypothermia and Wernicke's encephalopathy, *Br. Med. J.,* 2, 563, 1976.

98. **Koeppen, A. H., Daniels, J. C., and Barron, K. D.,** Subnormal body temperatures in Wernicke's encephalopathy, *Arch. Neurol.,* 21, 493, 1969.

99. **Victor, M., Adams, R. D., and Collins, G. H.,** *The Wernicke-Korsakoff Syndrome,* Blackwell Scientific, Oxford, 1971.

100. **Lipton, J. M., Payne, H., Garza, H. R., and Rosenberg, R. N.,** Thermolability in Wernicke's encephalopathy, *Arch. Neurol.,* 35, 750, 1978.

101. **Novak, D. J. and Victor, M.,** The vagus and sympathetic nerves in alcoholic polyneuropathy, *Arch. Neurol.,* 30, 273, 1974.

102. **Winship, D. H., Caflisch, C. R., Zbaralske, F. F., and Hogan, W. J.,** Deterioration of oesophageal peristalsis in patients with alcoholic neuropathy, *Gastroenterology,* 55, 173, 1968.

103. **Duncan, G., Johnson, R. H., Lambie, D. G., and Whiteside, E. A.,** Evidence of vagal neuropathy in chronic alcoholics, *Lancet,* 2, 1053, 1980.

104. **Johnston, L. C., Patel, S., Vankineni, P., and Kramer, N.,** Deficient slowing of the heart among very heavy social drinkers, *J. Stud. Alcohol,* 44, 505, 1983.

105. **Melgaard, B. and Somnier, F.,** Cardiac neuropathy in chronic alcoholics, *Clin. Neurol. Neurosurg.,* 83-4, 219, 1981.

106. **Ewing, D. J., Campbell, I. W., and Clarke, B. F.,** The natural history of diabetic autonomic neuropathy, *Q. J. Med.,* 49, 95, 1980.

107. **Amorin, D. S., Dargie, H. J., Heer, K., Brown, M., Jenner, D., Olsen, E. G. J., Richardson, P., and Goodwin, J. F.,** Is there autonomic impairment in congestive (dilated) cardiomyopathy?, *Lancet,* 1, 525, 1981.

108. **Amorin, D. S. and Olsen, E. G. J.,** Assessment of heart neurons in dilated (congestive) cardiomyopathy, *Br. Heart J.,* 47, 11, 1982.

109. **Tan, E. T. H., Johnson, R. H., Lambie, D. G., and Whiteside, E. A.,** Alcoholic vagal neuropathy: recovery following prolonged abstinence, *J. Neurol. Neurosurg. Psychiatry,* 47, 1335, 1984.

110. **Guilleminault, C., Briskin, J. G., Greenfield, M. S., and Silvestri, R.,** The impact of autonomic nervous system dysfunction on breathing during sleep, *Sleep,* 14, 263, 1981.

111. **Tan, E. T. H., Lambie, D. G., Johnson, R. H., Robinson, B. J., and Whiteside, E. A.,** Sleep apnoea in alcoholic patients after withdrawal, *Clin. Sci.,* 69, 644, 1985.

112. **Lambie, D. G., Tan, E. T. H., and Johnson, R. H.,** Respiratory responsiveness in male alcoholics after withdrawal, *Alcoholism Clin. Exp. Res.,* 11, 49, 1987.

113. **Tan, E. T. H., Lambie, D. G., Johnson, R. H., and Whiteside, E. A.,** Release of glucagon in alcoholics with vagal neuropathy, *Alcoholism Clin. Exp. Res.,* 7, 461, 1983.

114. **Tan, E. T. H., Lambie, D. G., Johnson, R. H., and Whiteside, E. A.,** Parasympathetic denervation of the iris in alcoholics with vagal neuropathy, *J. Neurol. Neurosurg. Psychiatry,* 47, 61, 1984.

115. **Vijayasenan, M. E.,** Alcohol and sex, *N.Z. Med. J.,* 93, 18, 1981.

116. **Tan, E. T. H., Johnson, R. H., Lambie, D. G., and Vijayasenan, E. A.,** Erectile impotence in chronic alcoholics, *Alcoholism Clin. Exp. Res.,* 8, 297, 1984.

117. **Noth, R. H. and Walter, R. M., Jr.,** The effects of alcohol on the endocrine system, *Med. Clin. North Am.,* 68, 133, 1984.

Chapter 14

# EFFECTS OF ETHANOL ON VASOPRESSIN, THIRST, AND WATER BALANCE

**Graeme Eisenhofer and Ralph H. Johnson**

## TABLE OF CONTENTS

# I. INTRODUCTION

Water balance is controlled by the relative rates of water excretion and water intake, the former primarily influenced by the action of vasopressin on free water clearance by the kidney and the latter affected by thirst. Vasopressin is secreted from the posterior pituitary gland under the control of neural input from the hypothalamus: this is primarily regulated by osmoreceptors, sensitive to changes in the tonicity of body fluids, and peripherally located baroreceptors, responsive to changes in blood volume or blood pressure. Thirst is regulated similarly by areas in the brain which appear to share in the control of vasopressin release.

The diverse physiological actions of ethanol include effects on both vasopressin release and thirst. These appear to differ according to the time course of intoxication and to whether the acute or chronic effects of ethanol are examined. Since many of the changes in water balance and its control reflect or result from homeostatic responses to the initial actions of ethanol, it is often difficult to distinguish the direct effects from the indirect effects of ethanol intoxication.

The following is an attempt to clarify the sequence of actions of ethanol on water balance and its control, and also to distinguish, where possible, the acute effects of ethanol from the chronic effects. In addition to the regulation of water balance by vasopressin and thirst, water balance is also influenced by changes in electrolyte balance, which is mainly regulated by the renin-angiotensin-aldosterone system. Although ethanol has important effects on the latter system, discussion is confined to the effects of ethanol on the control of water balance by thirst and vasopressin release. The clinical relevance of the effects of ethanol on vasopressin, thirst, and water balance is discussed in relationship to the mental manifestations of ethanol intoxication, tolerance, the dependence process, and the pathophysiology of alcoholism.

# II. ACUTE EFFECTS

## A. Vasopressin and Urine Production

Apart from the mental manifestations of ethanol intoxication, perhaps the most well-known effect of ethanol involves the increase in urine flow (diuresis) that follows drinking. It is well established that the diuresis is not simply due to the fluid load associated with the consumption of alcoholic beverages, but is secondary to a specific inhibitory action of ethanol on vasopressin release. This was first indicated in studies showing a greater diuresis after a given quantity of ethanol compared with a similar quantity of water, and complete blockade of the diuresis after administration of posterior pituitary extract before ethanol dosing.[1,2] Localization of the cause of the diuretic effect of ethanol to an action on posterior pituitary function, rather than to a direct action on the kidney, was indicated by studies showing increased urine flow independent of changes in urinary electrolyte and creatinine clearance,[3,4] and a lack of diuretic effect in animals or patients with defective vasopressin release associated with diabetes insipidus.[5,6]

Proof of the inhibitory action of ethanol on vasopressin release required the development and availability of sensitive and precise radioimmunoassay methods to measure the low plasma concentrations of vasopressin. Using these methods it was shown that plasma vasopressin concentrations were reduced after ethanol ingestion, but that this effect was only transient; vasopressin concentrations returned to baseline values within 3 h of drinking despite the continued presence of high blood ethanol concentrations (Figure 1).[7-10] These results were consistent with previous findings showing that the ethanol-induced diuresis was of short duration and could not be maintained by continued ethanol consumption.[2,3,11] With repeated doses of ethanol the diuretic response was successively reduced in volume;[2,5] also, the diuretic effect of ethanol could be blocked after stimulation of vasopressin release by prior administration of hypertonic saline.[3,12,13]

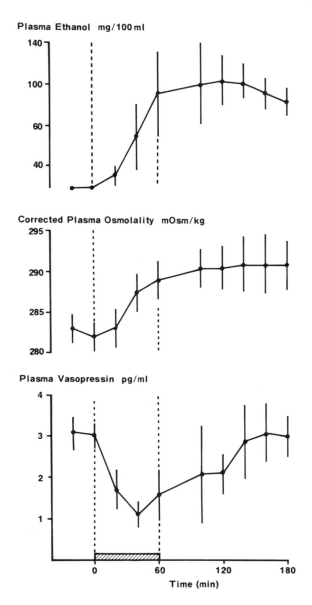

FIGURE 1. Plasma ethanol concentrations, plasma osmolality (corrected for the contribution of ethanol), and plasma vasopressin concentrations before, during (0 to 60 min), and after consumption of 70 ml ethanol (mean ± SD). (From Eisenhofer, G. and Johnson, R. H., *Am. J. Physiol.*, 242, R522, 1982. With permission.)

An explanation for the above findings and elucidation of the mechanism of action of ethanol on vasopressin release followed examination of the effects of ethanol on the vasopressin response to hypertonic saline infusion.[9] Infusion of hypertonic saline causes increased vasopressin release in response to an osmotic load. This enables positive relationships to be established between plasma osmolality and plasma concentrations of vasopressin and allows examination of the responsiveness and sensitivity of osmoreceptors controlling vasopressin release. After ethanol, relationships between plasma vasopressin and osmolality were shifted to higher osmolalities in a dose-dependent manner, with higher doses causing a greater shift

PLASMA VASOPRESSIN (pg/ml)

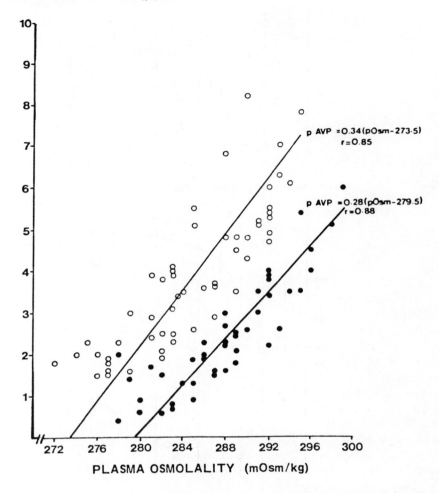

FIGURE 2.    Plasma vasopressin and plasma osmolality in five male subjects during i.v. hypertonic saline infusion (with ethanol, ●; without ethanol, ○). Plasma osmolalities in studies in which ethanol was ingested have been corrected for influence of ethanol. Ethanol shifted relationships between vasopressin and osmolality to higher osmolality values. (From Eisenhofer, G. and Johnson, R. H., *Am. J. Physiol.*, 242, R522, 1982. With permission.)

(Figure 2). Therefore, ethanol ingestion causes inhibition of vasopressin release by raising the threshold at which hypothalamic osmoreceptors respond to increases in plasma osmolality with stimulation of vasopressin release. Studies in dogs also showed that production of an antidiuretic response after ethanol required a higher level of electrical stimulation to neurons in the supraoptic nuclei[14] and a greater increase in plasma osmolality during hypertonic saline administration.[15]

The above findings explain why inhibition of vasopressin release and the diuresis after ethanol occur transiently despite the continued presence of ethanol in the blood. Increased plasma vasopressin concentrations in the latter period of intoxication after the initial decrease reflect a counterregulatory response to increased plasma osmolality and also possibly to contracted blood volume. Increased plasma osmolality reflects the state of dehydration induced by the initial diuresis; dehydration progressively becomes sufficient to overcome the depressant effects of ethanol on osmoreceptor responsiveness and return plasma vaso-

**Table 1**
**THIRST THRESHOLD: TIME AND PLASMA**
**OSMOLALITY (OSM) AT OCCURRENCE OF THIRST**
**DURING HYPERTONIC SALINE INFUSIONS WITH**
**AND WITHOUT ETHANOL**

| Subject | Control study | | Ethanol study | | Dose of ethanol (ml/kg) |
|---|---|---|---|---|---|
|  | Time (min) | Osm (mOsm/kg) | Time (min) | Osm (mOsm/kg) |  |
| 1 | 25 | 282 | 25 | 287 | 0.5 |
| 2 | 30 | 283 | 55 | 288 | 0.5 |
| 3 | 30 | 286 | 55 | 293 | 0.5 |
| 4 | 35 | 285 | 100 | 291 | 0.5 |
| 5 | 35 | 285 | 80 | 294 | 1.0 |
| 6 | 70 | 291 | 115 | 302 | 1.0 |
| 7 | 45 | 286 | 72 | 295 | 1.0 |
| 8 | 15 | 284 | 15 | 291 | 1.0 |
| 9 | 110 | 296 | >120 | >298 | 1.0 |

*Note:* Ethanol consumption caused thirst to be experienced at significantly ($p < 0.01$) higher plasma osmolalities and significantly ($p < 0.05$) later during hypertonic saline infusions.

From Eisenhofer, G. and Johnson, R. H., *Am. J. Physiol.*, 244, R568, 1983. With permission.

pressin concentrations back to baseline levels. As ethanol is metabolized, plasma vasopressin concentrations are further raised to the higher-than-baseline values observed later during intoxication and in the hangover period.[7,8]

Whether intracellular, extracellular, and intravascular fluid compartments become equally dehydrated by acute ethanol ingestion, or whether one compartment is more affected than the others is not well established. Some studies have suggested that it is the intracellular compartment that is most affected; there may be little effect or a minor increase in the volume of the extracellular and intravascular compartments.[11,16-18] An action of ethanol on fluid balance between intra- and extracellular compartments independent of the effect of ethanol on free water clearance remains a possibility. Confirmation of such an action requires further radioisotopic investigations following acute doses of ethanol.

## B. Thirst and Drinking

The use of hypertonic saline infusions also uncovered a previously unrecognized effect of ethanol on thirst mechanisms. Similar to the effect of ethanol on vasopressin release, it was found that ethanol also inhibited thirst by raising the plasma osmolality at which thirst was experienced (Table 1) and shifting the relationship between plasma osmolality and thirst score to higher osmolalities (Figure 3).[19] Thus, after ethanol, higher doses of salt were required to elicit thirst than without ethanol. Studies in rats supported the above findings by showing that ethanol dose-dependently suppressed the drinking response to s.c. injection of hypertonic saline (Figure 4), but not to intravascular volume depletion by i.p. injection of dextran.[20] These results challenged the widely held belief that ethanol is primarily a dipsogenic agent[21-23] by showing that the initial, although not obvious, effect of ethanol is to inhibit osmotically mediated thirst. The thirst and polydipsia that are commonly experienced or observed after ethanol could be blocked by administration of posterior pituitary extract or a vasopressin analogue, indicating that ethanol-induced thirst is caused by the state of dehydration resulting from inhibition of vasopressin release.[19,24,25] Furthermore,

THIRST SCORE

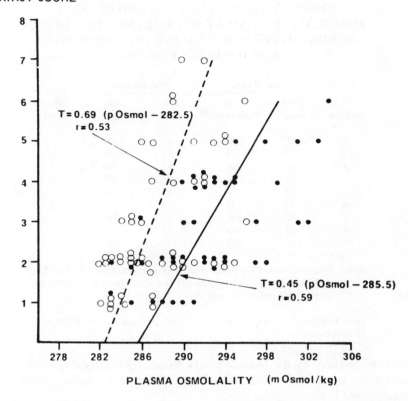

FIGURE 3.   Thirst score (T), plasma osmolality relationships from i.v. hypertonic saline infusions with (●) and without (○) ethanol consumption. Ethanol significantly ($p < 0.02$) shifted the relationship to higher osmolality values. (From Eisenhofer, G. and Johnson, R. H., *Am. J. Physiol.*, 244, R568, 1983. With permission.)

increased thirst and fluid consumption developed as ethanol was metabolized, presumably in response to release of the suppressant effects of ethanol on osmoreceptor responsiveness.[19] This offers an explanation for the commonly observed phenomenon that thirst after ethanol is most readily observed during the hangover period. During the initial stages of intoxication, thirst is suppressed despite the presence of dehydration and becomes most apparent as ethanol is metabolized.

## C. Vasopressin, Thirst, and Osmoreceptors

The end result of acute ethanol ingestion is a resetting of the homeostatic regulatory mechanisms controlling water balance so that a state of dehydration is induced. This state of dehydration is not actually perceived by the organism as such — with production of the appropriate dipsogenic and antidiuretic responses — until blood ethanol concentrations decline, returning to normal the setting of the water balance regulatory mechanisms (Figure 5).

It appears likely that inhibition of both thirst and vasopressin release may be secondary to a common action of ethanol on osmosensitive cells that subserve these responses. Such an action may be related to the selective inhibition by ethanol of the electrical activity of sodium-sensitive cells in the lateral hypothalamus.[26] Animal studies, showing no effect of ethanol on thirst and vasopressin release stimulated by intravascular volume depletion,[20,27] nicotine, hexamethonium,[28] and pain,[27] suggest that the inhibitory effect of ethanol on

CUMULATIVE FLUID CONSUMPTION (ml)

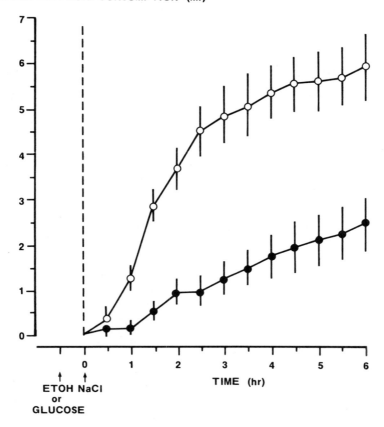

FIGURE 4. Cumulative fluid consumption (mean ± SEM) for 6 h after s.c. injection of hypertonic saline in rats administered ethanol (●) or an isocalorific and equivolumetric dose of dextrose in water (○). Ethanol significantly ($p < 0.005$) reduced the drinking response at all time intervals 2 h and more after hypertonic saline. (From Heyward, P. and Eisenhofer, G., *Pharmacol. Biochem. Behav.*, 22, 493, 1985. With permission.)

vasopressin release and thirst is confined to osmotic mechanisms. Ethanol in *vivo* affects the intra-extracellular distributions of potassium and sodium,[29] an action which may particularly influence the function of osmosensitive cells controlling thirst and vasopressin release. Characteristically, agents which alter membrane transport of sodium and potassium have parallel effects on both vasopressin release and thirst.[30]

An alternative hypothesis for the mechanism by which ethanol alters the responsiveness of osmoreceptors has been proposed by Oiso and Robertson.[31] They showed that administration of naloxone to rats partially reversed the effect of ethanol in elevating the set of the osmostat, suggesting that the effects of ethanol on osmoreceptor responsiveness may be mediated, in part, by endogenous opioids.

## D. Water Balance and Ethanol Consumption

A role of vasopressin and thirst has been implicated in tolerance and dependence to ethanol, and changes in water balance have been proposed to influence ethanol consumption. Ethanol-induced thirst may stimulate further ethanol consumption which could contribute to the common inability of ethanol-dependent individuals to curb or control their consumption of ethanol once drinking has begun ("loss-of-control drinking").[22] During saline loading or

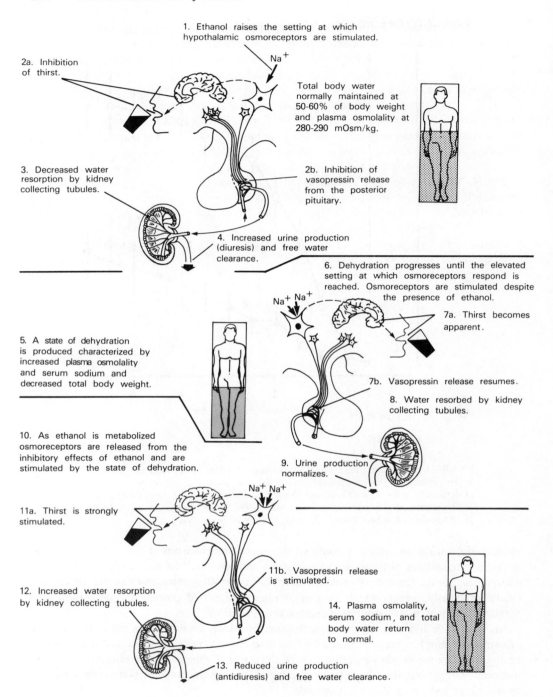

1. Ethanol raises the setting at which hypothalamic osmoreceptors are stimulated.

Na$^+$

2a. Inhibition of thirst.

Total body water normally maintained at 50-60% of body weight and plasma osmolality at 280-290 mOsm/kg.

3. Decreased water resorption by kidney collecting tubules.

2b. Inhibition of vasopressin release from the posterior pituitary.

4. Increased urine production (diuresis) and free water clearance.

6. Dehydration progresses until the elevated setting at which osmoreceptors respond is reached. Osmoreceptors are stimulated despite the presence of ethanol.

Na$^+$ Na$^+$

7a. Thirst becomes apparent.

5. A state of dehydration is produced characterized by increased plasma osmolality and serum sodium and decreased total body weight.

7b. Vasopressin release resumes.

8. Water resorbed by kidney collecting tubules.

10. As ethanol is metabolized osmoreceptors are released from the inhibitory effects of ethanol and are stimulated by the state of dehydration.

9. Urine production normalizes.

11a. Thirst is strongly stimulated.

Na$^+$ Na$^+$

11b. Vasopressin release is stimulated.

12. Increased water resorption by kidney collecting tubules.

14. Plasma osmolality, serum sodium, and total body water return to normal.

13. Reduced urine production (antidiuresis) and free water clearance.

FIGURE 5.   The effects (in sequence, numbered 1 to 14) of ethanol on water balance and its control during the initial stages of intoxication when blood ethanol concentrations are rising (1 to 4), during the intoxication phase after blood ethanol concentrations reach a plateau (5 to 9), and as ethanol is metabolized (10 to 14).

the administration of a high salt diet, mice and rats have been found to show an increased selection for solutions of ethanol compared with other available fluids.[32-34] Normally, experimental animals show a strong aversion to solutions of ethanol, but, in one study, administration of a high salt diet completely reversed the normal aversion to ethanol and resulted in a sustained preference for ethanol over other available fluids.[34] It was suggested that this effect resulted from the association of ethanol consumption with greater alleviation of osmotically induced thirst than consumption of water. If so, this may have relevance to loss-of-control drinking in alcoholism. According to this view the dehydrating effects of ethanol do not become apparent or uncomfortable until blood ethanol concentrations decline, lowering to normal the osmotic threshold at which thirst is experienced.[19] A person who exhibits loss-of-control drinking may be compelled to drink ethanol in order to maintain an elevated osmotic threshold for thirst, thus, postponing the discomfort of dehydration.

A modulatory influence of fluid balance changes on ethanol intake is supported by studies in dogs in which vasopressin administration after a period of forced ethanol consumption suppressed the increase in voluntary ethanol intake normally exhibited when forced ethanol consumption was discontinued.[35] Also, in the Brattleboro rat with diabetes insipidus, vasopressin administration reversed the abnormally high absolute ethanol intake observed under free-choice situations.[36,37] In the above situations vasopressin administration may have reduced the consumption of ethanol secondary to correction of polydipsia or dehydration. Differences in the mechanisms controlling water and electrolyte balance between ethanol preferring and avoiding strains of rats[38-41] have further indicated that fluid balance may be a factor influencing ethanol consumption. Ethanol preferring strains of rats showed lower urinary sodium excretion,[38,40] higher vasopressin excretion,[39] and greater dipsogenic responses to an osmotic load.[41] In human alcoholics, diuretic responses to a given dose of ethanol were greater than diuretic responses to the same dose in normal control subjects,[42,43] and it might be expected that the associated changes in water balance and thirst would also be greater.

## E. Tolerance

Animal studies have indicated that vasopressin may affect tolerance to and physical dependence on a number of drugs, including ethanol.[44] In mice made tolerant to ethanol and then subsequently withdrawn, s.c. injection of vasopressin sustained the tolerance to ethanol, as indicated by the duration of the loss of the righting reflex and development of hypothermia associated with acute ethanol administration (Figure 6).[45] A facilitatory effect of vasopressin and related peptides on the maintenance of tolerance to ethanol was confirmed by a number of subsequent animal studies.[46-49] In one study,[46] vasopressin was found to potentiate handling-induced withdrawal seizures in mice, suggesting that physical dependence processes may also be affected by vasopressin. Whether endogenous vasopressin has a modulatory influence on the development of tolerance or physical dependence is not established. In the Brattleboro rat with defective vasopressin release, impaired development of tolerance to ethanol suggests that endogenous vasopressin may affect tolerance to ethanol.[50,51] If so, however, it is unlikely that a facilitatory influence of endogenous vasopressin would be related to the acute inhibitory action of ethanol on vasopressin release. It remains a possibility that the increase in vasopressin release following the initial inhibition may have some influence on tolerance to ethanol. Vasopressinergic nerves are now established to extend from the hypothalamus into many areas of the central nervous system other than the neurohypophysis. Whether release of vasopressin into these areas of the central nervous system is affected in the same manner as peripheral release by ethanol has not been established.

## F. Cognitive Function

The effects of vasopressin and related peptides on tolerance to ethanol and other drugs may be related to the established positive effects of these peptides on memory processes.[44] Impairment of memory and changes in cognition after ethanol may be related to inhibition

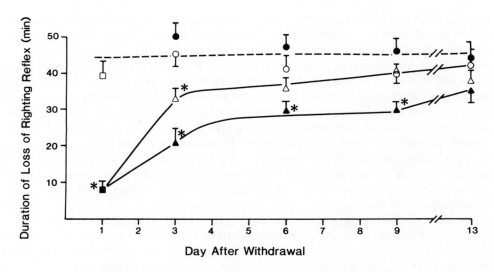

FIGURE 6.   Effect of arginine vasopressin on ethanol tolerance. C57B1 mice were fed a liquid diet containing ethanol ($\triangle$, $\blacktriangle$) or a control diet ($\blacksquare$, $\square$) for 5 d. Mice were given daily intracerebroventricular injections of artificial CSF ($\triangle$, $\bigcirc$) or 1 ng of arginine vasopressin ($\blacktriangle$, $\bullet$) beginning on the first day after ethanol withdrawal. Tolerance (measured as reduced duration of loss of righting reflex) to the hypnotic effect of an i.p. injection of ethanol (3.1 g/kg) was determined on the indicated days after ethanol withdrawal. $p < 0.05$ compared to corresponding control group (ANOVA and Tukey-Kramer test). (From Hung, C. R., Tabakoff, B., Melchoir, C. L., and Hoffman, P. L., *Eur. J. Pharmacol.*, 106, 645, 1984. With permission.)

of vasopressin release,[52,53] a possibility which is supported by the time course of the effects of ethanol on these variables. Greater cognitive impairment occurs during the ascending as opposed to the descending limbs of the blood ethanol curve,[54] which corresponds to the time course of the fall, then the rise, in plasma vasopressin concentrations after acute ethanol ingestion.[7,9] However, examination of the effects of intranasal administration of the vasopressin analogue 1-desamino-8-D-arginine vasopressin on memory and cognitive function has failed to show any improvement in the memory deficits produced by acute ethanol ingestion.[55] This suggests that ethanol-induced inhibition of vasopressin release is unlikely to be involved in cognitive impairment after ethanol. A role for ethanol-induced changes of vasopressin secretion into the central nervous system in the production of some of the cognitive and mental manifestations of ethanol intoxication remains to be established.

## III. CHRONIC EFFECTS

### A. Chronic Ethanol Consumption
   While acute ingestion of ethanol causes a diuresis with resulting dehydration, the majority of experimental and clinical evidence indicates that fluid retention and overhydration are the predominant features following chronic ethanol ingestion. This was first indicated by studies in which dogs that were administered ethanol daily for 8 weeks showed increases in total body water, extracellular volume, and plasma volume that were associated with retention of water and sodium.[56] Overhydration and reversal of ethanol-induced dehydration after chronic ethanol consumption were confirmed by subsequent animal studies.[57,58] In these studies, ethanol was required to be consumed over a prolonged time period before overhydration could be demonstrated, although in other animal studies fluid retention[23] and an increase in extracellular fluid volume[59] were observed 18 to 24 h after a single large dose (2.25 to 3.0 g/kg) of ethanol. Presumably, in the latter studies, increases in vasopressin

release[7-9] and thirst,[9] after their initial inhibition, may have resulted in a subsequent anti-diuresis and increase in fluid intake that caused water to be retained in excess of that lost during the diuretic phase. This may offer an explanation for some initial reports of increased extracellular water and blood volume after single doses of ethanol in man,[11,17] and may be one mechanism by which overhydration could follow repeated ethanol consumption.

Alternatively, overhydration during chronic ethanol consumption could result from a resetting of extracellular fluid volume regulatory mechanisms.[60] Hypertrophy of the neurohypophysis after long-term administration of ethanol to rats[61,62] may reflect a compensatory response of the pituitary to facilitate vasopressin release that could contribute to overhydration. Disturbances in the balance between vasopressin release and synthesis as determined after long-term ethanol administration to rats[63] may also be involved. Greater diuretic responses to acute doses of ethanol in rats chronically administered ethanol[64] and in alcoholics[42,43] indicate that sensitization rather than adaptation occurs to the acute inhibitory effect of ethanol on vasopressin release after prolonged exposure. Thus, water retention after chronic ethanol consumption does not appear to result from the development of resistance to the acute diuretic effects of ethanol; however, a greater diuretic response to acute ethanol consumption in chronic alcoholics could cause a subsequently greater increase in vasopressin release and thirst which could contribute to fluid retention and overhydration.

**B. Alcohol Withdrawal**

One situation where evidence suggests that increased vasopressin secretion may contribute to overhydration is during the alcohol withdrawal syndrome. Higher plasma and urinary vasopressin concentrations have been observed in withdrawing alcoholics with symptoms of withdrawal, as compared to alcoholics without symptoms and to control subjects (Figure 7).[65,66] These observations are consistent with findings in withdrawing alcoholics of raised plasma concentrations of another group of neurohypophyseal peptides, the neurophysins.[67] Neurophysins are peptides produced from the same prohormones as vasopressin or oxytocin and their secretion from the posterior pituitary often parallels that of vasopressin and oxytocin. The cause of increased neurohypophyseal activity during alcohol withdrawal is unknown. One possibility is that it reflects alterations in central nervous system function associated with the withdrawal syndrome. Dehydration is unlikely to be the cause of increased neurohypophyseal activity, since fluid retention and overhydration in alcoholism are particularly evident during the first few days of alcohol withdrawal, as indicated by findings of weight gain, of sodium retention,[60] and of increased total body water and plasma volume.[68,69] Use of nuclear magnetic resonance imaging has also shown increased cerebral free water in alcoholics during alcohol withdrawal[70,71] and decreased cerebral free water following an acute dose of ethanol.[72]

Findings of associations between periorbital edema and raised plasma vasopressin concentrations during alcohol withdrawal,[66] together with increased cerebral-free water after vasopressin administration,[72] support the possibility that changes in vasopressin secretion could at least be partly responsible for fluid retention and overhydration following chronic ethanol consumption and withdrawal. This may occur secondary to the actions of vasopressin on free water clearance by the kidney, but results showing that intraventricular infusion of vasopressin enhances brain water permeability[73] suggest that an increase in cerebral free water may also result from increased vasopressin release within the brain. Central and peripheral release of vasopressin appear to be independently controlled.[74] The chronic effects of ethanol consumption on central vasopressinergic neuronal activity, like the acute effects, have not been well established, although in one report reduced cerebrospinal fluid concentrations of vasopressin were observed in six alcoholic patients.[75] Unfortunately, the clinical state of these patients at the time of cerebrospinal fluid sampling was unclear, so it is uncertain whether these findings reflect a change associated with intoxication, withdrawal, and recovery from ethanol, or result from central nervous system damage.

PLASMA VASOPRESSIN (pg/ml)

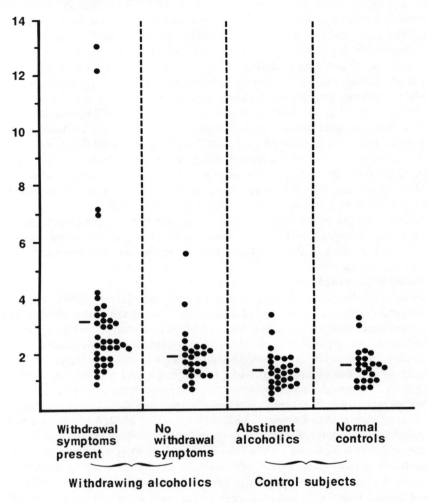

FIGURE 7.　Plasma vasopressin (individual results and means) in withdrawing alcoholics with and without withdrawal symptoms, abstinent alcoholics, and normal control subjects. Alcoholics with withdrawal symptoms had significantly higher plasma vasopressin concentrations than withdrawing alcoholics without withdrawal symptoms ($p < 0.005$), control subjects ($p < 0.001$), and abstinent alcoholics ($p < 0.001$). (From Eisenhofer, G., Whiteside, E., Lambie, D., and Johnson, R., *Br. J. Addict.*, 80, 195, 1985. With permission.)

Despite evidence for an overhydrating effect of chronic ethanol consumption, there remains a common presumption based on the acute effect of ethanol on water balance — that withdrawing alcoholics may particularly suffer from dehydration. Thus, some therapeutic regimens have advocated the use of fluid and electrolyte replacement therapy in the treatment of the detoxifying alcoholic.[76,77] While this may have a place for the patient presenting with vomiting and diarrhea, studies have shown that administration of diuretics[69] and fluid restriction[78] will accelerate the recovery of the majority of detoxifying alcoholics. Furthermore, early therapies for the treatment of delirium tremens included restriction of fluids in the belief that cerebral edema was an important etiologic factor.[79]

It is possible that i.v. administration of fluids, used in the correction of electrolyte imbalances, may actually exacerbate withdrawal symptomatology. Fluid loading after vasopressin administration is well established to provoke seizures in susceptible individuals,[80,81] and vasopressin administration to mice during alcohol withdrawal increases the incidence and severity of handling-induced withdrawal seizures.[46] Thus, restriction of fluids and care with i.v. administration of fluids may be particularly relevant for the patient with a past history of withdrawal seizures.

## C. Alcoholic Brain Damage

Overhydration and cerebral edema may contribute to alcoholic brain damage, adding to and possibly potentiating the effects of nutritional deficiency and traumatic head injury.[82] According to this view, certain cell bodies and regions in the brain may be particularly vulnerable to cellular edema caused by rapid changes in hydration status and intra-extra-cellular fluid and electrolyte shifts following ethanol consumption and withdrawal. Support for this view includes pathological findings of altered cerebral electrolyte balance[83] and the edematous nature of the alcoholic brain on post-mortem examination.[84,85] Also, numerous case reports of neurological damage in alcoholism attest to an association with disorders of fluid balance, including edema, hyponatremia, and inappropriate vasopressin secretion.[86-91] In one report, attention was drawn to cerebral edema as a causal factor in the production of central pontine myelinosis, the first descriptions of which coincided with the introduction of i.v. fluid replacement therapy for the treatment of alcohol withdrawal.[90]

Brain damage, associated with changes in hydration status, is generally attributed to rapid correction of hyponatremia.[92] In alcoholism, overhydration is generally iso-osmotic.[59,60,66,68] Thus, changes in hydration status as a cause of brain damage in alcoholism are unlikely to be solely related to overhydration as it commonly occurs after chronic ethanol consumption.

Apart from a possible role of disordered water balance in the production of alcoholic brain damage, it is also possible that alcoholic brain damage may affect the systems controlling water balance. Normal plasma vasopressin responses to insulin hypoglycemia have been shown to occur in brain-damaged alcoholics with Korsakoff's psychosis,[93] which indicate that peripheral release of vasopressin is unimpaired in alcoholics with this degree of central nervous system damage. Others have shown a reduction in cerebrospinal fluid concentrations of vasopressin in patients with Korsakoff's psychosis.[94] Thus, Korsakoff's psychosis may represent one neurological condition resulting from chronic ethanol consumption where central and peripheral release of vasopressin are differentially affected. Since vasopressin is involved in memory consolidation,[44] it is possible that the memory disorder characteristic of Korsakoff's psychosis may be partly related to damage of central vasopressinergic nerves.

## D. Beer Drinkers' Hyponatremia and Cirrhosis

Although overhydration in alcoholism is generally iso-osmotic, in certain instances hyponatremia is a presenting symptom, and evidence indicates that inappropriate vasopressin secretion may be the main etiologic factor. For instance, a number of reports have documented a neurological syndrome involving disorders of consciousness, coma, and seizures secondary to hyponatremia and inappropriate concentration of the urine following excessive beer drinking.[95-97] Hyponatremia may also be observed in cirrhosis,[98] and experimental evidence has indicated that nonosmotic stimulation of vasopressin secretion secondary to a decrease in effective blood volume is an important factor in the impaired water excretion characteristic of the patient presenting with cirrhosis.[99,100]

## IV. CONCLUSIONS

There are marked differences between the acute and the chronic effects of ethanol on water balance. In the short term (during acute intoxication), ethanol consumption causes dehydration, while in the long term (in alcoholism or following repeated ethanol consumption), overhydration is the predominant feature. Dehydration following acute ethanol consumption occurs secondary to the increase in urinary-free water clearance resulting from inhibition of vasopressin release. It is possible that intra-extracellular fluid shifts also contribute. The mechanism by which ethanol inhibits vasopressin release involves a depressant action on the responsiveness of osmoreceptors to changes in body fluid tonicity, an action which also suppresses the osmotic stimulation of thirst.

The cause of overhydration following chronic ethanol consumption is not well delineated. Although increased vasopressin release may contribute to water retention, overhydration is generally iso-osmotic, indicating that other factors such as sodium retention are likely to be involved and complicated by the presence of clinical features of chronic alcoholism such as liver disease and the withdrawal syndrome. Overhydration may reflect an adaptive response to repeated periods of dehydration after acute ethanol consumption. This, however, does not appear to involve tolerance to the inhibitory effects of ethanol on vasopressin release, since exaggerated diuretic responses to ethanol occur in alcoholics and experimental animals chronically administered ethanol.

The effects of ethanol on vasopressin, thirst, and water balance are potentially important for understanding some of the pathophysiological changes after ethanol. Effects on thirst and water balance may influence the dependence process, and effects on vasopressin may influence tolerance and cognition. In alcoholism, alterations in water balance may contribute to the withdrawal syndrome, hypertension, and brain damage.

## ACKNOWLEDGMENTS

We thank the Medical Research Council of New Zealand and the New Zealand Alcoholic Liquor Advisory Council for support for our personal work included in this article. Thanks are also extended to Joan Darcey for careful typing of this manuscript.

## REFERENCES

1. **Murray, M. M.,** The diuretic action of alcohol and its relation to pituitrin, *J. Physiol.,* 76, 379, 1932.
2. **Eggleton, M. G.,** The diuretic action of alcohol in man, *J. Physiol.,* 101, 172, 1942.
3. **Strauss, M. B., Rosenbaum, J. D., and Nelson, W. P.,** The effect of alcohol on the renal excretion of water and electrolyte, *J. Clin. Invest.,* 29, 1053, 1950.
4. **Rubini, M. E., Kleeman, C. R., and Lamdin, E.,** Studies on alcohol diuresis. I. The effect of ethyl alcohol ingestion on water, electrolyte and acid-base metabolism, *J. Clin. Invest.,* 34, 439, 1955.
5. **Van Dyke, H. B. and Ames, R. G.,** Alcohol diuresis, *Acta Endocrinol.,* 7, 110, 1951.
6. **Lamdin, E., Kleeman, C. R., Rubini, M., and Epstein, F. H.,** Studies on alcohol diuresis. III. The response to ethyl alcohol in certain disease states characterized by impaired water tolerance, *J. Clin. Invest.,* 35, 386, 1956.
7. **Helderman, J. H., Vestal, R. E., Rowe, J. W., Tobin, J. D., Andres, R., and Robertson, G. L.,** The response of arginine vasopressin to intravenous ethanol and hypertonic saline in man: the impact of aging, *J. Gerontol.,* 33, 39, 1978.
8. **Linkola, J., Ylikahri, R., Fyhrquist, F., and Wallenius, M.,** Plasma vasopressin in ethanol intoxication and hangover, *Acta Physiol. Scand.,* 104, 180, 1978.
9. **Eisenhofer, G. and Johnson, R. H.,** Effect of ethanol ingestion on plasma vasopressin and water balance in humans, *Am. J. Physiol.,* 242, R522, 1982.
10. **Gill, G. V., Baylis, P. H., Flear, C. T. G., Skillen, A. W., and Diggle, P. H.,** Acute biochemical responses to moderate beer drinking, *Br. Med. J.,* 285, 1770, 1982.
11. **Nicholson, W. M. and Taylor, H. M.,** The effect of alcohol on water and electrolyte balance in man, *J. Clin. Invest.,* 17, 279, 1938.
12. **Kleeman, C. R., Rubini, M. E., Lamdin, E., and Epstein, F. H.,** Studies on alcohol diuresis. II. The evaluation of ethyl alcohol as an inhibitor of the neurohypophysis, *J. Clin. Invest.,* 34, 448, 1955.

13. **Roberts, K. E.,** Mechanism of dehydration following alcohol ingestion, *Arch. Intern. Med.,* 112, 154, 1963.
14. **Millet, Y. A., Tieffenbach, A. L., and Tack, J. L.,** Influence de l'alcool sur le reponse antidiuretique du noyau supraoptique a la stimulation electrique chez le cobaye, *C. R. Soc. Biol.,* 162, 1437, 1968.
15. **Millerschoen, N. R. and Riggs, D. S.,** Homeostatic control of plasma osmolality in the dog and the effect of ethanol, *Am. J. Physiol.,* 217, 431, 1969.
16. **Nicholson, W. M. and Taylor, H. M.,** Blood volume studies in acute alcoholism, *Q. J. Stud. Alcohol,* 1, 472, 1940.
17. **Lolli, G., Rubin, M., and Greenberg, L. A.,** The effect of ethyl alcohol on the volume of extracellular water, *Q. J. Stud. Alcohol,* 5, 1, 1944.
18. **Huang, K. C., Knoefel, P. K., Shimomura, L., and Buren King, N.,** Some effects of alcohol on water and electrolytes in the dog, *Arch. Int. Pharmacodyn.,* 109, 90, 1957.
19. **Eisenhofer, G. and Johnson, R. H.,** Effects of ethanol ingestion on thirst and fluid consumption in humans, *Am. J. Physiol.,* 244, R568, 1983.
20. **Heyward, P. and Eisenhofer, G.,** Ethanol-induced inhibition of the drinking response to hypertonic saline in the rat, *Pharmacol. Biochem. Behav.,* 22, 493, 1985.
21. **Essig, C. F.,** Increased water consumption following forced drinking of alcohol in rats, *Psychopharmacologia,* 12, 333, 1968.
22. **Lawson, D. M.,** The dipsogenic effect of alcohol and the loss of control phenomenon, *Adv. Exp. Med. Biol.,* 85B, 547, 1977.
23. **Sargent, W. Q., Simpson, J. R., and Beard, J. D.,** Twenty-four-hour fluid intake and renal handling of electrolytes after various doses of ethanol, *Alcoholism,* 4, 74, 1980.
24. **Baisset, A. and Montastruc, P.,** Inhibition par la vasopressine de la soif provoquee par ingestion d'alcool, *C. R. Soc. Biol.,* 155, 1128, 1961.
25. **Montastruc, P., Faruch, M., and Montastruc, J. L.,** Essai de traitement par la vasopressine de l'acoolisme chronique, *Therapie,* 37, 103, 1982.
26. **Wayner, M. J.,** Effects of ethyl alcohol on lateral hypothalamic neurons, *Ann. N. Y. Acad. Sci.,* 215, 13, 1973.
27. **Tata, P. S. and Buzalkov, R.,** Vasopressin studies in the rat. III. Inability of ethanol anesthesia to prevent ADH secretion due to pain or hemorrhage, *Pfluegers Arch.,* 290, 294, 1966.
28. **Bisset, G. W. and Walker, J. M.,** The effects of nicotine, hexamethonium and ethanol on the secretion of the antidiuretic and oxytocic hormones of the rat, *Br. J. Pharmacol.,* 12, 461, 1957.
29. **Pierson, R. N., Wang, J., Frank, W., Allen, G., and Rayyes, A.,** Alcohol affects intracellular potassium, sodium, and water distributions in rats and man, in *Currents in Alcoholism,* Vol. 1, Seixas, F. A., Ed., Grune & Stratton, New York, 1977, 161.
30. **Rundgren, M., McKinley, M. J., Leksell, L. G., and Andersson, B.,** Inhibition of thirst and apparent ADH release by intra-cerebroventricular ethacrynic acid, *Acta Physiol. Scand.,* 105, 123, 1979.
31. **Oiso, Y. and Robertson, G. L.,** Effect of ethanol on vasopressin secretion and the role of endogenous opioids, in *Vasopressin,* Schrier, R. W., Ed., Raven Press, New York, 1985, 265.
32. **Iida, S.,** Experimental studies on the craving for alcohol. I. Alcoholic drive in mice following administration of saline, *Jpn. J. Pharmacol.,* 6, 87, 1957.
33. **Grupp, L. A., Perlanski, E., and Stewart, R. B.,** Dietary salt and DOCA-salt treatments modify ethanol self-selection in rats, *Behav. Neural Biol.,* 40, 239, 1984.
34. **Eisenhofer, G.,** Increased voluntary ethanol intake after dietary salt-loading in rats, *Pharmacol. Biochem. Behav.,* 24, 1825, 1986.
35. **Baisset, A. and Montastruc, P.,** Effet de l'hormone antidiuretique sur le besoin d'alcool cree par l'habitude, *C. R. Soc. Biol.,* 156, 945, 1962.
36. **Rigter, H. and Crabbe, J.,** Ethanol preference in homozygous diabetes insipidus (Brattleboro) rats: effect of vasopressin fragments, *Ann. N. Y. Acad. Sci.,* 394, 663, 1982.
37. **Myers, R. D., Critcher, E. C., and Cornwell, N. N.,** Effect of chronic vasopressin treatment on alcohol drinking of brattleboro HZ and DI rats, *Peptides,* 4, 359, 1983.
38. **Linkola, J.,** Urine sodium, potassium and osmolality in two rat strains selected for their different ethanol preferences, *Med. Biol.,* 54, 254, 1976.
39. **Linkola, J., Fyhrquist, F., and Forsander, O.,** Effects of ethanol on urinary arginine vasopressin excretion in two rat strains selected for their different ethanol preferences, *Acta Physiol. Scand.,* 101, 126, 1977.
40. **Linkola, J., Fyhrquist, F., Poso, A. R., and Tikkanen, I.,** Electrolyte excretion in alcohol preferring and alcohol avoiding rats, *Life Sci.,* 26, 103, 1980.
41. **Linkola, J., Tikkanen, I., Fyhrquist, F., and Rusi, M.,** Renin, water drinking, salt preference and blood pressure in alcohol preferring and alcohol avoiding rats, *Pharmacol. Biochem. Behav.,* 12, 293, 1980.
42. **Ogata, M.,** Clinical and experimental studies on water metabolism in alcoholism, *Q. J. Stud. Alcohol,* 24, 398, 1963.
43. **Kissin, B., Schenker, V. J., Schenker, A. C., Belanger, I. R., Stucker, M., and Woerner, R.,** Hyperdiuresis after ethanol in chronic alcoholics, *Am. J. Med. Sci.,* 248, 660, 1964.

44. **Van Ree, J. M. and DeWied, D.,** Involvement of neurohypophyseal peptides in drug-mediated adaptive responses, *Pharmacol. Biochem. Behav.,* 13, 257, 1980.
45. **Hoffman, P. L., Ritzmann, R. F., Walter, R., and Tabakoff, B.,** Arginine vasopressin maintains ethanol tolerance, *Nature,* 276, 614, 1978.
46. **Rigter, H., Rijk, H., and Crabbe, J. C.,** Tolerance to ethanol and severity of withdrawal in mice are enhanced by a vasopressin fragment, *Eur. J. Pharmacol.,* 64, 53, 1980.
47. **Hoffman, P. L.,** Structural requirements for neurohypophyseal peptide maintenance of ethanol tolerance, *Pharmacol. Biochem. Behav.,* 17, 685, 1982.
48. **Hung, C. R., Tabakoff, B., Melchoir, C. L., and Hoffman, P. L.,** Intraventricular arginine vasopressin maintains ethanol tolerance, *Eur. J. Pharmacol.,* 106, 645, 1984.
49. **Ritzmann, R. F., Colbern, D. L., Zimmermann, E. G., and Krivoy, W.,** Neurohypophyseal hormones in tolerance and physical dependence, *Pharmacol. Ther.,* 23, 281, 1984.
50. **Pittman, Q. J., Rogers, J., and Bloom, F. E.,** Arginine vasopressin deficient Brattleboro rats fail to develop tolerance to the hypothermic effects of ethanol, *Regul. Pept.,* 4, 33, 1982.
51. **Pittman, Q. J., Rogers, J., and Bloom, F. E.,** Deficits in tolerance to ethanol in Brattleboro rats, *Ann. N. Y. Acad. Sci.,* 394, 764, 1982.
52. **Gold, P. W., Goodwin, F. K., and Reus, V. I.,** Vasopressin in affective illness, *Lancet,* 1, 1233, 1978.
53. **Linnoila, M. and Mattila, M. J.,** How to antagonize ethanol-induced inebriation, *Pharmacol. Ther.,* 15, 99, 1981.
54. **Pishkin, V., Lawrence, B. E., and Bourne, L. E.,** Cognitive and electrophysiologic parameters during ascending and descending limbs of the blood-alcohol curve, *Alcoholism,* 7, 76, 1983.
55. **Eisenhofer, G., Lambie, D. G., and Robinson, B. J.,** No improvement in ethanol-induced memory deficits after administration of a vasopressin analog, *Life Sci.,* 37, 2499, 1985.
56. **Beard, J. D., Barlow, G., and Overman, R. R.,** Body fluids and blood electrolytes in dogs subjected to chronic ethanol administration, *J. Pharmacol. Exp. Ther.,* 148, 348, 1965.
57. **Crow, L. T.,** Water metabolism with prolonged ethanol consumption in the rat, *Psychol. Rep.,* 27, 675, 1970.
58. **Wright, J. W. and Donlon, K.,** Reversal of ethanol-induced body dehydration with prolonged consumption in rats, *Pharmacol. Biochem. Behav.,* 10, 31, 1979.
59. **Sargent, W. Q., Simpson, J. R., and Beard, J. D.,** Extracellular volume expansion after ethanol in dogs, *J. Stud. Alcohol,* 36, 1468, 1975.
60. **Sereny, G., Rapoport, A., and Husdan, H.,** The effect of alcohol withdrawal on electrolyte and acid-base balance, *Metabolism,* 15, 896, 1966.
61. **Karlsson, L., Hirvonen, J., Salorinne, Y., and Virtanen, K.,** Functional status of the hypothalamo-neurohypophyseal system after two weeks' administration of adrenalin and insulin in rats under continuous influence of ethanol, *Ann. Med. Exp. Fenn.,* 45, 72, 1967.
62. **Zeballos, G. A., Basulto, J., Munoz, C. A., and Salinas-Zeballos, M. E.,** Chronic alcohol ingestion: effects on water metabolism, *Ann. N. Y. Acad. Sci.,* 273, 343, 1976.
63. **Midoux, C., Burlet, A., Boulange, M., and Legait, E.,** Modifications du systeme hypothalamo-neuro-hypophysaire du rat apres administration repetee d'ethanol, *C. R. Soc. Biol.,* 169, 151, 1975.
64. **Marquis, C., Marchetti, J., Burlet, C., and Boulange, M.,** Secretion urinaire et hormone antidiuretique chez des rats soumis a une administration repetee d'ethanol, *C. R. Soc. Biol.,* 169, 154, 1975.
65. **Eisenhofer, G., Whiteside, E., Lambie, D., and Johnson, R.,** Brain water during alcohol withdrawal, *Lancet,* 1, 50, 1982.
66. **Eisenhofer, G., Lambie, D. G., Whiteside, E. A., and Johnson, R. H.,** Vasopressin concentrations during alcohol withdrawal, *Br. J. Addict.,* 80, 195, 1985.
67. **Legros, J. J., Deconinck, I., Willems, D., Roth, B., Pelc, I., Brauman, J., and Verbanck, M.,** Increase of neurophysin II serum levels in chronic alcoholic patients: relationship with alcohol consumption and alcoholism blood markers during withdrawal therapy, *J. Clin. Endocrinol. Metab.,* 56, 871, 1983.
68. **Beard, J. D. and Knott, D. H.,** Fluid and electrolyte balance during acute withdrawal in chronic alcoholic patients, *J.A.M.A.,* 204, 135, 1968.
69. **Knott, D. H. and Beard, J. D.,** A diuretic approach to acute withdrawal from alcohol, *South. Med. J.,* 62, 485, 1969.
70. **Besson, J. A. O., Glen, A. I. M., Foreman, E. I., MacDonald, A., Smith, F. W., Hutchison, J. M. S., Mallard, J. R., and Ashcroft, G. W.,** Nuclear magnetic resonance observations in alcoholic cerebral disorder and the role of vasopressin, *Lancet,* 2, 923, 1981.
71. **Smith, M. A., Chick, J., Kean, D. M., Douglas, R. H. B., Singer, A., Kendell, R. E., and Best, J. J. K.,** Brain water in chronic alcoholic patients measured by magnetic resonance imaging, *Lancet,* 1, 1273, 1985.
72. **Mander, A. J., Smith, M. A., Kean, D. M., Chick, J., Douglas, R. H. B., Rehman, A. U., Weppner, G. J., and Best, J. J. K.,** Brain water measured in volunteers after alcohol and vasopressin, *Lancet,* 2, 1075, 1985.

73. **Raichle, M. E. and Grubb, R. L.,** Regulation of brain water permeability by centrally-released vasopressin, *Brain Res.,* 143, 191, 1978.
74. **Luerssen, T. G., Shelton, R. L., and Robertson, G. L.,** Evidence for separation of plasma and cerebrospinal fluid vasopressin, *Clin. Res.,* 25, 14A, 1977.
75. **Noto, T., Kato, N., Inoue, K., Kitabayashi, M., and Nakajima, T.,** The levels of vasopressin in cerebrospinal fluid of patients with alcoholism, *Endocrinol. Jpn.,* 29, 121, 1982.
76. **McNichol, R. W., Cirksena, W. J., Payne, J. T., and Glasgow, M. C.,** Management of withdrawal from alcohol, *South. Med. J.,* 60, 7, 1967.
77. **Victor, M. and Wolfe, S. M.,** Causation and treatment of the alcohol withdrawal syndrome, in *Alcoholism: Progress in Research and Treatment,* Bourne, P. G. and Fox, R., Eds., Academic Press, New York, 1973, 137.
78. **Balsano, N. A. and Reynolds, B. M.,** Fluid restriction in the management of acute alcoholic withdrawal, *Surgery,* 68, 283, 1970.
79. **Steinbach, R.,** Treatment of delirium tremens by spinal drainage, *Dtsch. Med. Wochenschr.,* 41, 369, 1915.
80. **McQuarrie, I. and Peeler, D. B.,** The effects of sustained pituitary antidiuresis and forced water drinking in epileptic children: a diagnostic and etiologic study, *J. Clin. Invest.,* 10, 915, 1931.
81. **Clegg, J. L. and Thorpe, F. T.,** Induced water retention in diagnosis of idiopathic epilepsy, *Lancet,* 1, 1381, 1935.
82. **Lambie, D. G.,** Alcoholic brain damage and neurological symptoms of alcohol withdrawal — manifestations of overhydration, *Med. Hypoth.,* 16, 377, 1985.
83. **Shaw, D. M., Camps, F. E., Robinson, A. E., Short, R., and White, S.,** Electrolyte content of the brain in alcoholism, *Br. J. Psychiatry,* 116, 185, 1970.
84. **Alexander, L.,** Neuropathological findings in brain and spinal cord of chronic alcoholic patients, *Q. J. Stud. Alcohol,* 2, 260, 1941.
85. **Umiker, W. O.,** Pathology of acute alcoholism, *U. S. Navy Med. Bull.,* 49, 744, 1949.
86. **Seitelberger, F. and Berner, P.,** Uber die marchiafavasche krankheit, *Virchows Arch.,* 326, 257, 1955.
87. **Norenberg, M. D., Leslie, K. O., and Robertson, A. S.,** Association between rise in serum sodium and central pontine myelinolysis, *Ann. Neurol.,* 11, 128, 1982.
88. **Hazratji, S. M. A., Kim, R. C., Lee, S. H., and Marasigan, A. V.,** Evolution of pontine and extrapontine myelinolysis, *J. Comput. Assist. Tomogr.,* 7, 356, 1983.
89. **Hamilton, D. V.,** Inappropriate secretion of antidiuretic hormone associated with cerebellar and cerebral atrophy, *Postgrad. Med. J.,* 54, 427, 1978.
90. **Messert, B., Orrison, W. W., Hawkins, M. J., and Quaglieri, C. E.,** Central pontine myelinolysis: considerations on etiology, diagnosis and treatment, *Neurology,* 29, 147, 1979.
91. **Cooles, P. E. and Borthwick, L. J.,** Inappropriate antidiuretic hormone secretion in Wernicke's encephalopathy, *Postgrad. Med. J.,* 58, 173, 1982.
92. **Sterns, R. H., Riggs, J. E., and Schochet, S. S.,** Osmotic demyelination syndrome following correction of hyponatraemia, *N. Engl. J. Med.,* 314, 1535, 1986.
93. **Eisenhofer, G., Johnson, R. H., and Lambie, D. G.,** Growth hormone, vasopressin, cortisol and catecholamine responses to insulin hypoglycaemia in alcoholics, *Alcoholism,* 8, 33, 1984.
94. **Mair, R. G., Langlais, P. J., Mazurek, M. F., Beal, M. F., Martin, J. B., and McEntee, W. J.,** Reduced concentrations of arginine vasopressin and MHPG in lumbar CSF of patients with Korsakoff's psychosis, *Life Sci.,* 38, 2301, 1986.
95. **Demanet, J. C., Bonnyns, M., Bleiberg, H., and Stevens-Rocmans, C.,** Coma due to water intoxication in beer drinkers, *Lancet,* 2, 1115, 1971.
96. **Gwinup, G., Chelvam, R., Jabola, R., and Meister, L.,** Beer drinker's hyponatraemia, inappropriate concentration of the urine during ingestion of beer, *Calif. Med.,* 116, 78, 1972.
97. **Hilden, T. and Svendsen, T. L.,** Electrolyte disturbances in beer drinkers, *Lancet,* 2, 245, 1975.
98. **Summerskill, W. H. J., Barnardo, D. E., and Baldus, W. P.,** Disorders of water and electrolyte metabolism in liver disease, *Am. J. Clin. Nutr.,* 23, 499, 1970.
99. **Linas, S. L., Anderson, R. J., Guggenheim, S. J., Robertson, G. L., Berl, T., and Dickmann, D. C.,** Role of vasopressin in impaired water excretion in conscious rats with experimental cirrhosis, *Kidney Int.,* 20, 173, 1981.
100. **Bichet, D., Szatalowicz, V., Chaimovitz, C., and Schrier, R. W.,** Role of vasopressin in abnormal water excretion in cirrhotic patients, *Ann. Intern. Med.,* 96, 413, 1982.

*Index*

# INDEX

## A